TREE STATION. SIERRA LEONE TRIANGULATION.

[*All Rights Reserved.*

# TEXT BOOK

OF

# TOPOGRAPHICAL AND GEOGRAPHICAL SURVEYING

BY

MAJOR C. F. CLOSE, C.M.G., R.E.

FELLOW OF THE ROYAL ASTRONOMICAL SOCIETY, MEMBER OF COUNCIL OF THE ROYAL GEOGRAPHICAL SOCIETY, HEAD OF THE
TOPOGRAPHICAL SECTION OF THE GENERAL STAFF, LATE INSTRUCTOR IN SURVEYING AT THE SCHOOL OF
MILITARY ENGINEERING, LATE BRITISH COMMISSIONER NYASA-TANGANYIKA BOUNDARY
COMMISSION, FORMERLY OF THE ORDNANCE SURVEY AND OF
THE SURVEY OF INDIA.

LONDON:
PRINTED FOR HIS MAJESTY'S STATIONERY OFFICE,
BY HARRISON AND SONS, ST. MARTIN'S LANE,
PRINTERS IN ORDINARY TO HIS MAJESTY.

And to be purchased, either directly or through any Bookseller, from
WYMAN AND SONS, LTD., FETTER LANE, E.C.; or
OLIVER AND BOYD, EDINBURGH; or
E. PONSONBY, 116, GRAFTON STREET, DUBLIN.

1905.

*Price Three Shillings and Sixpence.*

DEPARTMENT OF ARCHITECTURE,
HARVARD UNIVERSITY.

146

(Wt. 11975  5000  5 | 05—H & S  5455)

TA
590
C62

# PREFACE.

THE methods described in this book are chiefly based on the actual experience of Royal Engineer Officers who are, or have been, engaged in carrying out geographical and topographical surveys.

Chapter IV., on "Traverses," is mainly written by Captain W. J. Johnston, Captain H. D. Pearson and Captain C. J. Heath.

Chapter VII., "Survey under Active Service Conditions," is from the writings of Colonel Sir Thomas Holdich, K.C.M.G., K.C.I.E., Colonel Gore, C.S.I., late Surveyor-General of India, Colonel the Hon. M. G. Talbot, Director of Surveys in the Sudan, and Colonel Wahab, C.I.E.

Chapter VIII., "Levelling," is due to Captain H. L. Crosthwait and Captain C. Russell-Brown.

A portion of Chapter X., the use of the theodolite for contouring, is taken from a report by Major P. H. du P. Casgrain.

Chapter XII., "The Reproduction of Maps in the Field," has been written by Captain B. A. G. Shelley, with some additions by Lieutenant F. V. Thompson.

Chapter XIV., "Field Astronomy," is based on the 'Notes on Astronomy,' written for the School of Military Engineering, by Colonel A. C. MacDonnell, re-arranged by Captain C. J. Heath.

In Chapter XVI. the description of "Photo-longitudes" is taken from the account published by Major E. H. Hills, C.M.G., in the 'Memoirs of the Royal Astronomical Society,' Vol. LIII., 1897.

The section on "Occultations" has been written by Captain C. H. Foulkes.

The "Graphic Method of predicting the circumstances of an Occultation" is due to Colonel S. C. N. Grant, C.M.G., and permission to use the diagrams explanatory of this method has been kindly granted by the Royal Geographical Society. Appendix VIII. is also due to Colonel Grant.

The extension of the "Graphic Method" is from a paper by Mr. A. C. D. Crommelin.

Tables I., II., IX., XI. and XII. were computed by Major C. H. Enthoven.

Tables XVII. and XVIII. were computed by Lieutenant G. F. Evans.

The diagrams for rapidly determining the Time and Altitude of a star on the Prime Vertical were devised by Captain C. J. Heath.

The Star Charts were drawn by Captain H. E. Winsloe.

Mr. A. R. Hinks, M.A., of Cambridge Observatory, has very kindly read and criticised the chapters which deal with field astronomical methods; these chapters have benefited greatly in consequence.

The reproductions of British and Foreign Topographical Maps were executed under the direction of Colonel D. A. Johnston, C.B., Director-General of the Ordnance Survey. Most of the other plates were drawn in the Topographical Section of the War Office under the direction of Major E. H. Hills, C.M.G.; a few, however, have been produced by H.M. Stationery Office.

Plate. V. is from a plane-table sheet of Kroonstad by Lieut.-Colonel H. M. Jackson, Surveyor-General of the Transvaal.

Plate VI. is from a plane-table survey executed in Somaliland under Captain G. A. Beazeley.

Plate VII. is from an original field sheet lent by Lieut.-Colonel F. B. Longe, Surveyor-General of India.

The field methods described are, for the most part, those in use by the Survey of India; but advantage has been taken of recent experience in mapping and exploring various territories in Africa and elsewhere to include useful methods which are not commonly employed in India.

# CONTENTS.

|  |  | Page. |
|---|---|---|
| CHAPTER I. | Introduction. General Principles | 1 |
| II. | Bases | 5 |
| III. | Triangulation | 10 |
| IV. | Traverses | 34 |
| V. | Subtense Methods and Tacheometry | 51 |
| VI. | Plane-Tabling | 57 |
| VII. | Survey under Active Service Conditions | 68 |
| VIII. | Levelling | 74 |
| IX. | Heights (Barometer and Hypsometer) | 86 |
| X. | Contouring and Hill Sketching | 89 |
| XI. | Map Projections and Graticules | 92 |
| XII. | Reproduction of Maps in the Field | 112 |
| XIII. | British and Foreign Government Maps and Surveys | 118 |
| XIV. | Field Astronomy | 123 |
| XV. | Latitudes and Azimuths | 182 |
| XVI. | Longitude | 184 |
| XVII. | Theory of Errors as applied to Topographical Work | 204 |

| TABLE I. | Values of P, Q, R, S, T (Terr., Lat., Long., and Reverse Azimuths) | 209 |
|---|---|---|
| II. (a) & (b). | Directions for applying signs | 222 |
| III. | Heights. Value of K for coefficient of refraction ·07 | 223 |
| IV. | Computation of value of K | 223 |
| V. & VI. | Barometer Heights | 224 |
| VII. & VIII. | Hypsometer Heights | 226 |
| IX. | Graticules for Maps, scale ¼″=1 m. (Sides and diagonals for 15′ Lat. and Long.) | 227 |
| X. | Length in miles of 15′ Lat. and Long. | 232 |
| XI. | Linear value in feet of 1″ arc and its logarithm, along Meridian | 234 |
| XII. | ,, ,, ,, ,, Parallel | 246 |
| XIII. | Spherical Excess | 258 |
| XIV. | Astronomical Refraction | 259 |
| XV. | Parallax of the Sun | 261 |
| XVI. | Reduction to Meridian (for Latitude) | 262 |
| XVII. | Correction of Moon's Equatorial Parallax (for Occultation) | 264 |
| XVIII. | Reduction of Latitude (for Occultation) | 265 |

| APPENDIX I. | Explanation of Table I. | 267 |
|---|---|---|
| II. | Computation of Distances and Azimuths | 268 |
| III. | Alternative Formulæ for Lat., Long., and Reverse Azimuth | 270 |
| IV. | Explanation of Table IV. | 271 |
| V. | Formula for correction for Sag in steel tapes | 272 |
| VI. | List of Instruments, Tools and Camp Equipment | 273 |
| VII. | Sketching in Dense Forest | 275 |
| VIII. | Photographic Surveying | 279 |
| IX. | Useful Trigonometrical Formulæ | 280 |
| X. | Length of the Metre | 282 |

# LIST OF PLATES.

|  |  | To face page |
|---|---|---|
| *Frontispiece.* | Tree Station, Sierra Leone Triangulation. | |
| I. | Port Elizabeth Base Extension | 8 |
| II. | Trigonometrical Signals for Short Distances | 10 |
| III. | Five-inch Micrometer Transit Theodolite | 12 |
| IV. | Service Plane-Table | 57 |
| V. | Example of Small-Scale Plane-Table Work in Undulating Country (Kroonstad) | 58 |
| V. (*b*). | Example of Two-inch Plane-Table Work (Jersey) | 60 |
| VI. | Specimen of Small-Scale Plane-Table Work. (A portion of a survey in Somaliland during the operations, 1903–4) | 62 |
| VII. | Form-Lines for Plane-Table Work. Plane-Table Field Sheet (Baluchistan) | 62 |
| VIII. | Diagram of Afghan Boundary Triangulation | 71 |
| IX. | Cooke's Reversible Level | 75 |
| X. | "Y" Level | 78 |
| XI. | Dumpy Level | 79 |
| XII. | Water Level, Ordnance Survey Pattern | 90 |
| XIII. | Ordnance Survey 1-inch, black with brown hills | 91 |
| XIV. | Gnomonic Projection | 104 |
| XV. | Clarke's Perspective Projection | 105 |
| XVI. | Conical Projection with Rectified Meridians | 109 |
| XVII. | Field Lithographic Press | 112 |
| XVIII. | Ordnance Survey 1 inch to 1 mile (coloured) | 118 |
| XVIII. (*b*). | ,,    ,,    2 miles to 1 inch   ,, | 118 |
| XIX. | ,,    ,,    4   ,,    to   ,,    ,, | 120 |
| XX. | French, $\frac{1}{50,000}$ (Bizerta) | 122 |
| XXI. | United States, $\frac{1}{62,500}$ | 122 |
| XXII. | Germany, $\frac{1}{100,000}$ | 122 |
| XXIII. | France, $\frac{1}{80,000}$, Engraved black | 122 |
| XXIV. | Switzerland, $\frac{1}{50,000}$, Coloured | 122 |
| XXV. | Italy, $\frac{1}{100,000}$, Engraved black | 122 |
| XXVI. | St. Helena, 2½ inches to 1 mile | 123 |
| XXVII. | Occultation Diagram No. 1 | 192 |
| XXVIII. | ,,     ,,    2 | 192 |
| XXIX. | Star Chart | End of Book. |
| XXX. | ,,    ,, | ,, ,, |
| XXXI. | ,,    ,, | ,, ,, |
| XXXII. | ,,    ,, | ,, ,, |
| XXXIII. | Diagram for Determining the Hour Angle of a Heavenly Body when on Prime Vertical | ,, ,, |
| XXXIV. | Diagram for Determining the Altitude of ditto | ,, ,, |

# TEXT BOOK OF TOPOGRAPHICAL SURVEYING.

## CHAPTER I.

### INTRODUCTION. GENERAL PRINCIPLES.

**Surveying** is the process of accurately determining the relative positions of the features of a portion of the Earth's surface.

**Topography** (τοπογραφία, a description of a place) is the delineation of the natural and artificial features of a locality.

**Topographical Surveying** is thus the process of accurately determining the relative positions of the surface features of the earth, and of delineating them on a map. The surface features are of two kinds: artificial, that is to say those produced by the agency of man, such as towns, roads or railways; and natural features, such as rivers, lakes, forests, hill features and the undulations of the ground.

**A Topographical Map** should depict all the surface features which can be legibly drawn; all the accidents of the surface should be shown which distinguish it from a level, featureless plain. The amount of information presented on a topographical map is thus conditioned by two principal factors, the complexity of the surface features, and the scale of the map.

Since it is not customary to construct maps in the field on scales smaller than about $\frac{1}{500,000}$, this scale will be taken as the lower limit for topographical maps. Maps on scales smaller than this are compilations or atlas maps.

It is not the primary intention of a topographical map to show property boundaries; this is the function of a large-scale *Cadastral Plan*. Nor, on the other hand, has a topographical survey for its object the investigation of any scientific questions connected with the shape and dimensions of the earth; this is the domain of a *Geodetic Survey*.

The production of a *map*, in the ordinary sense of the word, is, in fact, the aim of a topographical survey.

The *scales* of topographical maps may vary from 6 inches to 1 mile to $\frac{1}{4}$ inch to 1 mile, or from about $\frac{1}{10,000}$ to $\frac{1}{500,000}$; but for military and administrative purposes, the most useful scales are 2 inches, 1 inch, $\frac{1}{2}$ inch, and $\frac{1}{4}$ inch to 1 mile, or say from $\frac{1}{30,000}$ to $\frac{1}{250,000}$, and this book is to be considered as dealing mainly with maps on such scales.

Topographical maps naturally vary in accuracy, according to the conditions under which they have been produced. Thus a plane table survey made rapidly on active service will be wanting in that refinement of accuracy which is to be expected of a map deliberately executed in peace time. In a broad sense, even a military sketch may be considered a rough topographical map; but military sketches will not be considered in this book, which will deal with:—

(1) Deliberate topographical surveys;
(2) Topographical surveys executed on service;
(3) Geographical surveys, such as are required for the delimitation of an international boundary;
(4) Exploratory surveys.

Topographical, geographical and exploratory surveys merge into one another, and there is no very definite dividing line between them. Generally speaking, there should be no detectable error in a topographical map, whilst geographical surveys are somewhat rough and ready. The value of a survey on active service depends on the extent to which it assists the operations in progress, and minute accuracy must often, in such a case, be sacrificed.

The extent of the field before the topographer may be realised from the fact that, excluding Canada, Australia, New Zealand and India, the total unmapped area of the British Empire amounts to about 3,700,000 square miles.

MAIN PRINCIPLE TO OBSERVE IN CARRYING OUT A TOPOGRAPHICAL SURVEY.

The main principle to observe in a map making is to work from the whole to the part, and not from the part to the whole. It would be entirely wrong, for instance, to map a block of four English counties by making independent surveys of each county and then joining the surveys together. It is still more incorrect to attempt to compile a map of a large country by piecing together farm surveys. Such a proceeding, which has several times been attempted, results in a thoroughly untrustworthy map. The technical reasons for this will be better understood later on; meanwhile it must be accepted as an axiom that however large or however small may be the area to be surveyed, it must be treated as a whole, and that all over the area a number of carefully determined points must be fixed, forming, as it were, an accurate framework or skeleton, on which the detail is to depend. It is clear that such a method prevents the accumulation of any system of errors, except that of the accurate framework.

The accuracy of a map will thus ultimately depend on the accuracy of the framework, which will consist of points fixed by one of the following methods:—

(1) **Triangulation;**
(2) **Traverses;**
(3) **Astronomical determinations;** or
(4) **Combinations of the above.**

Of these, triangulation is always to be preferred, and should be adopted whenever possible.

**Triangulation.**—It is an elementary proposition in trigonometry that, given the angles of a triangle and the length of one side, the lengths of the other sides can be calculated.

If, then, we have a chain of triangles, as below, and know all the angles and the length of *one* side, we can compute the lengths of all the other sides.

Fig. 1.

The same obviously applies to a network, or to any figure built up of triangles.

Fig. 2.

Now imagine that these points are objects on the earth's surface and that at each object horizontal angles have been observed to those surrounding it, and suppose that the horizontal distance between any two of these points has been measured, it is clear that the distance between any two connected points is easily computed and plotted on paper.

The measured side is called the *base*.

### CLASSES OF TRIANGULATION.

(1.) **Principal Triangulation** or **Geodetic Triangulation** is the main triangulation of a country carried out with every refinement of care, of which the sides vary in length from about 10 to 50 miles (though rays well over 100 miles have been used), in which no precaution is neglected, and of which the computations are most exact and laborious. The triangular error of a principal triangle, *i.e.*, the algebraic sum of the errors of the three angles of the triangle, would not be expected to exceed 1 second.

(2.) **Secondary Triangulation.**—In a regularly surveyed country this would break up the principal triangles into sides of 5 to 10 miles in length, or else be carried out in the form of secondary chains connecting the principal chains. Triangular error, say, 2 to 5 seconds.

The observations are not usually reduced so rigorously as in principal work.

(3.) **Tertiary Triangulation.**—The minor triangulation of a regularly surveyed country, covering the whole surface of the land with triangles the sides of which are from 1 to 5 miles in length. Triangular error, say, 5 to 15 seconds.

In the Ordnance Survey of the United Kingdom—

The sides of the Principal Triangles average 35½ miles.
,, ,, Secondary ,, ,, 5 ,,
,, ,, Tertiary ,, ,, 1¼ ,,

In the Survey of India the average length of the sides of the principal triangles is 32 miles.

(4.) **Topographical Triangulation.**—Though it may be considered certain that in course of time the whole land surface of the world will be systematically surveyed and covered with principal, secondary, and tertiary triangles, still for some time to come there will be, in many cases, the need for topographical maps produced rapidly and with a minimum of expense without the aid of rigorous triangulation. As an example of a geographical survey carried out rapidly without waiting for a rigid framework, the survey of Burma may be mentioned. This country, which was annexed at the end of 1885, was rapidly surveyed on the scale of 4 miles to 1 inch. The triangulation for this survey was carried out during the execution of the topography, the sides were of very varying lengths, the points observed were mainly pagodas or well defined hill-tops; the sole object was to provide points for the plane-tablers to work on, and provided there was no plotable error on the particular scale used, no further accuracy was demanded. This sort of triangulation, of which the sole intention is to provide a framework sufficiently accurate for the scale in use, and which will never be used for geodetic or cadastral work, or for larger scale topographical maps, may be called a *topographical triangulation*.

(5.) **Rapid Triangulation.**—A topographical triangulation carried out during an expedition, or when marching with a column on active service; in this many devices, which would be objectionable in more deliberate triangulation, are used to expedite the work.

### OTHER METHODS OF FORMING THE FRAMEWORK.

As has been mentioned on p. 2, the other methods of forming a framework of accurately fixed points, on which the detail of the map is to depend, are, *Traverses, Astronomical Determinations*, or combinations of these methods with each other or with triangulation.

**Traverses** are described in Chapter IV.

**Astronomical Determinations** are dealt with in Chapters XIV, XV, and XVI.

### NOTE.

The framework should always be trigonometrical where the circumstances permit. The two chief hindrances to triangulation are, the absence of hills or the existence of forest, and the most difficult country to triangulate is a flat, forest-clad region. In such a region the expense of triangulation may render it necessary to execute a framework of traverses.

Generally speaking, astronomical determinations are to be avoided; a map, of which the framework is entirely astronomical, falls far short of the standard of a good topographical map. In exploratory and geographical surveys, however, it may be necessary to make a free use of astronomical methods.

## CHAPTER II.

### BASES.

IF the angles of a triangulation are correctly observed, any error in the measurement of the base will be reproduced in the triangulation. Thus, if the measurement makes the base $\frac{1}{500}$ longer than it really is, and the furthest point of a triangulation is really 500 miles distant from the base, the triangulation will place that point at a distance of 501 miles, or 1 mile too far.

Bases are of various degrees of accuracy, according to the class of triangulation which is to depend on them, viz. :—

(1) Base-lines of a Geodetic Triangulation with a probable error of $\frac{1}{700.000}$ to $\frac{1}{1,500,000}$
(2) Base-lines for accurate Secondary work with a probable error of $\frac{1}{100,000}$ to $\frac{1}{500,000}$.
(3) Base-lines for Topographical work with a probable error of $\frac{1}{10,000}$ to $\frac{1}{50,000}$.
(4) Rough base-lines.

The measurement of a Geodetic Base-line is a very special operation which occupies some weeks and costs much money. There is a large literature on the subject which will not be further referred to in this book.

**Topographical Bases.**—The *length* of a topographical base will depend on the stretch of level or evenly sloping ground available. Such bases have varied in length from ½ mile to 3 or 4 miles. *It is more important to be able to extend the base by well-conditioned triangles than to measure a long base.*

The base at Mandalay measured in 1886 for the Topographical Survey of Upper Burma was 1 mile long.

The base for the Bloemfontein 2-inch Survey in 1900 was 5,300 feet.

Base for Bermuda was 2,890 feet.

Base for Guernsey was 3,165 feet.

Base for the Sierra-Leone-Liberia Boundary was 3,073 feet.

**Site for a Base.**—The ideal site is one on level, even ground, from which good views of the surrounding country can be obtained.

**Measurement of Base.**—A topographical base should always be measured with a steel tape. This should be carefully compared with a standard steel tape before and after using. The error of the standard at a given temperature should be known. This can be ascertained at home at the Ordnance Survey Office, Southampton, in India at the offices of the Survey of India, in South Africa at the Cape Observatory.

The standard tape itself should not be used.

Steel tapes can be bought of the following lengths—100, 300 or 400 feet or 25 or 100 metres.

The line of the base should first be cleared of trees, bushes or high grass; ant heaps and such small mounds should be removed.

Each terminal point of the base should, if time allows, be marked by burying a mark stone 2 feet under ground; accurately centred over this should be built a small masonry pillar, surface flush with the ground. A fine mark should be made in a metal plug sunk into the surface of the pillar.

Having marked the terminal points of the base, set a theodolite over one of them and direct it on the other end, over which a pole should have been fixed; with the aid of the theodolite align banderoles at intervals of about 300 yards. By means of this alignment hammer into the ground square headed pickets, accurately in the line of the base, and at tape lengths from each other, starting from one end of the base. The tops of the pickets should be just flush with the ground; on these pickets nail strips of wood or zinc.

In measuring the base, stretch the tape from terminal point to first picket, and from picket to picket; the tension should be taken on a spring balance and should be about 15 lb. The end of the tape should be marked by a fine pencil line on the wood strip, or by a cut with a knife. Take the temperature of the tape at frequent intervals by letting the bulb of the thermometer come in contact with the under side of the tape.

At the other terminal, the small space between the last graduation and the end of the base should be measured with a pair of dividers.

Now measure the inclinations to horizontal from peg to peg with level or theodolite.

If the section consists of one or two fairly uniform slopes, it is sufficient to take the inclinations of these only, noting where the changes of slope occur.

Minor undulations crossed by the tape are to be disregarded.

The base should be thus measured *at least once in each direction*: on a dull day, if possible, or in early morning or late afternoon.

**Base Computations.**—The measured length of the base is subject to five corrections* :—

    i. For *standard*;
    ii. ,, *temperature*;
    iii. ,, *inclination*;
    iv. ,, *sag*;
    v. ,, *height above sea*.

The following case may be taken as an example :—

One tape was kept as a reference tape and not used for measuring. This, when tested at Southampton, had been found to be 0·05 feet short at 62° F.

The base tape (supposed to be 100 feet long) used was 0·02 feet long on reference tape at 71° F.

The temperature of base tape during measurement was 82° F.

Length, as measured, 5301·24 feet.

Height of base line above sea, 4,520 feet.

For 2,300 feet, base had a slope of 1° 15′, remainder level.

    i. *Standard* :—

Reference tape was 0·05 short at 62° F.

Now steel expands ·00000625 of its length for 1° F.

Therefore, at 71° it was $\dfrac{9 \times 100 \times 62\cdot5}{10{,}000{,}000}$ longer, or ·0056 feet longer.

    Reference tape was therefore 0·0444 short at 71° F.
    Base   ,,   ,,  0·0200 long on reference tape at 71°.
      ,,   ,,   ,,  0·0244 short at 71° . . . . . . . . . . (*a*).

ii. *Temperature* :—

Temperature of base tape during measurement was 82°, when compared was 71°.

$$\text{Increase of length } \dfrac{11 \times 100 \times 62\cdot5}{10{,}000{,}000}, \text{ or } 0\cdot0069 \quad . \quad . \quad . \quad . \quad . \quad (b).$$

Subtracting (*b*) from (*a*), we find that the base tape during measurement was 0·0175 short; correction for standard and temperature is therefore

$$-\cdot0175 \times \dfrac{5301}{100} = -0\cdot928 \text{ feet}.$$

* There is also a correction for *elastic extension*, which may be disregarded if the tape is always stretched at the same tension.

iii. *Inclination* : —

For 2,300 feet the base had a slope of 1° 15′.
Correction = − 2,300 (1 − cos 1° 15′) = − 0·550.

iv. *Sag* :—

In the measurement of a base it is desirable that the whole length of the tape should as far as possible be supported by the ground, so that no correction is necessary for the sag of the tape. Where the ground is uneven, it is customary to support the tape at intervals by pickets; but it may happen in the measurement of a base that a nullah or ravine has to be crossed. In such a case the sag, *i.e.*, the difference between the length when suspended and the length when laid on a plane surface, must be determined and corrected for.

If $s$ = the correction for sag,
$l$ = the length of the tape suspended between two supports (in feet).
$w$ = the weight of the tape in lbs.,
$t$ = the tension applied in lbs.,
$s = l \dfrac{w^2}{24 t^2}$ approximately.

See Appendix V.

*Height above sea* : —

All measurements are supposed to be made at sea-level. If the base is measured at a mean height $h$ above sea-level, it will require a correction of $-\dfrac{h}{R} \times$ length of base, where R, the radius of the earth, may be taken as 20,900,000 feet.

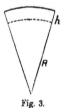

Fig. 3.

In this case $h$ = 4,520.

Then correction $= -\dfrac{4,520 \times 5,301}{20,900,000} = -1\cdot146$.

Collecting these corrections :—

|  | Feet. |
|---|---|
| i. and ii. Standard and temperature | −·0·928 |
| iii. Inclination | − 0·550 |
| iv. Sag | − 0·000 |
| v. Height above sea | − 1·146 |
| Total correction | − 2·624 |
| Measured length | 5301·240 |
| Reduced length of base | 5298·62 |

It is clear that corrections iii. and iv. must always be negative; i. and ii. may be either positive or negative.

Where there is time, topographical bases would always be measured as carefully as possible, and the corrections would be worked out as above; but in rapid surveys for military or exploration purposes, where time is of the first importance, many of these refinements must be given up, and a double measurement with tape or chain must suffice, the chain lengths being marked with arrows, the data for the computations being, if possible, noted, but the computations being left to a more convenient season.

**Base of Verification.**—In topographical or exploratory surveys, it is as well, after the triangulation has reached its intended limit, to measure another base connected with the triangulation. Two values will be obtained for this base, one by direct measurement, and one by computation from the original base. This affords a check on the accuracy of the work.

In all bases for topographical work the governing condition is, that their inaccuracy should not be sufficient to introduce any error capable of being shown on the scale of the survey in the furthest extension of the work depending on them.

**Base-line Figures.**—It is important that the extension of a triangulation from a base should be composed of well-conditioned triangles. In topographical work where the base is about 2 miles long and the sides of the triangles will eventually be 20 miles long or so, this requires some care.

Plate I shows the extension from one of the geodetic bases in South Africa. The principle of base extension is just the same in topographical work.

**Measurements of Greater Precision.**\*—The comparatively rough methods above described will give an accuracy of from about $\frac{1}{10,000}$ to $\frac{1}{50,000}$. When greater precision is required the following method may be adopted:—The base should be measured with an Invar tape (see next paragraph). The tape should be allowed to hang in a natural catenary, being supported on fine steel edges fixed to very firm tripods. The tension is given by means of a weight (say of 20 lbs.) attached to the end of the tape which passes over a pulley on a tripod beyond the knife-edge tripod. Attached to the other extremity of the tape is a weight of 30 lbs. resting on the ground, and the tape at this end also passes over a pulley. Convenient lengths of tape are 25 and 50 metres. The tape is graduated in the neighbourhood of the knife edges, and the readings of the graduations resting on the edges are read simultaneously. The tape should then be taken off, replaced, and another measurement made.

The point on the tripod is transferred to the ground (at the ends of the base) by a simple optical arrangement.

The *sag* of a tape varies as the cube of the span of the wire, hence it is easy by comparing the readings of the complete catenary, and those resulting when the tape is supported at 3 or 4 equidistant points, to deduce the sag for any given space. (Appendix V.)

On an incline ($\theta$) the sag varies as $\cos^2 \theta$.

The usual corrections for standard, inclination, and height above sea, must of course be applied.

There is no difficulty in obtaining a probable error of $\frac{1}{100,000}$ by this method.

Before use the tapes should be tested at the National Physical Laboratory, Teddington, or at the Ordnance Survey Office, Southampton.

**Invar.**—This is an alloy of steel and nickel containing 36 per cent. of nickel. It is due to the researches of the late Dr. J. Hopkins, of Mr. Benoit, and of Mr. C. E. Guillaume, especially of the latter. Mr. Guillaume has published an account of "Nickel-Steel Alloys and their Application to Geodesy," in the 'Report of the Thirteenth Conference of the International Geodetic Association, 1901.' The particular alloy we are dealing with has been called *invar* by Mr. Thury. Its importance consists in the fact that it has the smallest known

---

\* This method is used in Australia and New Zealand and a similar system has been employed very successfully by Mr. E. P. Cotton, Commissioner of Crown Lands in Lagos. See also a pamphlet by Mr. C. W. Adams, of Blenheim, New Zealand; Wellington, John Mackay, Government Printer.

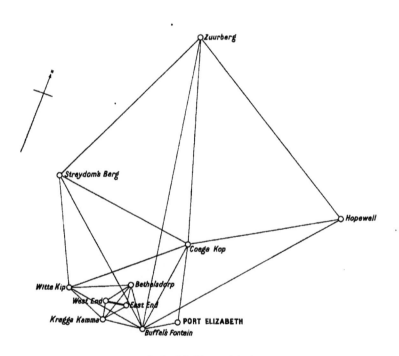

expansion of any metal or alloy, *i.e.*, $\frac{1}{10}$ that of platinum, or $\frac{1}{20}$ that of brass. (Less than $\frac{1}{1,000,000}$ for 1° Centigrade.)

Tapes of this metal are now available, 25 metres long, or of special lengths if required.

The agent for the sale of Invar in England is Mr. J. H. A. Baugh, 92, Hatton Garden, London. The metal requires special treatment in manufacture and is liable to slow residual variations of length; it is therefore necessary that tapes used for the measurement of accurate bases should be tested by exact standards before and after measurement.

The use of Invar for the measurement of good topographical bases renders negligible the errors caused by uncertainty of tape temperatures.

## TRIANGULATION.

**Selection of Stations.**—Before the measurement of the angles of the triangulation of a regular survey is commenced, the site of each station is selected and the exact spot marked in some permanent manner. The positions of these stations are so chosen that the triangles shall be *well conditioned*, i.e., no angle of a triangle should be less than 30° or greater than 120°. The nearer each triangle approaches to being equilateral the less error will be introduced into the calculations by any inaccuracy in the observed angles.

**Marks and Mark Stones.**—For geodetic work in India it is customary to build an isolated pillar and under this to bury a mark stone, on top of this another mark stone is plumbed with the surface flush with the ground; over this a masonary pillar is built, and the whole surrounded with a cairn of stones.

For topographical work less elaboration is necessary; it is sufficient, if the surface is rock to cut a small dot surrounded by a ring; otherwise a large stone is buried with its surface flush with the ground and a similar dot and ring are cut on it. In places where this is liable to be disturbed, it is a wise precaution to bury a tile, stone, or brick about 2 feet down, as is done in the tertiary triangulation of the Ordnance Survey.

**Signals to observe to.**—For geodetic work these are always luminous, either heliotropes, paraffin lamps with parabolic reflectors, lime-light, acetylene lamps, or electric light; the latter was used in connecting Spain and Algeria, one of the rays being over 200 miles long.

For good secondary work also the signals would be luminous.

For topographical and geographical triangulation the signals would usually be opaque, with perhaps an occasional use of a heliostat or heliotrope.

Where the sides are only a few miles long the following signals may be used:—

(1) Pole and brush, as used in India;
(2) Ordnance Survey signal;
(3) West African basket signal.

For longer rays, say those of 10 to 30 miles, large tripods or quadripods must be erected. These should be 30 feet high, and the upper part down to within 7 feet of the ground should be filled in with brushwood or covered with matting, so as to present a good mark at a long distance. The quadripod is slightly better than the tripod, as it is a symmetrical mark for intersection. (See Plate II.)

In forest-clad country it is sometimes possible to cut down all the trees on a hill top save one; a solitary tree on a hill top is an excellent mark.

For short rays of 7 miles or so, a white flag, or white covered signal is a great help in picking up the point.

In rapid triangulation, the tops of hills, rocks and conspicuous trees are used, as nature may provide.

**Size of Triangles.**—In geodetic triangulation in hilly country the sides would average about 30 to 35 miles; in the plains, about 11 miles.

The secondary triangulation, if independent, about the same size as for geodetic work; if for breaking up geodetic work, about 5 to 8 miles.

In tertiary triangulation the sides would be a mile or two long (average in United Kingdom, 1¼ miles).

In rapid triangulation, try and keep the sides from 10 to 20 miles long.

For a topographical triangulation a convenient rule is that the trigonometrical stations

## TRIGONOMETRICAL SIGNALS
## FOR SHORT DISTANCES

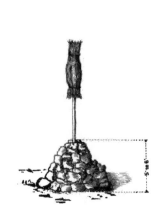

INDIAN POLE AND BRUSH SIGNAL

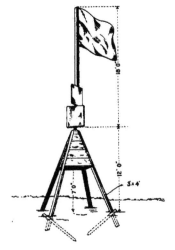

QUADRIPOD SIGNAL

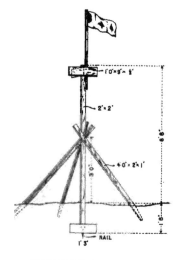

TRIGONOMETRICAL POLE

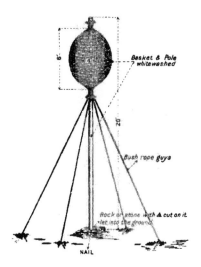

WEST AFRICAN BASKET SIGNAL

should be some 4 or 5 inches apart on paper. Thus for triangulation for a 1-inch map the sides would be 4 or 5 miles long, for a ¼-inch map some 16 to 20 miles long. But so much will depend on the nature of the country that this rule must only be treated as a rough guide.

**Chain or Net.**—When the extension of the base is completed in a large country, such as India or South Africa, the geodetic triangulation is carried out in the form of chains of triangles, running for preference N. and S., and E. and W., thus :—

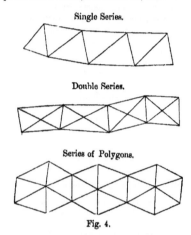

Fig. 4.

Secondary chains are run to fill up the spaces between the principal chains.

In small countries, such as the United Kingdom, the country is covered with a network of principal triangles.

Triangulation executed rapidly during a march, or along a frontier during a boundary survey, will usually take the form of a chain, but, of course, less symmetrical and less well conditioned than a principal or secondary chain.

A topographical triangulation, which is essentially required to provide points for the detail surveyors, will necessarily cover with a net the whole country to be surveyed.

**Intersected Points.**—In tertiary, topographical, and rapid triangulation, angles are often observed to objects which are not visited ; such objects are called intersected points, and should always be fixed by at least three rays ; the extension of the triangulation should not, except in the case of rapid triangulation, depend on these points, which are, however, very often most useful to the detail surveyors ; such objects will be spires, tall chimneys, pagodas, or inaccessible rocks or trees.

**The Theodolite.**—The angles of a triangulation are invariably observed with a theodolite, which is an instrument devised for the measurement of horizontal and vertical angles.

Theodolites are designated by the diameter in inches of their engraved horizontal circles.

In times past in geodetic triangulation theodolites have been used with circles as much as 36 inches in diameter ; but improvements in graduation of the circles have made it possible to obtain even better results with far smaller instruments. In Burma, for instance, the geodetic triangulation was executed with a 12-inch, and in South Africa with a 10-inch theodolite.

For ordinary topographical work it may be assumed that the triangulation will be carried out with one of the following instruments :—

    (1) 5-inch micrometer transit theodolite.
    (2) 6-inch vernier    ,,    ,,
    (3) 5-inch   ,,    ,,    ,,

A transit theodolite is one in which the telescope can be rotated vertically end for end, and in which the vertical circle is graduated throughout its circumference.

Verniers and micrometers are different means of reading graduated circles.

**Verniers.—**

*Principle of the Vernier.*—If the length of, say, three divisions of a primary scale be taken and divided into four (fig. 5) it is evident that the length of each of the latter is ¾ that of a primary division and that the difference between the length of one of the four divisions and

Fig. 5.

the length of a primary division is ¼ of a primary division, and generally if $(n - 1)$ primary divisions be divided into $n$ divisions to form a vernier, then the difference in length between a primary and a vernier division is $\frac{1}{n}$th of a primary division.

The same result obtains if we take, say, five divisions of a primary scale and divide it into four for a vernier (fig. 6) each of the vernier divisions is then $\frac{5}{4}$ of a primary division,

Fig. 6.

and the difference between their lengths is, as before, ¼ of a primary division, and generally if $n + 1$ primary divisions be divided into $n$ vernier divisions, the difference between the length of a primary and a vernier division is $\frac{1}{n}$th of a primary division.

For the application of this principle see fig. 7 below, which represents a portion of the graduated arc of a theodolite.

Fig. 7.

In this example the lower set of numbers represents degrees; these in turn are divided into six divisions, each representing 10 minutes.

The upper set of figures represents minutes; these are also divided into six divisions, each being equal to 10 seconds. To read the present example it will be seen that the arrow of the vernier plate comes between the second and third divisions of the lower plate, which indicates the reading to be between 20′ and 30′.

PLATE III.

5" MICROMETER TRANSIT THEODOLITE.

To obtain the correct reading, look along the vernier plate and see where a division on that plate coincides with a division on the horizontal circle. In this instance a coincidence occurs at the second line to the left of the figure 8 on the vernier. Thus the reading would be 20′ on horizontal plate + 8′ on vernier + 20″ on vernier, and the reading would be entered in the observation book as :—

0° 28′ 20″.

Always note the direction in which the numbers increase. The largest number on the vernier shows the number of parts into which the scale is divided.

Thus, if a theodolite vernier has 30 for its greatest number, then it reads to $\frac{1}{30}$ of a division of the scale, and if this scale is divided into 30-minute spaces, the vernier reads to minutes.

If no divisions exactly coincide, split the difference between those which most nearly agree. Thus if the 20″ and 30″ nearly coincide but each is a little out of coincidence in opposite directions, the reading will be 25″.

**Micrometer Microscopes.**— Another means of reading the angle shown on a graduated circle is the micrometer microscope. Having focussed the eye-piece of the microscope so that the wires and limb graduations are quite distinct (by pushing in or pulling out the eye-piece), the field of the microscope will present the appearance shown in fig. 8 (a). The firm upright

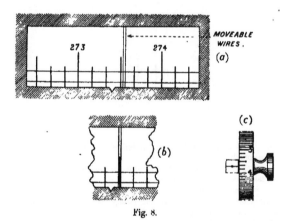

Fig. 8.

lines are the divisions of the limb, the numbers are degrees, and the limb in this example is divided into 10-minute spaces. At the bottom of the field will be seen a V-shaped notch in the frame of the diaphragm. It is the position of this notch which it is required to measure.

It will be at once seen that the notch is somewhere between 273° 20′ and 273° 30′. Two fine vertical lines will be noticed; these can be moved laterally by turning the graduated micrometer head, of which a separate sketch is shown in fig. 8 (c). These should be moved laterally until they occupy a position one on each side of a graduation on the limb, as in (b), and the graduation selected should be that which is nearest to the notch; it is immaterial which side. The minutes and seconds can now be read off the graduated head of the micrometer (c); in this example the graduated head reads 4′ + between 10″ and 20″, by estimation 4′ 17″. The complete reading is therefore 273° 24′ 17″.

**Taking Runs.**—It is part of the principle of this apparatus that a lateral movement of the parallel wires from one division of the limb to the next should be accomplished in a complete revolution of the micrometer head. This should be tested occasionally before beginning work; the test is called "taking runs." If the mean of two or three runs does not differ from the proper amount by more than 2 or 3 seconds, the error, for topographical work, may be neglected. Should there be a large error, it means that the microscope is not properly adjusted.

**Adjustment for "Run."**—The adjustment will be described as applied to a 5" Micro. microscope theodolite by Messrs. Troughton & Simms. (See Plate III.)

Take runs four times each on each microscope. Determine the error of the mean of the two microscopes, and correct this by altering one microscope twice the amount. Thus, supposing that the mean of two microscopes, instead of reading 10' in traversing the wires from one division of the limb to the next, reads 10' 5", correct one of the microscopes so that it shall read 10" less than it did in a run, *i.e.*, the image of the limb in the field of the microscope must be diminished in size.

Now, if it is wished to increase or diminish the power of the microscope, it will be necessary to increase or decrease the distance between the object-glass and the wires.

The microscope has at the end a screw in the body, *b*, a clamping ring embracing the

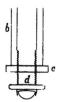

Fig. 9.

object-glass cell, *c*, and a cell *d* carrying the object-glass, the cell having a long screw cut on it for the purpose of adjustment.

First release the ring *c*, then move *d* towards the limb if it is desired to increase the readings, then take care to make fast again.

A little adjustment for focus will be necessary after every alteration; this is done by releasing the screw in the aluminium socket and sliding the whole microscope towards or from the graduations as may be necessary.

In practice, it is found that the microscopes very rarely need adjustment. Instruments of this class have recently been much used on boundary commissions and even on active service. Their advantages over vernier instruments are, that they are more accurate and more quickly read; no one accustomed to micrometers would willingly go back to verniers

**Examination of Theodolite.**—On receiving a theodolite, examine it to see if it is in good order.

Inspect the stand and see that there is no shakiness in any part of it. When the feet are shod with metal it sometimes happens that shrinkage of the wood allows a little play in one of the feet, which is quite enough to vitiate the observations.

The theodolite box should be opened, the positions of the object and eye ends of the telescope marked on the box, and the places of the tangent screws of the upper and lower plates carefully noted.

**Setting up at a Station.**—Supposing, now, that the observer has reached the station of observation with his instrument.* The first thing to do is to place the stand with its legs wide apart, so as to give a firm base, and with its centre exactly over the station dot, its head being as nearly level as the eye can judge. Dust the grooves of the tribrach, and, having taken out the lower portion of the theodolite—remembering always to hold it by the strongest parts—place it gently on the stand, and clamp it to the tribrach.

It is well at this stage to level the upper plate approximately, as it is quite possible that it may be necessary to alter the position of the legs in order to avoid bringing undue strain on the foot-screws. If the legs are moved, care must be taken to keep the instrument centred over the station dot.

Having centred and approximately levelled the lower portion of the theodolite, open the cleats and dust the Y supports of the transit axis of the telescope. Take out the telescope, wipe the transit axis and lower it gently into the Y's : close the cleats and screw them down. Take the cap off the object-glass and focus the telescope. Dust with a soft camel's hair brush the horizontal and vertical circles. Set the vernier or microscope lenses. Make quite sure that there is no shake in any part of the theodolite or stand, and that the feet of the latter rest upon firm ground, into which there is no fear of their sinking during the observations. In loose or marshy soil, hammer in pickets on which the feet of the theodolite may rest.

See that the tangent screws are in about the centre of their runs.

### THE ADJUSTMENTS OF THE THEODOLITE.

**Adjustments of the Telescope, Focus and Parallax.**—To focus the telescope, first remove the cap from the object-glass. If this moves on a hinge, no remark is necessary ; but if it pulls off, care must be taken to draw it with a screwing-up motion, to avoid any chance of the object-glass being partially unscrewed.

The telescope should now be pointed to the sky, and the eye-piece pushed in or pulled out, also with a screwing-up motion, until the wires are seen with perfect distinctness.

The telescope is now directed on to some distant and well-defined object, and the object-glass moved in or out until distinct vision is obtained. When this has been done, the foci of the object-glass and the eye-piece should coincide on the plane of the wires.

To test the focus, move the eye slightly to one side and note if the object appears to move relatively to the wires. If it does not move, the focus is perfect; but if there is movement, it is due to imperfect focus or *parallax*, which must be eliminated before observing is commenced. To effect this, note whether the object moves with the eye or against it—*i.e.*, if when the eye is moved to the left, the object moves to the left or to the right of the wires. If the former, it shows that the focus of the object-glass is beyond the wires, and the object-glass must therefore be brought nearer to the observer. If the latter, the focus is too near, and the object-glass must be moved further out.

From what has been said above, it will be seen that the proper position of the eye-piece varies with the eyesight of the individual observer, but not with the distance of the object ; while the position of the object-glass is unaffected by the eyesight of the observer, but varies with the distance of the object.

The object-glass, however, will be set at the same position, called *stellar or solar focus*, for all objects so far distant that the rays falling on it from them may be taken as practically parallel to one another.

In the case of a theodolite of the class described, this applies to all objects not less than half a mile from the observer ; but for nearer objects at any distance over 250 yards the change required is very slight.

* For very accurate work with large instruments isolated pillars are built. In this case it is assumed that the legs of the stand rest upon the natural soil or rock.

When the proper adjustment of the eye-piece and of the object-glass for the solar focus have been found, it is convenient to mark the tubes, so that on subsequent occasions focus may be found at once.

**Adjustment for Collimation in Azimuth.**—A theodolite is said to be collimated in azimuth, when the *line of collimation*, i.e., the imaginary straight line joining the intersection of the wires and the optical centre of the object-glass, cuts the transit axis or axis of *vertical* motion at right angles.

In fig. 10 let $tt_1$ denote the transit axis, then when the line of collimation, which for shortness call $c$, coincides with the perpendicular to $tt_1$, or the line $pp_1$, the instrument is in collimation, and if the telescope be made to describe a semicircle in its Y's, its line of collimation produced will pass through the zenith, describing a great circle. But if $c$ be represented by $ow$, then the semicircle traced out will be represented by $ow_1$, the false zenith being $z_1$, while the arc described will be that of a small circle. In other words, by changing face, the object end, or $o$, will arrive at $o_1$, while the wires, $w$, will describe a similar small circle and reach $w_1$. Whence if any object on the horizon give a horizontal reading at face left of $o^o$, and then the pivots of the instrument be changed and the telescope be turned sufficiently in azimuth to intersect the object, the two readings thus obtained will, instead of exactly coinciding, disagree by $ow_1$.

The error in collimation will, therefore, be $\dfrac{ow_1}{2} = op.$*

*Description of the Diaphragm.*—Before entering into the methods of collimating, a short description of the diaphragm is necessary.

The wires, the arrangement of which varies in different theodolites, are attached to a metal ring or diaphragm, whose position in the tube of the telescope can be regulated by

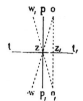

Fig. 10.

means of four capstan-headed screws, $a$, $b$, $c$, $d$ (see fig. 11), which pass loosely through slots, $S$, in the tube, so as to allow the diaphragm to be moved as required. Each of these slots is covered by a small plate to exclude dirt.

If it is necessary to move the diaphragm towards $b$, loosen the screw $a$ and slacken the screws $c$ and $d$; then by screwing $b$ into the tube, the diaphragm is drawn in its direction. When it has been moved sufficiently, tighten up the three screws $a$, $c$, $d$.

If an ordinary eye-piece is used with the usual inverting telescope, with the result that the image is seen upside down, it must be remembered that, if the object that has to be intersected appears to the left of the wire, the wire must be shifted to the left, and the diaphragm moved accordingly.

---

* A theodolite is said to be face left or face right when the vertical circle is on the observer's left or right. Face is changed by turning the telescope through 180 degrees vertically so as to bring the object-glass where the eye-piece was before. If now the instrument is turned in azimuth so as to bring the eye-piece to the observer's eye, the vertical circle will be found to be on the opposite side to that it occupied previously with referen to the observer.

**Methods of Collimating in Azimuth.**—There are three methods of collimating in azimuth:—

(1) By changing pivots. This process would be used for all small theodolites, the telescope of which cannot revolve through a complete vertical circle.
(2) By changing face. This is applicable to those theodolites only whose telescope can revolve through a complete vertical circle (*i.e.*, transit theodolites).
(3) By Gauss's method. This is applicable to all theodolites, but as it requires two other telescopes fitted with cross wires it is comparatively seldom available, and will, therefore, not be treated of further.

*Collimating by Changing Pivots.*—Throw open the cleats and see that the clip screws of the index arm are barely biting their shoulder and no more. Intersect the object azimuthally, and note the reading, taking care that the clamps are holding firmly. Next, release the clip screws, raise the telescope gently and revolve the transit axis through 180 degrees, so that the level comes underneath, leaving the index arm and clip screws upwards. Replace the telescope gently with the pivots changed and with the object-glass still pointing towards the object. If the wires in this second position of the telescope still intersect the object, the instrument is in collimation; if not, complete the intersection by the tangent screw of the upper plate and note the second reading. Set the instrument to the mean of the two readings, and then, looking into the telescope, move the diaphragm by its collimating screws $a$, $b$, $c$, $d$, fig. 11, until the

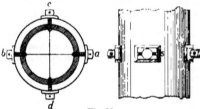
Fig. 11.

wires again intersect the object. Repeat the process until the intersection remains constant on changing pivots, remembering that half the apparent is the real error.*

*Collimating by Changing Face.*—Intersect and note the reading of any fixed object, change face, intersect and read again. The two readings should differ by exactly 180 degrees. If they do not, the error is corrected as above.

In collimating by changing face, any errors that may exist in the graduation of the limb are involved in the process.

In collimating an instrument, it is desirable that the object selected for intersection should be at such a distance that the object-glass be set to the focus which will be employed during the subsequent observations, as the line of collimation is liable to be distorted when the focus is altered.

**To make the Vertical and Horizontal Wires respectively Vertical and Horizontal.**—As the correct positions for these wires are cut on the diaphragm by the maker so that the lines cut each other at right angles, it follows that to adjust one wire is to adjust both, and this may be done by the following method:—

Level the instrument with care, intersect any small well-defined object with the vertical wire, and see if it continues bisected along the wire when the telescope is moved in a vertical plane. If this is not the case, the collimating screws must be slackened sufficiently to allow the diaphragm to be revolved until this condition is secured, when they must be again

---

* The same result can be obtained by estimating half the error as it appears in the telescope and dispensing with the readings.

tightened. It will now be found that the horizontal wire, if properly placed by the maker, will continue to bisect an object on which it has been placed, when the instrument is turned in azimuth.

**Levelling the Theodolite.**—This consists in making the vertical axis truly perpendicular to the horizon.

Place the longer of the levels, $o$, of the horizontal plate parallel to any two foot screws, $c$, $c$, and by suitably turning these latter bring the bubble into the centre of its tube. In doing this, the observer must remember always to turn the foot screws in opposite directions, that is to say, to work both hands either towards or away from himself. Then turn the upper plate 180 degrees in azimuth. Supposing the bubble to occupy a different place from what it did before, move it by means of the same two foot screws into the mean of these two places, and it is evident that if the upper plate be turned another 180 degrees until it occupies its original position, the bubble will continue to occupy that mean place, called the *zero of the level*. Now turn the upper plate 90 degrees in azimuth, and by means only of the third and hitherto unused foot screws bring the bubble into its mean place.

The instrument is now approximately level, and the above described process must be continued until the level remains steady in whatever direction the telescope may be pointed. The mean place is not necessarily in the centre of the tube, but if the deviation is considerable the bubble can be brought into the centre by means of the two drawing screws which secure the level to the horizontal plate. The length of the bubble varies inversely as the temperature, and consequently the positions of the ends of the bubble alter with a change of temperature. It is convenient to have one end of the level marked, and to watch that end only while levelling.

The levelling is now as perfect as it can be made by the body level. As the level attached to the microscope arms is usually more delicate, the final levelling can be completed with this, by bringing the bubble approximately into the centre of its run by the clip screws, and then proceeding exactly as above described.

**Level on Microscope Arms.**—If the level is on the vernier or microscope arms, it is necessary to have it well adjusted.

To do this, level the instrument; bring the bubble into the centre by means of the clip screws, $ZZ$; intersect the object and note the reading; change face and set the bubble by the clips as before, and intersect again. The mean of the two readings will be the true elevation. Set the vertical circle to this reading and intersect the object by the clip screws, and, keeping the instrument clamped, bring the bubble of the level into the centre by means of its drawing screws. It is clear that where the error is small the discrepancy between the two faces may be eliminated in a similar way to that used for collimation in azimuth. But the method just described is better.

When a level is carried on the vernier arms of the vertical limb, it is desirable to know the value of the level scale, for the purpose of correcting the observed vertical angles. See p. 19.

*To Level the Transit Axis.*—Level the upper plate. Place the striding level on the pivots, and, by tapping it gently sideways, bring the bubble of the cross level into the centre of its run.

When the bubble of the striding level is steady, read and register the number of the level scale at which one end of it stands, say the end nearer to the cross level.

Now reverse the level end for end, cross level, read and register the same end of the bubble as before. If the transit axis is level, the same end of the bubble will read the same on both occasions. If it does not read the same, half the apparent is the real error, which can be corrected by the capstan-headed screw, $k$, which raises or lowers the Y above it.

Unless the error of collimation and the dislevelment of the transit axis are very large, it is better not to try to correct them, as their effect can be eliminated by a suitable method of observation.

DETERMINATION OF VALUE OF ONE DIVISION OF LEVEL SCALE.

The level is usually graduated, but, if not, make a paper scale of equal parts, reading outwards from the centre towards both ends, and paste it on to the side of the level, so that the centre of the scale corresponds with the centre of the run of the bubble, assuming that the level is on the vernier arms (as it is only in that position that it affords a means of correcting the vertical angles); by means of the clip screws move the bubble up to one end of its run, say towards the object end, so that the object end of the bubble corresponds approximately with the extreme reading of the scale. Intersect with the horizontal wire some convenient object for observing. Read and record one end of the bubble, say the object end, and the vertical angle. Now, by means of the clip screws, bring the bubble back towards the eye as far as you can, taking care that it is really floating and within the graduations of the scale. Re-intersect the same object as before, and record the vertical angle and the reading of the object end of the bubble in its new position. The difference between the two readings of the object end of the bubble gives the dislevelment in terms of divisions of the scale, and the difference between the two vertical angles gives the same dislevelment in minutes and seconds of arc. Dividing this angular measurement by the number of divisions of dislevelment, you obtain the value in arc of one division of the scale.
Thus:—

|  | Elevation. | Object end of bubble. |
|---|---|---|
| 1st observation | 7  3  28 | 18 |
| 2nd       ,, | 7  0   0 | 5 |
| Difference | 0  3  28 | 13 |
| Value of one division = $\frac{208''}{13} = 16''$. | | |

This operation must be repeated several times in order to get a good mean value. The bubble of a level is very susceptible to changes of temperature, so care must be taken that it is not exposed to such changes while this operation is being performed. Should there be any chance of the bubble altering its length while you are determining the value of the divisions of the scale, it will be necessary to read and record both ends of the bubble. In observing, as described previously, for each vertical angle taken, the readings of both ends of the bubble must be recorded. *To apply the correction*, the rule is as follows:—

Divide the difference between the sums of the readings of the object end and eye end by the total number of readings, and the result will be the dislevelment in terms of divisions of the scale. Multiply this result by the angular value of one division of the scale, and you obtain the angular correction for dislevelment to be applied to the mean vertical angle. Supposing two observations are taken to a point one face left and one face right, and the readings are as follows:—

|  | O. | E. |
|---|---|---|
| F.L. | 5 | 8 |
| F.R. | 7 | 6 |
|  | 12 | 14 |

In this case the sum of the readings of the eye ends exceeds that of the object end by 2. The number of readings of ends of the bubble is 4. So to get the dislevelment in terms of division of the scale we must divide 2 by $4 = \frac{1}{2}$. Suppose the value of one division of the level scale is 16 seconds, then to get the correction we must multiply $\frac{1}{2}$ by 16 seconds = 8 seconds. The 8 seconds of arc must be applied to the mean of the two observed angles. With regard to its sign, the eye end being in excess, the correction must be subtracted from an elevation and added to a depression. If the object end were in excess, the process would, of course, be reversed.

Generally correction to altitude $= + \dfrac{O - E}{\text{number of readings}} \times$ value of 1 division.

If the level graduations read from one end and not outwards from the centre, book the readings of the zero end as negative, and then apply the above rule.

**Method of Observing Horizontal Angles.**—Observations taken to elevated objects are subject to two great sources of error, viz., dislevelment of the transit axis and want of adjustment of the line of collimation. These causes of error are generally large in small instruments, and, although capable of rectification, adjustments do not long remain perfect, hence the system of observation should be such as to cancel these errors. This is attained by changing face, *i.e.*, by observing the same angle with the vertical circle or face alternately to the left and to the right. This system should be adopted in all theodolite observations, for it is as effectual as it is simple.

The angles at a station are taken thus : supposing the observer to be at B, and that it is required to observe the angles to the signals at C, D, E.

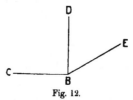

Fig. 12.

Clamp the upper plate to the lower, having set A vernier or microscope to about 0° on the horizontal limb. Loosen the lower clamp and turn the whole instrument till station C comes into the field of the telescope; now tighten the lower clamp, and intersect C with the lower tangent screw ; see that the vernier still reads about 0°. Now loosen the upper clamp and turn the upper plate and telescope completely round till C comes into the field of view, taking care not to overshoot it ; clamp the upper plate and intersect.

Read off both verniers or microscopes and enter in angle book. The reading of A vernier should be identical with the setting. If the difference is greater than can be accounted for by a slight difference in the intersection, there must be a defect in the instrument, which should be examined.

Now loose the upper clamp, and turn the telescope on to D, clamp, intersect with the upper tangent screw without overshooting. Read the verniers or microscopes, and book the readings. Repeat the same process with E, and so on, with a continuous movement from left to right ; this is termed **"swinging right."**

It is generally desirable to intersect the zero station again at the end of the swing, so as to detect any movement of the instrument which may have occurred. If such has occurred, the observations must be taken again. In some surveys the starting point is called the R.O., or Referring Object ; there is no necessity to use this term in topographical work.

Suppose that during these observations the vertical circle had been to the right of the telescope; this set would be described as one round of angles, **face right**, swing right, on zero 0°. The **zero** or **arc** is the setting of the first station or signal read during a round of angles. In the geodetic triangulation of Burma, for instance, the successive rounds of angles were taken on the following zeros: 0° 0′, 20° 1′, 40° 2′, 60° 3′, and so on, 9 zeros in all. The object of changing zero is to minimize the effects of errors of graduation, and to prevent any bias due to reading the same angle constantly on the same portion of the arc.

The number of changes of face, of swings, and the number of zeros will depend on the class of the work.

In geodetic triangulation the number of zeros will probably be at least 6, with 2 faces on each zero, 2 swings on each face.

In secondary work, probably 2 or 3 zeros each of 2 faces and 2 swings.

In tertiary, 2 zeros, 2 faces, 1 swing on each face.

In topographical work the numbers will naturally vary with the time available. The following system is generally suitable:—

    Zero 0°, F.R., swing R.,
      „   0°, F.L.,    „  L.,
      „  35°, F.L.,    „  L.,
      „  35°, F.R.,    „  R.,

or 4 rounds of angles at each station.

But if time does not allow of so many, then

    Zero 0°, F.R., swing R.,
      „  35°, F.L.,    „  L.

The reason for taking 35° as the second zero is that, whether there are 2 or 3 microscopes or verniers, they will not fall on changing zero over the same part of the arc. But any arbitrary change—not 60° nor 180°—will do.

**Method of Observing Vertical Angles.**—To observe vertical angles, see that the instrument is perfectly levelled. Direct the telescope on the object, leaving the horizontal clamp free; clamp the vertical circle when the object comes into the field, and intersect with the tangent screw. It is generally convenient to intersect with a portion of the horizontal wire at one side of the intersection of the cross-wires. If this is done, care must be taken to intersect with exactly the same spot on the wire when the telescope is reversed on changing face, remembering that if the spot appeared to be on the left of the centre F.L., it will be on the right when the object is observed F.R.

Read and record both the verniers, noting if the angle is an elevation or a depression. If there is a level on the vernier arms, enter the readings of both ends of the bubble under the headings of O or object end, and E or eye end.

If the object was intersected from *above* the first time, it should be intersected from *below* when a second observation is taken.

Now change face and observe the object F.R. in exactly the same manner. The mean of the F.L. and F.R. observations should give the true elevation or depression. If, however, there is a level on the vernier arms, and the sum of the readings of the two ends do not exactly agree, a correction will have to be applied. See p. 20.

The height above their station dots of the instrument and of the objects observed should be entered.

A vertical angle is of very small value if it has not been observed on both faces.

Generally, when taking horizontal angles leave the vertical arc unclamped, and when taking vertical angles leave the horizontal arc unclamped. This implies that the horizontal and vertical angles should not be taken simultaneously, and that the object should not be observed on intersection of the cross-wires, but separately on vertical and horizontal wires. If time presses, however, this restriction may be ignored, and vertical and horizontal angles may be taken together.

**General Remarks on Observations.**—It is a general rule of all regular surveys that all the observations should be entered in ink before the observer leaves his station. A description of each station should also be entered in the angle book, in order to enable any other observer or the detail surveyor to recognise it without difficulty.

This description should contain not only such information as would enable anyone to identify both the general locality and the exact spot used as a station, but should also state the most convenient camping ground in the neighbourhood, and the best direction from which to ascend the hill in the case of a hill station. The nature of the mark left should also be given, and its height above the station dot.

**Satellite Stations.**—It sometimes happens that angles are required from some point, such as a tree, steeple, temple, or pagoda, where it is impossible to erect the theodolite. This case never occurs in geodetic or good secondary triangulation, but it is not very uncommon in rapid work, where the stations have not been laid out beforehand, and the observer is dependent on natural objects for his marks. In such cases the theodolite is set up at the nearest convenient point, called a *satellite station*, at a short distance from the object, and

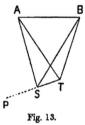

Fig. 13.

the angles taken from this station can afterwards be reduced to the true values at the object itself.

Supposing, now, that the triangulator has observed the trunk of a tree on the summit of a hill, and that on his arrival he finds that it is impossible for him to observe with his theodolite nearer to the tree than several feet; he would put up his theodolite in the most convenient place, as near to the tree as possible, and observe his round of angles as usual, taking care to include the centre of the trunk of the tree. He must also measure accurately the distance from his station to the trunk of the tree. He can now correct his angles in either of two ways :—

(1) If a number of angles have been taken from the station, and only two or three have been taken to the tree, and if it is possible to raise upon the station a mark sufficiently conspicuous to be visible from future stations, it is preferable to correct the two or three angles already taken to the tree.

(2) If, on the other hand, only a few angles have been taken at the station, and if it is impossible to erect on it as useful a mark as the tree, then it is advisable to reduce the angles taken at the station to what they would have been had they been taken on the spot occupied by the tree. This is called reduction to the centre.

Let AB be two stations from which the tree, T, has been observed, and let S be the station or spot where the theodolite was put up.

(1) In the first case the angles SAB and ABS are required, the angles TAB and ABT being known. Evidently SAB = TAB + SAT and $\sin SAT = \sin TSA \frac{ST}{AT}$. AT can be computed approximately or measured with sufficient accuracy from the working chart of the triangulation.

(2) In the second case, which is the usual one, ATB is required. Produce TS to P.
Then
$$BSP = BTS + BTS \quad \ldots \ldots \ldots \quad (1)$$
$$ASP = ATS + SAT \quad \ldots \ldots \ldots \quad (2).$$

Subtracting (2) from (1)

ASB = ATB + SBT − SAT, whence ATB = ASB + SAT − SBT.

Now $\sin SAT = \sin TSA \frac{ST}{AT}$, AT being found in any approximate way, such as measurement from the chart of triangulation, and $\sin SBT = \sin TSB \frac{ST}{BT}$, BT being found in any approximate way.

## COMPUTATIONS.

**Lengths of the Sides.**—Suppose AB to be the measured base, C a point in the triangulation, the angles A, B, C, having been observed.

Now since the surface of the earth is a curved surface, the horizontal angles at A, B, and C have been observed in planes which are very slightly inclined to each other, and, if perfectly observed, the sum of A, B, and C will exceed 180° by a small quantity known as the **spherical excess**. In geodetic and secondary triangulation this quantity is always computed and subtracted from the sum of the three angles, and any remaining excess or defect on 180° is an error to be adjusted by an elaborate process which will not here be described.

But in tertiary and topographical triangulation the procedure is very simple, viz.: to divide the excess or defect on 180° equally between the three angles. The angles thus corrected are the *corrected angles*.

By elementary plane trigonometry, we have

$$b = c \times \frac{\sin B}{\sin C} \quad \text{and} \quad a = c \times \frac{\sin A}{\sin C},$$

$c$ being the known base.

The most convenient arrangement for logarithmic computation is shown on p. 24; the ends of the known side being written first and the stations entered in order in the direction of motion of the hands of a watch.

The other triangles are solved in succession in the same way.

It is sometimes, though not often, necessary to compute the third side of a triangle of which two sides, $a$, $b$, and the included angle C are known. The following formula is used:—

$$\tan \frac{A - B}{2} = \frac{a - b}{a + b} \cot \frac{C}{2},$$

whence $\frac{A - B}{2}$ is found, and

$$\frac{A + B}{2} = 90 - \frac{C}{2}.$$

From these A and B are found, and the triangle can be solved in the ordinary way.

It is desirable, whenever possible, to obtain more than one value for the length of a side, the mean being taken in subsequent computations.

FORM A.—(S.M.E. Topographical Survey, Form I.)

## Computation of Sides of Triangles.

Formula: $\dfrac{\sin A}{a} = \dfrac{\sin B}{b} = \dfrac{\sin C}{c}$.

| Triangle. | Letter of station. | Name of station. | Mean of observed angles;—or satellite angles reduced to centre. | Correction. | Corrected angles. | Logarithmic computation. | Sides (log. feet). |
|---|---|---|---|---|---|---|---|
| | | | ° ′ ″ | ″ | ° ′ ″ | | |
| 14 | A<br>B<br>C | Darnet Fort<br>Callum Hill<br>Stoke Ch. | 96 50 26<br>29 38 26<br>53 30 59 | + 3<br>+ 3<br>+ 3 | 96 50 29<br>29 38 29<br>53 31 2 | log AB = 4·3937724<br>L sin = 9·9968967<br>L sin = 9·6942276<br>L cosec = 10·0947247 | 4·4353938 (BC)<br>4·1827247 (AC) |
| | | | 179 59 51 | + 9 | 180 0 0 | | |
| 15 | D<br>B<br>C | Naval Hospital<br>Callum Hill<br>Stoke Ch. | 56 28 36<br>55 16 26<br>68 15 26 | −10<br>− 9<br>− 9 | 56 28 26<br>55 16 17<br>68 15 17 | log DB = 4·5323728<br>L sin = 9·9209741<br>L sin = 9·9147976<br>L cosec = 10·0320580 | 4·4854049 (BC)<br>4·4792277 (CD) |
| | | | 180 0 28 | −28 | 180 0 0 | | |

**Co-ordinates or Latitudes and Longitudes.**—The next step in the computations is to put the results in a form which permits of accurate plotting, and which enables a record to be kept of the position of each trigonometrical point.

This can be done in two ways :—

I. By rectangular co-ordinates;
II. By latitudes and longitudes.

**I. Rectangular Co-ordinates.**—If the distance east or west of the starting point is not likely to exceed 100 miles, and there is no probability of geographical positions being required accurately, nor of there being any further extension of the triangulation, and if the azimuths of the sides are not likely to be wanted, proceed as follows :—

(a) Calculate by plane trigonometry the lengths of all the sides of the triangles.

(b) Select one station as the initial station and the meridian through that station as the initial meridian, then, treating the earth's surface as a plane, calculate the *co-ordinates* of all the stations, the co-ordinates being measured along and at right angles to the initial meridian, thus—

If $\alpha$ is the azimuth, or true bearing of the first side, AB, A the initial point, then $x_1$ (the co-ordinate at right angles of the meridian) = AB sin $\alpha$,

$$y_1 = AB \cos \alpha.$$

$$\left.\begin{array}{l} x_2 = BC \sin \beta \\ y_2 = BC \cos \beta \end{array}\right\} \text{where } \beta = ABC - \alpha.$$

Then co-ordinates of A are 0, 0.

    ,,   ,, B ,, $x_1$,   $y_1$.

    ,,   ,, C ,, $x_1 + x_2$, $y_1 - y_2$, &c

All these points can now be plotted with reference to the initial point, A and its meridian

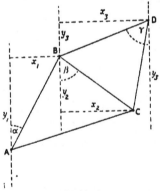

Fig. 14.

It is important to note that once the initial meridian has been left, the angles $\beta$, $\gamma$, &c., formed by the sides with lines parallel to the meridian, are no longer the true bearings of these lines, and should never be used as such.

It is clear that if a chain of triangles runs in an oblique direction, it may be convenient to plot along, and at right angles to, a line of known bearing, running generally in the same direction.

**II. Latitudes and Longitudes.**[*]—If the distance east or west of the starting point is likely to exceed 100 miles, or if the geographical positions of the stations are of importance, or if the azimuths of the sides are likely to be wanted, or if the triangulation is likely to be extended in the future, then calculate the latitude and longitude of each trigonometrical station, having regard to the true shape of the earth, as follows :—

It is first necessary to know the latitude and longitude of some trigonometrical station, and the azimuth (*i.e.*, true bearing) of some ray from that station. The latitude, longitude, and azimuth are either determined astronomically (see Chapter XIV.) or have been brought up by former triangulation. In the case of astronomical determination there is no difficulty in obtaining the required latitude and azimuth, but the longitude may present some difficulty, and an approximate value may have to be assumed. This will not, in any way, affect the accuracy of the work in itself.

Suppose now that A and B are two stations, and that we know :—

$\lambda$ = latitude of A,
L = longitude of A,
A = azimuth of B from A,
$c$ = distance in feet at mean sea-level between A and B,

then we require $\lambda_b$ = latitude of B,
$L_b$ = longitude of B,
B = azimuth of A from B.

If we put $\lambda_b = \lambda + \Delta\lambda$,
$L_b = L + \Delta L$,
$B = \pi + A + \Delta A$,

and if $\Delta\lambda = \delta_1\lambda + \delta_2\lambda$,
$\Delta L = \delta_1 L + \delta_2 L$,
$\Delta A = \delta_1 A + \delta_2 A$,

then it can be shown that

$\delta_1\lambda = P \cos A \cdot c$, $\qquad\qquad\delta_2\lambda = \delta_1 A \cdot R \cdot \sin A \cdot c$,
$\delta_1 L = \delta_1\lambda \cdot Q \cdot \sec \lambda \tan A$, $\qquad\delta_2 L = \delta_2\lambda \cdot S \cdot \cot A$,
$\delta_1 A = \delta_1 L \sin \lambda$, $\qquad\qquad\delta_2 A = \delta_2 L \cdot T$,

which are the expressions used in the form for computation.

It will be noticed that each term is found consecutively from the preceding one.

The quantities P, Q, R, S, T will be found by interpolation from Table I. The proper signs to be given to $\delta_1\lambda$, $\delta_1 L$, &c., will be found in Tables II. (*a*) and (*b*).

**Examples of the Computation of Latitude, Longitude, and Reverse Azimuth.**— This is shown on Form B. In this example Sugar Loaf is the station of which the latitude and longitude are known, Blorenge that of which they are required. The distance Sugar Loaf—Blorenge is known and also the azimuth of the ray Sugar Loaf—Blorenge. *Azimuth for this computation is always to be reckoned from south by west.*

In the example given, the longitude of the known station was *west* of Greenwich, and was therefore booked as a negative quantity; east longitude is booked as a positive quantity. The signs of the quantities $\delta_1\lambda$, $\delta_2\lambda$, $\delta_1 L$, $\delta_2 L$, $\delta_1 A$, $\delta_2 A$ are to be booked exactly as found in Tables II. (*a*) and (*b*).

Suppose that we start with a triangle ABC, of which the latitude and longitude of A are known, the azimuth of AB, and the lengths of all the sides :—

First compute the latitude and longitude of B and reverse azimuth BA. Then, by adding

[*] See also Appendix III.

## Form B.—(S.M.E. Topographical Survey Form III.)
### Latitudes, Longitudes, and Reverse Azimuths.

| | | | | |
|---|---|---|---|---|
| Station A | | Sugar Loaf. | | ° ′ ″ |
| Lat. of A ($\lambda$) | | 51 51 44·00 | | |
| Long. of A (L) | | W. 3 3 23·00 | | |
| Station B | | Blorenge. | | |
| Az. B from A (A) | | 1 23 58 | | ° ′ ″ |
| log AB (c) in feet | | 4·3556240 | | |
| | | | | |
| P | | 3·9939483 | | |
| log cos A | | 1·9998704 | | |
| log c | | 4·3556240 | | |
| | | | | |
| log $\delta_1 \lambda$ | | 2·3494427 | | |
| Q | | 1·0988671 | | |
| log sec $\lambda$ | | 0·2093240 | | |
| log tan A | | 2·3879194 | | |
| | | | | |
| log $\delta_1$ L | | 0·9455538 | | |
| log sin $\lambda$ | | 1·8957142 | | |
| | | | | |
| log $\delta_1$ A | | 0·8412680 | | |
| R | | 8·37848 | | |
| log sin A | | 2·38779 | | |
| log c | | 4·35562 | | |
| | | | | |
| log $\delta_2 \lambda$ | | 5·96317 | | |
| S | | 0·50922 | | |
| log cot A | | 1·61208 | | |
| | | | | |
| log $\delta_2$ L | | 2·06447 | | |
| T | | 0·01267 | | |
| | | | | |
| log $\delta_2$ A | | 2·09714 | | |
| | | | | |
| $\lambda$ | | 51 51 44·00 | | ° ′ ″ |
| $\delta_1 \lambda$ | | − 0 3 43·58 | | |
| $\delta_2 \lambda$ | | − 0 0 0·00 | | |
| | | | | |
| Lat. of B | | 51 48 0·42 | | ° ′ ″ |
| | | | | |
| *L | | W. − 3 3 23·00 | | ° ′ ″ |
| $\delta_1$ L | | − 0 0 8·82 | | |
| $\delta_2$ L | | + 0 0 0·01 | | |
| | | | | |
| Long. of B | | − 3 3 31·81 | | ° ′ ″ |
| | | | | |
| $\pi$ + A | | 181 23 58 | | ° ′ ″ |
| $\delta_1$ A | | − 0 0 6·9 | | |
| $\delta_2$ A | | + 0 0 0·01 | | |
| | | | | |
| Az. of A from B | | 181 23 51 | | ° ′ ″ |

* When A is W. of Greenwich, give L the negative sign.

the angle at B to this last obtain the azimuth of BC, and, knowing now the latitude and longitude of B and azimuth BC, calculate the latitude and longitude of C.

Now, as a check, calculate the latitude and longitude of C direct from A.

And generally, to avoid arithmetical errors, compute the latitude and longitude of each point from at least two rays.

Latitudes and longitudes should be worked out to two places of decimals of a second.

The latitudes and longitudes of all the stations of a triangulation are thus computed successively from those of previously computed stations. These latitudes and longitudes are definite and sufficient records of the positions of the stations, and the positions can be plotted on any scale required.

**Trigonometrical Heights.**—Having computed the lengths of the sides and the co-ordinates or latitudes and longitudes, the next step is to calculate the heights of the stations above mean sea-level. For this purpose some initial height is necessary, which may have been determined by former triangulation, or may be determined by direct observation of the mean sea-level or by careful barometer readings at sea-level and at the point in question (see p. 87).

We will assume that the height above mean sea-level of one station of the triangulation is known, and that vertical angles have been taken at each station to all stations visible therefrom, and that these vertical angles have been observed with a theodolite on both faces, and that the means have been corrected for level error.

All vertical angles require a correction for refraction. The ray of light from a distant object lies, as a rule, in such a curve as to make the object appear higher than it really is. The amount of this deflection varies with the place, season, and time of day. Sometimes, but rarely, it makes an object appear lower than it really is.

In fig. 15 A and B are the summits of two hills, AR, BR' the apparent directions of the distant stations (due to refraction). As AR, BR' are portions of an arc of small curvature, BAR and ABR' are sensibly equal. Make OC = OA. Then it is required to find BC the difference of height of A and B.

Let the observed depressions at A and B be $D_a$ and $D_b$.

BAC = OCA − ABO  and

BAC = BAO − CAO,  and since  OCA = CAO,

BAC = ½ (BAO − ABO) = ½ (RAO − R'BO) (since BAR = ABR') = ½ ($D_b$ − $D_a$).

Now as AC is a very small part of the earth's surface,

BC = AC tan BAC (sensibly);  or

$$h = C \tan \frac{D_b - D_a}{2},$$ or if one angle is an elevation

$$h = C \tan \frac{E_a + D_b}{2}, \text{ or } C \tan \frac{E_b + D_a}{2},*$$

where

$h$ is the required difference in height
$C$ the distance between the stations
$D_a$, $D_b$ the angles of depression at A and B

It has been assumed above that the reciprocal observations were taken simultaneously. This is usually impracticable, so in order to attain the same result as far as possible, vertical angles should be taken (when time allows) about the same time of day, that time being between noon and 3 p.m., when refraction is fairly constant.

* See Form C, p. 31.

Where reciprocal angles have not been observed, it is necessary to adopt some approximate value for the refraction (BAR).

From fig. 15 it is clear that, calling the refraction $r$,

$$r = \frac{c - D_a - D_b}{2},$$

where $c$ is the arc of the earth's surface between A and B.

So that whenever reciprocal angles have been observed the amount of the refraction can be found.

The result of a very large number of observations all over the world is to show that the

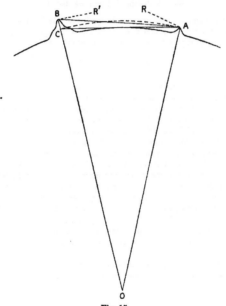

Fig. 15.

quantity $\frac{r}{c}$, called the **coefficient of refraction**, is fairly constant for a given country and climate.

| | | | | | |
|---|---|---|---|---|---|
| The mean coefficient of refraction for South Africa is | | | | ... ... | ·064 |
| ,, | ,, | ,, | ,, Burma... | ... ... ... | ·067 |
| ,, | ,, | ,, | ,, the United Kingdom (for rays not crossing the sea) | ... | ·075 |
| ,, | ,, | ,, | ,, the United Kingdom (for rays crossing the sea) | ... | ·081 |

Generally the coefficient of refraction may be taken as ·070.

Having adopted a suitable coefficient of refraction, it is clearly easy to compute the effect produced by curvature and refraction combined. The labour saving table on p. 31 has been computed for a coefficient of ·070. Its use is as follows :—

If a vertical angle has been observed at only one of the stations, if this is a depression call it D, if an elevation E.

Then look out the value of K in the Table, corresponding to the distance between the stations.

Let
$$S = K - D \text{ or } K + E,$$
then the difference in height between the stations
$$= \tan S \times \text{distance between the stations.}$$

To get the exact height, a correction must be applied for the heights in feet above their station dots of the instruments at A and B, $i_a$ and $i_b$, and of the signals at A and B, $g_a$ and $g_b$.

This correction, to be applied to the already obtained difference of height,
$$= + \frac{g_a - g_b}{2} + \frac{i_a - i_b}{2}.$$

Two examples of the computation of heights are given on Form C; one where reciprocal angles have been observed, and the other where only one angle has been observed.

**Datum Level.**—All heights on a map should be referred to mean sea-level. The mean sea-level for the Ordnance Survey of the United Kingdom was determined at Liverpool. The datum level of the Survey of India is the mean sea-level at Karachi.

**Mean Sea-Level.**—This is the mean height of the surface of the sea during an indefinitely long period. As determined by the Survey of India, the mean sea-level is the mean of the hourly heights extending over a period of 5 years. Five years is chosen so as to cover the cycle of recurrence of astronomical phenomena producing tidal effect. This is the mean sea-level which should be used in all civilized countries having an organized survey.

It will often, however, be impossible on a topographical survey to determine mean sea-level so accurately. In such a case, if there are any tidal records at the port chosen, the mean of high and low-water heights throughout a year will give a good value.

If there are no tidal records, fix a graduated pole in the sea and observe the heights of high and low water during the progress of the survey, the period to extend over a definite number of lunations.

Under any circumstances the period must extend over at least one lunation.

This latter method will give only a rough value for mean sea-level.

### TRIGONOMETRICAL INTERPOLATION.

In rapid triangulation during military operations, or on an expedition, or with a boundary commission, many artifices have to be employed which it would be improper to use on a regular topographical survey, in the execution of which rapidity is not a governing condition. A very valuable expedient in such cases is **interpolation**, or the determination of the observer's position by means of observations taken to previously fixed points.

There are three methods of interpolation :—

(a) By observing the horizontal angles from 1 unfixed point to 3 known points.
(b) ,,   ,,   ,,   2 ,, points to 2 ,,
(c) ,,   ,,   ,,   1 ,, point to 2 ,,
with the help of an azimuth.

FORM C.—(S.M.E. Topographical Survey Form II.)
## Computation of Trigonometrical Heights.

DATA.

A is the Station of which the height is known.
B   "    "    "    "    required.
$c$ = Distance in feet between A and B.
$D_a$ = Vertical angle of depression at A.
$D_b$ =   "    "    "    "    B.
$E_a$ =   "    "    elevation  A.
$E_b$ =   "    "    "    "    B.
$i_a$ = Height of instrument above ground at A.
$i_b$ =   "    "    "    "    B.
$g_a$ = Height of signal    "    "    A.
$g_b$ =   "    "    "    "    B.
$K$ = Correction for curvature and refraction (see Table below).

GENERAL FORMULÆ.

Height of B = Height of A + $c$. tan S + $\frac{g_a - g_b + i_a - i_b}{2}$.

(I.) Where S or Subtended Angle = $\frac{D_b - D_a}{2}$.

If either angle is an elevation, it is to be treated as a *negative* quantity in the above formulæ.
Signs must be carefully attended to.
Where reciprocal vertical angles have been taken, no corrections for curvature and refraction are required.

(II.) If one angle only is observed—
S = K − D  or  K + E.

| DATA. | | | DATA. | | |
|---|---|---|---|---|---|
| Station (A): | Pabiya. | | Station (A). | Chiltan. | |
| Known height of A | | 642·2 | Known height of A | = | 453 |
| Station (B) | Sung. | | Station (B). | Murdah. | |
| Length $c$ (in feet) | = | 50,540 | Length $c$ (in feet) | = | 53,420 |
| | | | | | |
| I. +$E_a$ or −$D_a$ | = | − 1° 17′ 20″ | I. +$E_a$ or −$D_a$ | = | ° ′ ″ |
| +$D_b$ or −$E_b$ | = | + 0  57  30 | +$D_b$ or −$E_b$ | = | |
| 2 | | 0  19  50 | 2 | | |
| ½ Algebraic sum (± S) | = | − 0  9  55 | ½ Algebraic sum (± S) | = | |
| | | | | | |
| II. +$E_a$ or −$D_b$ | = | ° ′ ″ | II. +$E_a$ or −$D_b$ | = | 0° 5′ 10″ |
| K (from Table) | = | | K (from Table) | = | 0  3  48 |
| Algebraic sum (± S) | = | | Algebraic sum (± S) | = | + 0  8  58 |
| | | | | | |
| Log tan S | = | 3·4600930 | Log tan S | = | 3·4168581 |
| Log $c$ | = | 4·7036352 | Log $c$ | = | 4·7277039 |
| Log $h$ | = | 2·1637282 | Log $h$ | = | 2·1440620 |
| $h$ (±) | = | −145·79 | $h$ (±) | = | +139·93 |
| $\frac{g_a - g_b + i_a - i_b}{2}$ * | = | + 8·3 | $\frac{g_a - g_b + i_a - i_b}{2}$ * | = | + 4·25 |
| Known height of A | = | 642·2 | Known height of A | = | 453·0 |
| Deduced height of B | = | 504·71 | Deduced height of B | = | 596·61 |

TABLE.—Value of K.

These values are obtained by using a coefficient of refraction of ·07.

| Distance ($c$) in feet. | K. | Distance ($c$) in feet. | K. | Distance ($c$) in feet. | K. |
|---|---|---|---|---|---|
| 1,000 | 0′ 4″ | 8,000 | 0′ 34″ | 60,000 | 4′ 15″ |
| 2,000 | 0  9 | 9,000 | 0  38 | 70,000 | 4  58 |
| 3,000 | 0  13 | 10,000 | 0  43 | 80,000 | 5  40 |
| 4,000 | 0  17 | 20,000 | 1  25 | 90,000 | 6  23 |
| 5,000 | 0  21 | 30,000 | 2  8 | 100,000 | 7  5 |
| 6,000 | 0  26 | 40,000 | 2  50 | | |
| 7,000 | 0  30 | 50,000 | 3  33 | | |

* Where only one angle is observed at A to the *ground line* at B, the correction will be ÷ $i_a$.

(a) **Fixing the position of a point by angles taken from it to 3 previously fixed points**; in this there are three cases, as below.

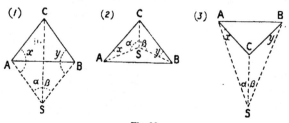

Fig. 16.

Let S be the observer's position, $\alpha$ and $\beta$ observed angles, then

$$x + y = 360 - (\alpha + \beta + C) \text{ in (1) and (2)};$$
$$x + y = C - (\alpha + \beta) \text{ in (3)}.$$

Let

$$\tan \phi = \frac{\sin \alpha}{\sin \beta} \times \frac{a}{b},$$

then

$$\tan \frac{x - y}{2} = \tan (\phi - 45°) \tan \frac{x + y}{2}.$$

NOTE.—In case (1) if S, A, B, C, are on the circumference of the same circle the problem is insoluble.

(b) **Fixing the position of 2 points by angles taken from them to 2 previously fixed points.**

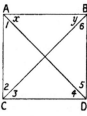

Fig. 17.

A and B are known points. The angles 2, 3, 4, 5 are observed. 1 6 are deduced. Then, if

$$P = \frac{\sin 1 \sin 3 \sin 5}{\sin 2 \sin 4 \sin 6},$$

$$\tan \tfrac{1}{2} (y - x) = \tan \tfrac{1}{2} (3 + 4) \frac{1 - P}{1 + P}.$$

(*c*) **Fixing the position of a point by angles taken from it to 2 known points and an azimuth.**

Fig. 18.

Let A and B be 2 known points. At S observe the angle ASB and the azimuth of SA or SB (by the sun will do. See p. 148).

Get some approximate value for the position of S from the plane-table, or by traverse.

Then compute reverse azimuths of AS and BS, as below

$$dA = dL \sin \lambda,$$

where   $dA$ = increment of azimuth in seconds,

$dL$ = difference of longitude in seconds of arc,

$\lambda$ = mean latitude.

The way to apply the quantity $dA$ can easily be seen by considering in which direction the meridians converge. Then knowing the azimuths of AS, BS, and that of AB (which is a known line), the differences give the angles A and B, and in the triangle ASB we know the three angles and one side (AB).

NOTE.—Method (*a*) and method (*b*) are much strengthened if an azimuth is observed. The use of an azimuth enables two solutions to be found in each case, and a check is thus obtained. It is a general maxim in interpolation that azimuths should be observed when possible.

## CHAPTER IV.

### TRAVERSES.

**Definition.**—A traverse consists of a connected series of straight lines (on the earth's surface) of which the lengths and bearings have been determined.

**Different Forms of Traverse.**—Traverses may be carried out in different ways, according to the accuracy required.

Where great precision is aimed at, the linear measurements are made with a chain or steel tape, and the included angles are measured with a theodolite.

Where rapid work is desired and where accuracy is not so important, the angles may be measured with a prismatic compass, and the distance either chained, paced, or taken by means of a perambulator, or by time.

These two classes of traverses are known as:—

(1) Theodolite traverses;
(2) Compass traverses.

The latter are dealt with in Appendix VII.

**Theodolite Traverses.**—The principal varieties of theodolite traverses are:—

A. Traverses between points already fixed trigonometrically.
B. ,, used on the line of march during an expedition.
C. ,, as a substitute for triangulation in a country unsuited to the latter.

#### A. TRAVERSES BETWEEN TRIGONOMETRICAL POINTS.*

This class of traverse is used:—

(*a*) On the Ordnance Survey of the United Kingdom, for large-scale town surveys.
(*b*) In India for cadastral surveys.
(*c*) For certain large-scale engineering surveys.
(*d*) For forest surveys.
(*e*) For the detailed survey of special routes, roads, railways, river banks, &c., &c.
(*f*) To fix points for detail survey which cannot be provided by triangulation.

The starting and finishing points of these traverses are trigonometrical points already fixed.

**Errors.**—Traverses are liable to errors of two kinds:—

(1) Errors of direction. These are due partly to inaccurate observations of the angles, and partly to bad centering of the theodolite, and also of the signals over the points marking the traverse stations.
(2) Linear errors, due to faulty chaining.

**Traverse Stations.**—These stations should be as far apart as possible consistent with the ultimate object of the traverse; that is to say, in the case of traverses along winding roads the positions of the stations are governed to a great extent by the fact that off-sets from the traverse line to detail along the road should not exceed a certain length.

All stations should be carefully marked, and where possible a picket of about 2" × 2" section should be driven in flush with the ground. A tintack should then be driven into the picket to show the exact centre point. The further the stations are apart, the less will be the effect produced by bad centering of the theodolite, or signal, over these points.

---
* Including the case where there are no trigonometrical points, but the traverse closes on the starting point.

**Signals.**—The sight vane commonly in use consists of a staff about 6 feet long and about 3" × 1" section, having a cross vane (a black cross on a white ground) which can be clamped at any height. The difficulty is to hold this staff vertically over the centre point. It is best, therefore, in order to minimise the errors in direction when taking horizontal readings, to make the cross-wires of the theodolite bisect the bottom of the staff at the ground level. In this case the horizontal reading should first be taken and recorded, after which the vertical angle to the cross vane, clamped to the height of the instrument, may be read. Where the distance between the traverse points is small, it increases the accuracy of the work, if the horizontal readings are taken on the tintack itself, if visible, or else on an arrow or nail held over the same.

There is an excellent sight vane in use on the Ordnance Survey, consisting of a tripod and staff with cross vane attached. This last can be adjusted so as to be vertical over the centre points.

**Observations.**—We will assume that we have to run a traverse along a certain defined route, starting at A and finishing at B, where A and B are stations mutually visible (fig. 19).

Proceeding to A, the theodolite is carefully centred over the station mark in the ground, and then levelled. The horizontal plates are now clamped together so that the reading on the A vernier is anything between 0° 0′ 0″ and 0° 10′ 0″ (it is best not to try to set it to an exact reading), and with the plates so clamped the theodolite is directed on point B (the R.O.) and the signal bisected by the cross hairs. The bottom plate is clamped and the horizontal verniers are read and recorded.

Now release the top plate and direct the theodolite on to traverse station No. I, whither a man has been sent on with a sight vane clamped to the height of the theodolite. The intersection of the cross hairs must then be carefully directed on the sight vane. The horizontal verniers are now read, and the verniers of the vertical arc, also the vertical level.

A complete observation has now been taken on one face of the instrument.

Now reverse the instrument and, setting the A vernier at about 35° 0′, proceed in precisely the same manner as just described.

This completes the observations at point A. The theodolite is then taken to traverse point I, where, after careful centering and levelling, a similar set of observations is made. Here the trigonometrical point A becomes the R.O. Each traverse station is then visited in turn, the station the observer has just left becoming the R.O. in each case. On arriving at the point B, the R.O. will be the last traverse station, and forward readings taken on point A (vide fig. 19).

Form D shows how the readings at traverse point VI would be recorded in the field book. It is unnecessary to observe the vertical angles to the back station. The details at VI correspond with those on Form E.

**Linear Measurements.**—The linear measurements are made with a chain or steel tape, preferably the latter. The distance between traverse stations should be taped twice, to ensure accuracy. A pull of 12 lbs. is to be applied when taping; this is done by means of a spring balance.

Where the slopes between stations are practically continuous, surface chaining should be employed. The reduction to the horizontal is made afterwards by means of the vertical angles observed.

Where the ground is broken, horizontal chaining must be resorted to. In this case the observation of vertical angles is unnecessary.

**Computations of Co-ordinates.**—In order to plot a traverse correctly, it is necessary to work out the co-ordinates of the station points referred to some origin. For the class of traverse under consideration, rectangular co-ordinates are always used. It is found convenient to calculate the ordinates ($y$) for north and south, and the abscissæ ($x$) for east and west.

Form D.

**Traverse Book.**

|  |  | (Horizontal.) |  |  | (Vertical.) |  |  |  |
|---|---|---|---|---|---|---|---|---|
|  |  | A. | B. |  | C. | D. | E. | O. |
| F.R. | V | 0° 3′ 20″ | 3′ 30″ |  | — — | — — | — | — |
|  | VII | 122° 15′ 20″ | 15′ 20″ | VI | 4° 30′ 0″ | 30′ 10″ | 7 | 5 |
| F.L. | V | 35° 5′ 10″ | 5′ 10″ |  | — — | — — | — | — |
|  | VII | 157° 17′ 5″ | 16′ 55″ |  | 175° 29′ 40′ | 29′ 50″ | 4 | 8 |

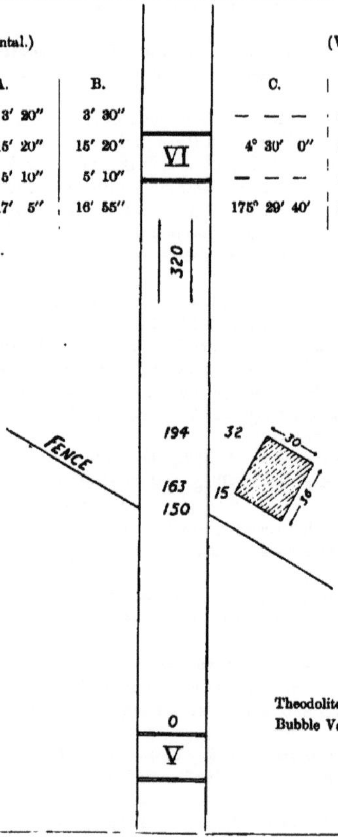

320

194  32
163  15
150

FENCE

Theodolite No. 101.
Bubble Value. 1 Div. = 20″

0

V

Form E is the form used for the computation of these co-ordinates. The traverse is that shown in fig. 19, the starting and finishing points, A and B, being mutually visible. The observations taken at point VI are shown on Form D.

Now A and B being points of a triangulation, the horizontal distance AB and also the true bearing of AB will be known.

The legs of the traverse together with the line AB form an enclosed figure, and the sum of all the interior angles should be equal to twice as many right angles as the figure has sides, less four right angles. We thus have a check on the observed horizontal angles.

It should be carefully noted that the included angles at the stations are always measured, either all to the right or all to the left of the back station.

This is an invariable rule. In the case of fig. 19 they are all measured to the left of the back station. Take point VI, for instance. Form D shows that the mean of the differences of the readings on V and VII is 122° 11′ 52″·5. The horizontal plate of a theodolite, however, is always graduated in the same direction as a watch. Hence we must deduct our result from

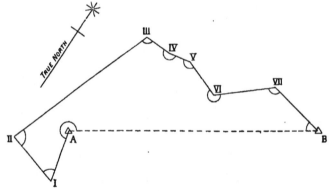

Fig. 19.

360°, which gives 237° 48′ 7″·5. The same must be done with all the angles before entering them in column 2, Form E. It is obvious that this last operation is unnecessary where the required angles are all measured to the right of the back station.

**Angular Error.**—The sum of all these interior angles then should be equal to $(2n \times 90 - 360)$ degrees, where $n$ is the number of sides to the figure. Any discrepancy between these two is called the "total angular error." This error must then be distributed equally among the observed angles. For example, if there are 6 sides to the figure, and the sum of the observed angles comes out to 719° 59′ 0″, then the error to be distributed is 1′. Hence 10″ must be added to each observed angle.

Columns 2, 3, and 4, Form E, are self-explanatory.

In this class of traverse it is quite unnecessary to take into account the convergence of the meridians, owing to the distance AB being short. Hence the bearing of each leg is a "false bearing," as computed in column 5.

It may here be stated that, before making any computations at all, it is best to draw a rough diagram of the traverse, such as in fig. 19.

Column 5 is computed thus:—the bearing A to I is 360° − (249° 17′ 14″·2) + (59° 4′ 33″),

## Form E.—(S.M.E. Form N.)

**Traverse.**

**Computation of Co-ordinates.**

Starting Point △ (A) ⎫ Mutually visible.
Closing Point △ (B) ⎭

*Data.*—Hor. Dist. AB = 1995·78 ft.
True Bearing of AB = 59° 4′ 33″.

| 1. | 2. | 3. | 4. | 5. | 6. | 7. | 8. | 9. | 10. | 11. | 12. | 13. | 14. | 15. | 16. | 17. |
|---|---|---|---|---|---|---|---|---|---|---|---|---|---|---|---|---|
| Stations. | Angles observed at Stations. | Angular Error. | Corrected Angles. | False Bearing of Traverse Lines. | Vertical Angles. | Distances in Feet. | Horizontal Distance in Feet. | \multicolumn{4}{c|}{Deduced Co-ordinates.} | Errors of Co-ordinates of Closing Point. | \multicolumn{4}{c|}{Corrected Co-ordinates from Origin.} | |
| From | To | ° ′ ″ | ′ ″ | ° ′ ″ | ° ′ ″ | ° ′ ″ | | | North-ings. | South-ings. | East-ings. | West-ings. | | N. | Corr. | S. | Corr. | E. | Corr. | W. | Corr. |
| A | B | 0 0 0 | | | 59 4 33 | | | 1995·78 | 1025·64 | | 1712·08 | | | 1025·64 | — | | | 1712·08 | — | | |
| A | I | 240 16 7·5 | +1 6·7 | 240 17 14·2 | 169 47 19 | E. 3 16 11 | 429·9 | 429·2 | | 422·4 | 76·90 | | | | | 422·555 | +·155 | 76·43 | +·35 | | |
| I | II | 60 21 0 | +1 6·7 | 60 23 6·7 | 288 22 12 | E. 0 10 20 | 476·0 | 475·8 | 157·87 | | | 449·86 | | | | ·31 | 264·94 | | | | 372·25 |
| II | III | 86 23 39·2 | +1 6·7 | 86 23 39·2 | 22 58 12 | E. 0 41 40 | 1362·1 | 1351·9 | 1254·0 | | 531·32 | | | 868·005 | ·465 | | | 159·63 | 1·08 | | ·72 |
| III | IV | 111 49 7·5 | +1 6·7 | 111 50 14·2 | 91 7·50 | E. 0 5 0 | 210·1 | 210 | | 4·13 | 209·92 | | | 864·7 | ·62 | | | 369·91 | 1·44 | | |
| IV | V | 191 24 7·5 | +1 6·7 | 191 25 14·2 | 79 42 44 | E. 0 3 20 | 164·0 | 164 | 29·29 | | 161·36 | | | 1013·335 | ·775 | | | 531·43 | 1·8 | | |
| V | VI | 144 6 43 | +1 6·7 | 144 7 51·7 | 113 34 53 | D. 2 1 33 | 259·0 | 258·43 | | 138·17 | 238·43 | | | 815·31 | ·93 | | | 820·61 | 2·15 | | |
| VI | VII | 237 48 7·5 | +1 6·6 | 237 49 14·1 | 57 43 39 | D. 4 30 20 | 501·5 | 500 | 266·73 | | 422·91 | | | 1142·385 | 1·085 | | | 1243·87 | 2·5 | | |
| VII | B | 133 47 0 | +1 6·6 | 133 48 6·6 | 103 57 32 | E. 4 56 36 | 483·9 | 482·2 | | 116·29 | 467·85 | | | 1025·64 | 1·24 | | | 1712·08 | 2·96 | | |
| B | A | 44 31 52·5 | +1 6·6 | 44 52 59·1 | | | | | 1707·89 | 681·01 | 2158·28 | 449·06 | Closing errors are + 1·24 N. and − 2·96 E. ∴ corrections S. { +·E and W. { } − ·E. | | | | | | | |
| Total | | 1256 50 0 | | 1260 0 0 | | | | B (deduced) | 681·01 | | 449·06 | | | | | | | | | | |
| Correct | | 1250 0 0 | | | | | | | 1026·88 | | 1709·22 | | | | | | | | | | |
| | | 0 10 0 | | | | | | Error + 1·24 | | | − 2·96 | | | | | | | | | |
| | | 0 1 6·6 | to be added. | | | | | | | | | | | | | | | | | |

NOTE.—There are 9 sides to the figure ∴ sum of included angles = 2 × 90 × 9 − 360 = 1260°.

or 169° 47' 18"·8. The next bearing is got by adding 180° to this last, and then subtracting the corrected angle at II; and similarly for all the rest. An examination of the figure will always reveal the correct method of computing the bearing.

Column 6 gives the vertical angles of elevation or depression; column 7 the distance chained along the surface. The horizontal distances in column 8 are obtained thus: —

Hor. dist. = surface dist. × cosine E (or D).

The co-ordinates, Form E, are computed with A taken as the origin. From the triangulation data we know the length AB to be 1995·78 feet, and the bearing of AB 59° 4' 33". The co-ordinates of B will then be 1995·78 × cos 59° 4' 33", or a north co-ordinate of 1025·64; and 1995·78 × sin 59° 4' 33", or an east co-ordinate of 1712·08.

The next step is to compute in this manner, by using the bearings in column 5 and the distances in column 8, the co-ordinates of each point of the traverse referred to the last point as origin.

We have now a set of deduced co-ordinates referred to parallel axes throughout, in columns 9, 10, 11, and 12. Thus the co-ordinates of III are 1254·0 N. and 531·52 E., referred to II as origin. Also the co-ordinates of II are 157·87 N. and 449·06 W., referred to I as origin.

It is easily seen that where the bearing of the line lies between certain values, the deduced co-ordinates must be measured from the axes in certain directions, thus:—

| Bearings. | Deduced co-ordinates. |
| --- | --- |
| 0 to 90 | North and East |
| 90 ,, 180 | South and East |
| 180 ,, 270 | South and West |
| 270 ,, 360 | North and West |

Now we see from the form that the co-ordinates of I referred to A are 422·4 S. and 76·09 E.; and the co-ordinates of II referred to I are 157·87 N. and 449·06 W. Hence the co-ordinates of II referred to A are—

$$\begin{array}{ccc} 422\cdot 4 & & 449\cdot 06 \\ 157\cdot 87 & & 76\cdot 09 \\ \hline 264\cdot 53 \text{ S.} & \text{and} & 372\cdot 97 \text{ W.} \end{array}$$

Thus it is evident that the co-ordinates of any point referred to the primary origin (in this case A) are obtained by taking the algebraic sum of the deduced northings and southings, and of the eastings and westings, up to the point in question.

These co-ordinates from the origin are entered in columns 14, 15, 16 and 17, with certain corrections to be now described.

On computing the co-ordinates of B by the algebraic sum of the deduced co-ordinates of the points throughout the traverse, we find that they come to 1026·88 N. and 1709·22 E. Now, from the data, the co-ordinates of B are 1025·64 N. and 1712·08 E; hence the total northing at B is too large by 1·24, and the total easting too small by 2·86.

In this class of traverse a rigorous adjustment of such errors is unnecessary. The simplest way to distribute the errors of the co-ordinates is as follows:—

If

$n$ = number of legs of traverse,

$y$ = error to be adjusted N. and S.,

$z$ = error to be adjusted E. and W.,

then the several corrections to be applied to the points of the traverse in order will be

$$\frac{y}{n}, \frac{2y}{n}, \frac{3y}{n} \ldots \frac{ny}{n},$$

$$\frac{x}{n}, \frac{2x}{n}, \frac{3x}{n} \ldots \frac{nx}{n}.$$

In the example on Form E the corrections to northings are all negative, and, therefore, to southings all positive. Similarly the corrections to all eastings are positive, and to westings negative.

For example, the corrections to be applied to the computed co-ordinates of VI are

$$\frac{6}{n} \times 1\cdot 24 = \cdot 93 \text{ on northing,}$$

and

$$\frac{6}{n} \times 2\cdot 86 = 2\cdot 15 \text{ on easting.}$$

The algebraic sum of the deduced co-ordinates up to this point gives 876·44 N. and 818·46 E. The corrections are entered in the columns provided, on the left of columns 14 and 16, and applied to above results. The corrected co-ordinates are then entered in columns 14 and 16.

Thus the computed traverse co-ordinates of B have been made to agree with the trigonometrical values; and those of the other points have been adjusted in proportion.

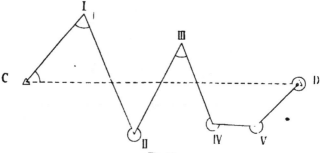

Fig. 20.

It oftens happen that the traverse crosses and re-crosses the line joining the two terminal points, as in fig. 20.

In such a case, where the traverse is started at C, the angle required at C must be under 180°. In fig. 20 this angle is measured to the left of the back station D. As stated before, all the other angles must be so measured.

Then, as before,

$$\Sigma\theta = 2n \times 90° - 360°$$

if the traverse crosses CD an *even* number of times.

Where, however, the traverse crosses CD an *uneven* number of times,

$$\Sigma\theta = 2n \times 90°.$$

The co-ordinates are computed in the manner already explained

Where it is desired to start and finish a traverse on the same point, the method is identical with that described, care being taken that only the interior angles of the figure are used in the computation, and that at the starting point the angle between the first and last legs of the traverse is observed. If the starting point be taken as origin, whose co-ordinates are 0, 0, then the co-ordinates carried throughout the traverse round to this point again must be 0, 0. If they do not come out to this value, the final co-ordinates must be made to have this value, and the remaining points adjusted, as shown in Form E.

**Chainage Errors.**—In reducing the surface measurements to the horizontal, the computer should bear in mind the errors involved in chaining. The larger the scale to which the traverse is plotted, the smaller must be the chainage error.

Thus for the Ordnance Survey maps of scales $\frac{1}{500}$ and $\frac{1}{1000}$, the error in ordinary chaining allowed is 1 in 1,000. For the $\frac{1}{2500}$ map, an error of 1 in 500 is permitted.

In traverses measured with steel tapes much greater accuracy is to be expected.

### B. TRAVERSES ON THE LINE OF MARCH (SUBTENSE TRAVERSES).

(*See* also p. 53.)

Often during an expedition in a tropical country, the nature of the ground and the time available render it impossible to execute a chain traverse. Under these circumstances, subtense methods may be employed.

Even without proper subtense instruments and subtense bars fitted with discs to be observed to, very fair accuracy can be obtained with the ordinary 5-inch micrometer theodolite and improvised signals placed at right angles to the forward ray.

With subtense work the subtended angle should be from 8 to 12 minutes, the base would therefore be some 30 feet for a ray of 10,000 feet. It is very important that the exact intersection of the signal posts with the ground should be measured, and it is consequently necessary that the ground round the signal posts should be systematically cleared of all grass; if the background is dark, a white sheet or piece of cloth will be found very useful to show up the darkened signal post. Where it is impossible to observe the actual ground line, great care should be exercised in selecting signal posts as straight as possible and carefully plumbing them before measurement is made.

It is immaterial what the exact distance between the posts is, so long as the subtended angle is somewhere within the limits specified above, and any suitable high ground or mounds should be selected in the vicinity of the traverse station, care, of course, being taken that the base is truly at right angles to the required ray.

This method of traverse was almost exclusively used in the Anglo-Liberian Boundary Commission of 1902–03, its advantages being that the computations were of the simplest and could be readily worked out in the field, enabling the necessary beacons to be put up at once.

**The procedure** is as follows :—

An assistant with a small clearing party and a compass is sent on ahead to a suitable hill as near the line of march as possible, and proceeds to clear the hill and fix his subtense posts at right angles to the direction of the post he has just left; this is done by any elementary method. A little practice will enable him to judge the distance apart the signals should be and to select a suitable place with a good background. He may or may not put up a separate signal as the traverse station proper, but it is preferable, perhaps, as on the appended sketch, to have a third signal traverse (theodolite) station.

As soon as his signals are fixed, he goes on to the next hill, selecting a hill by his compass as near the line as he can judge. He should be kept constantly informed how much he is distant from the required direction, so that he may select suitable hills.

The observer behind will take, say, three sets of 10 readings to the subtense base, see p. 42, and if they accord, having taken the angle between forward and back stations, and altered the subtense signals at his own station so as to be at right angles to the new ray, and carefully measured the distance in between them, will proceed to the forward station. Here he will at once, before there is any chance of the signals being disturbed, measure the length of base he has observed to. He will then set up the instrument over the new traverse station and take the subtense angle, on the traverse station he has just left, so that he will have two values for the length of each ray, and if the two readings have been taken under equal conditions of light and clearness, the mean of the two should be his accepted length for the ray.

Owing to the difficulty of measuring the exact distance between the signals without the help of another white man, and to the shimmer caused by the sun and heat, except at certain times of the day, great accuracy with small instruments cannot be expected. To measure the distance in between the posts it was found that the best method was to drive a nail into the middle of one of the posts and fix the ring at the end of the steel tape on it, pulling the tape tight to the middle of a pencil mark on the other signal post. This was more satisfactory than trusting the other end of the tape to a native.

The posts should be invariably guyed with rope or wire to make them as firm as possible.

As regards the accuracy of the work, it may be noted that during the Anglo-Liberian Boundary Delimitation above quoted a difference of 20 feet in a ray of 10,000 feet was thought good, i.e., a discrepancy of 1 in 500. At the end of the march, when rays as long as 10 miles had to be taken, the discrepancy in a length of 21 miles between the amplitudes of the latitudes deduced respectively from the traverse and from astronomical observations was 1".

With these long rays as many as 60 or even 100 readings were taken of the subtended angle.

As regards rate of working, from 2 to 3 legs can be measured in a day, each leg being 2 or 3 miles long.

Fig. 21. Subtense Station.

**Computations.**—In the instance quoted, the latitude and longitude of each station of traverse was computed from the known length and azimuth of the ray by means of Form B, p. 27.

### C. Traverses as a Substitute for Triangulation in Countries Unsuited to the Latter, and where Subtense Traverses are not Sufficiently Accurate.

It is not suggested that traverses should ever be substituted for triangulation where the latter is possible, but there are localities, such as a dense forest or a flat grass-covered country, where the difficulties and expense of triangulation are such as to render it practically impossible. In these circumstances, **long traverses** will be found extremely useful in obtaining a framework of a high degree of accuracy on which topographical work can be based.

In considering the relative accuracy of these two forms of survey, it must be remembered that, after the triangles have been subjected to a rigorous treatment for the elimination of errors, no side of a geodetic triangulation is likely to have an error of $\frac{1}{100,000}$.

The unsurmountable difficulty with regard to traverses, in countries unfurnished with a Geodetic Survey, is the impossibility of arriving at an accurate conclusion as to what errors have actually occurred. In the absence of fixed geodetic points, we are obliged to fall back on astronomical latitudes and telegraphic longitudes as a possible means of checking our measurements.

**Local Attraction.**—Unfortunately, all astronomical observations are affected by "local attraction." The effect of this is that the plumb bob and the normal to the assumed geometrical surface of the earth do not coincide. The determination of the value of "local attraction" is one of the problems of a geodetic survey.

Those interested in the subject are referred to the "Account of the Principal Triangulation, Ordnance Survey;" Vol. II., "Great Trigonometrical Survey of India," or "Geodesy," by Col. Clarke, C.B., F.R.S., R.E. Chapter X of the Account of the Principal Triangulation, Ordnance Survey, gives an account of the deflections produced by masses of different density and form. It may be taken that "the probable error of an observed latitude as due to local disturbance of gravity is certainly not less than $1''\cdot 5$." Several instances, however, exist where "local attraction" far exceeds this amount. At a station in Elginshire, determinations show it to be something like $8''$, and at a station in the plains of Lombardy—a fact all the more remarkable owing to the absence of any large range of hills—it is over $20''$.

Despite the most careful observations we cannot, therefore, say with any certainty that an astronomical latitude is correct within about $2''$.

This represents 200 feet on the earth's surface, and, dealing with 2 latitude stations, the error of amplitude probably amounts to $2''\sqrt{2}$, or say 300 feet.

We cannot, therefore, put confidence in latitudes as a check on accurate traverses.

We may, perhaps, lay down the rule, that the latitude stations shall be so far apart, that the errors of the traverse and latitude amplitudes shall be about equal.

**Distance Apart of Astronomical Stations.**—Suppose that a traversing error of $\frac{1}{1000}$ may be expected. We have seen above that 300 feet is the probable error of two latitude stations due to "local attraction." The distance at which latitude observations should be made should, therefore, not be closer than $300 \times 1000$ feet—say 60 miles (of latitude). The distances between longitude stations determined telegraphically in a traverse running east and west should be even greater.

**Latitude Observations.**—Latitude observations should be made with a zenith telescope* —the mean of two night's work, 6 pairs on each night, being taken. This should give a probable error due to observing of about $0''\cdot 2$.

**Clearings.**—The principal aim of long traverses is to establish points within the area to be surveyed on which topographical surveys can be based. In most countries where traverses will be used this involves clearing the bush or forest in order to obtain as long legs as possible and the preparation of the surface of the ground for the chain measurement.

It is probable that native labour only will be available for this class of work, but it is desirable, if possible, to put a European in charge of the "clearing party." The traverse direction having been decided on, it is best to keep as much as possible in the same straight line, and it is advisable to let the traverse touch the villages falling near the proposed line. These form good closing points for further topographical work.

**Rate of Clearing.**—The rate of clearing depends on the nature of the bush and the number of men employed. On the Anglo-German Boundary Commission on the Gold Coast of 1901, 16 men, under a native clerk, cleared on the average about 2 miles a day. The country was, however, not very difficult for clearing and differed totally from that in the Ashanti forest, where the Gold Coast Survey experienced the greatest difficulty in getting long straight clearings.

The tools required for a party of, say, 20 natives for six months would be—

    50 matchets or billhooks,
    5 shovels,
    3 cross cut saws,
    2 tapes,
    1 magnetic compass.

* Or a theodolite suitably fitted.

It is best, where possible, to make the clearing fairly wide. This permits of a fair scope for the selection of the theodolite stations. About 10 to 12 feet is a useful width.

The "clearing party" should always be a mile ahead of the survey party which follows it.

**Detail of Traverse.**—The detail in carrying out these traverses is generally similar to that already described for traverses between fixed points. Owing, however, to the impossibility of obtaining any check on the results, it is important to use the greatest care in the measurement of the distance between stations. The angular measurements, too, must be made with every possible care, although they are to a certain extent checked by means of azimuth observations taken at every 30 or 35 stations apart.

**Causes of Error.**—The chief causes of errors in traverses are given in the account on Trigonometrical Traverses, but may be repeated here. They are due to—

(1) Inaccurate centering of the instrument over the station mark ;
(2) Inaccuracy in reading the verniers or micrometers and in the bisection of the station marks ;
(3) Chainage errors ;
(4) Insufficient length of traverse legs.

Of these (1) has been dealt with and requires no further explanation beyond emphasizing the fact that, having no fixed points to work between, it is of even greater importance in these long traverses to see that the centering is accurate.

(2) This can best be minimised by taking a sufficient number of arcs. The probable error of a mean varies inversely as the square root of the number of observations, if all are equally good. As regards the bisection of the station marks, it must be borne in mind that the shorter the leg the more refined must be the method of bisection.

When the leg is over 1000 feet, it is sufficient to bisect a flag (at its junction with the ground).

When the leg is short, say under 200 feet, a nail or even a fine needle driven into a wooden picket forms the best station mark. Between these distances an arrow stuck in the ground can be observed to.

(3) Chaining should be done by steel tapes, either of 100-feet or 300-feet lengths. All steel tapes should be carefully standardized before they leave England, and it is advisable to have them tested at the same "pull," say 12 lbs., with which it is intended to use them in the field. The mean coefficient of expansion of steel* may be used, and temperatures should be taken about every hour.

If available "Invar" tapes should be used, on account of their low coefficient of expansion and of their smaller liability to rust. See p. 8. The distance between stations should be measured twice, once in each direction. One tape should be kept solely for comparison purposes.

The measured lengths must be corrected for expansion of the tape as explained in Trigonometrical Traverses.

(4) The length of a traverse leg should not be less than 350 yards, but it is not always an easy matter to arrange this—depending as it does on the nature of the country and character of the bush. On the Anglo-German Gold Coast Boundary Commission the average length was considerably over 2000 feet, whereas in the Gold Coast Survey an average of 500 feet was extremely difficult to get.

**Azimuth Stations.**—Mention has been made of azimuth observations, and these must be taken frequently in order to check any errors of angular measurement. They should be taken, with short legs, at every 30th or 35th station. Much, however, depends on local circumstances. In the Gold Coast Survey it is found almost impossible to adhere to this owing to the amount of clearing it would involve in order to see the sun or stars in a convenient

* ·00000625 of its length for 1° Fahrenheit.

position near the prime vertical. Under these circumstances one has to be contented with taking observations for azimuth as native clearings in the vicinity of villages present themselves.

As a rule it is sufficient to take the mean of three sets of observations to the sun. The probable error should not exceed 20". When, however, it is found impossible, for reasons above stated, to get azimuth stations as close together as 35 legs, precautions should be taken to ensure a higher degree of accuracy, and east and west stars, or better still, circumpolar stars at elongation should be observed.

The difference found to exist between an observed azimuth and the azimuth deduced through the traverse should be proportionally divided amongst the angles. This difference is composed partly of the angular errors of the traverse and partly of the convergency of the meridians. The proportional adjustment is sufficiently exact for practical purposes.

**Example.**—An example is given on p. 48 of a portion of a traverse carried out by the British Party of the Togo–Gold Coast Boundary Commission. The headings at the top of each column explain sufficiently what is required in adjusting the various errors in order to obtain the final true bearing of each leg.

**Computations.**—The computations of the co-ordinates are similar to those already explained in the Trigonometrical Traverse, but in this case the sine and cosine of the *true bearing* are used to compute the co-ordinates.

**Latitude Deduced through Traverse.**—Having now obtained the value of the co-ordinates of each point, it is an easy matter to deduce the latitude corresponding to the value of the co-ordinates in the north and south direction.

In doing so it is better to calculate this for each azimuth station, and to add the amplitudes in order to obtain the final difference of latitude.

To determine the latitudes and longitudes of the azimuth stations of a long traverse :—

If the legs are short—

Let $n$ be the total northings or southings between two adjacent azimuth stations in feet, let M be the value in feet of 1" of arc at the mean latitude, then the difference of the latitudes of the two stations is $\frac{n}{M}$ seconds of arc.

The sum of all the amplitudes of the arcs between the azimuth stations will give the amplitude of the arc between the terminal stations of the traverse.

Similarly, if $w$ is the total westings or eastings between two adjacent traverse stations in feet, P the value in feet of 1" of arc of longitude at the mean latitude, the difference of the longitudes of the two stations is $\frac{w}{P}$ seconds of arc.

If the legs are long (over 1 mile) the latitudes and longitudes should be determined by the method described on p. 26.

When the final latitude has been calculated it can be compared with that obtained astronomically; but, as previously remarked, owing to our ignorance of the value of "local attraction," this is but an indifferent test of the accuracy of the traverse.

To give a comparison between two latitude stations, and to show the method of deducing latitude through traverses, the following table is taken from the results of the British Party of the Togo–Gold Coast Boundary Commission (see p. 47) :—

*Column* 1 gives the names of the azimuth stations.

*Column* 2 the distance in feet between these stations, measured along the meridian. These are the summation of the co-ordinates in the north and south direction.

*Column* 3 gives the amplitudes of arc corresponding to these distances.

*Column* 4 gives the values of the latitudes obtained astronomically at the two stations.

*Column* 5 gives the latitudes deduced through the traverse.

It will be observed that the difference between the final observed and deduced latitudes is 0·84".

The mean of the deduced and observed values is taken as the true latitude.

Form H, p. 49, gives a further illustration of work of this nature. It is taken from the same Boundary Commission and shows the total differences between all the latitude stations observed and deduced in the two independent traverses carried out by both divisions of the Commission.

**Errors to be expected.**—To recapitulate, we see that the errors to which a traverse is subject are :—

Errors of chainage and angular measurement, called traverse errors.

Errors of observed latitude, longitude, and azimuth.

Of these, traverse errors vary considerably (from $\frac{1}{1000}$ to $\frac{1}{50,000}$), depending chiefly on the character of the country.

On the Boundary Commission already quoted the following results were obtained as the measurement between two co-incident points by two quite independent traverses.

|  | Feet N. | Feet W. |
|---|---|---|
| British | 279,335 | 24,698 |
| German | 279,347 | 24,679 |
| Difference | 12 | 19 |

If the first of these be taken as representing principally the chainage errors (the traverses running almost due north), the resulting discrepancy was $\frac{1}{23,280}$.

Astronomical observation errors (omitting "local attraction") should not exceed :—

Latitude by zenith telescope ... ± 0".5
Longitude by telegraphic exchange of signals ... ± 3".0
Azimuth ... ± 20"

**Reduction to Mean Sea-Level.**—When the length of each traverse line has been finally computed (but not adjusted), it remains to reduce it to mean sea-level. This is done by the same method as used in the reduction of measured bases :—

Correction to length of traverse = length × $\frac{h}{R}$,

R is the mean radius of the earth,
h is the mean height of the line.

**General Scheme for carrying out a Traverse Survey.**—The following is a general scheme :—

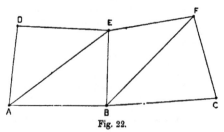

Fig. 22.

Each of the lines here represents a traverse of about 60 to 70 miles.

FORM F.

| Name of Station. | Distance in Feet of y Co-ordinate. | Amplitude of Arc. | Observed Latitude. | Deduced Latitude. |
|---|---|---|---|---|
|  |  | ′ ″ | ° ′ ″ | ° ′ ″ |
| Padjai |  |  | 8 10 30 ·19 |  |
|  | 32171 ·13 | 5 19 ·20 |  |  |
| Camp |  |  |  | 8 15 49 ·39 |
|  | 47309 ·54 | 7 49 ·40 |  |  |
| Kurupi |  |  |  | 8 23 38 ·79 |
|  | 16860 ·14 | 2 47 ·29 |  |  |
| Peg 90 |  |  |  | 8 26 26 08 |
|  | 39723 ·85 | 6 34 ·13 |  |  |
| Salaga |  |  |  | 8 33 1 ·21 |
|  | 20183 ·60 | 3 20 ·25 |  |  |
| Kalandé |  |  |  | 8 36 21 ·46 |
|  | 38060 ·80 | 6 17 ·6 |  |  |
| Daka |  |  |  | 8 42 39 06 |
|  | 39899 ·64 | 6 35 ·88 |  |  |
| Kadengé |  |  |  | 8 49 14 ·94 |
|  | 28361 ·93 | 4 41 ·39 |  |  |
| Gyègi |  |  |  | 8 53 56 ·33 |
|  | 21656 ·44 | 3 34 ·85 |  |  |
| Djonayire |  |  |  | 8 57 31 ·18 |
|  | 15372 ·88 | 2 32 ·54 |  |  |
| Pillar (near 9th degree) |  |  | 9 0 2 ·88 | 9 0 3 ·72 |

48

## Form G.

### Road Traverse 9th Degree N.L. Pablyá.

| Stations. | Included angle from angle book. | Deduced bearing. | Azimuth observed. | Error divided proportionally. | True bearing. | Horizontal distance in feet. | Co-ordinates. | | Co-ordinates from origin. (Station 1.) | |
|---|---|---|---|---|---|---|---|---|---|---|
| | ° ′ ″ | ° ′ ″ | ° ′ ″ | ″ | ° ′ ″ | | x. | y. | E. + W. − | N. + S. − |
| 1–2 | 223 40 6 | 342 18 44 | 342 18 44 | … | 342 18 44 | 1356·0 | − 411·69 | + 1290·9 | − 411·69 | + 1290·0 |
| 2–3 | 193 59 53 | 356 18 37 | | − 2 | 356 18 35 | 2020·4 | − 130·04 | + 2016·2 | − 541·73 | + 3307·1 |
| 3–4 | 179 58 18 | 356 16 55 | | − 4 | 356 16 61 | 980·8 | + 60·37 | + 928·82 | − 602·10 | + 4235·9 |
| 4–5 | 126 48 9 | 303 5 4 | | − 6 | 303 4 58 | 1847·9 | −1129·4 | + 734·06 | −1731·5 | + 4970·0 |
| 5–6 | 238 48 44 | 1 53 46 | | − 8 | 1 53 40 | 3260·7 | + 108·03 | + 3258·8 | −1623·47 | + 8228·6 |
| 6–7 | 179 43 26 | 1 37 14 | | − 10 | 1 37 4 | 2377·7 | + 67·18 | + 2376·7 | −1556·34 | + 10605·3 |
| 7–8 | 180 7 1 | 1 44 15 | | − 12 | 1 44 3 | 2905·9 | + 87·93 | + 2904·25 | −1468·41 | + 13509·5 |
| 8–9 | 179 58 43 | 1 42 58 | | − 14 | 1 42 44 | 3452·3 | + 104·05 | + 3450·7 | −1364·36 | + 16960·2 |
| 9–10 | 252 20 55 | 74 3 53 | | − 16 | 74 3 37 | 2330·4 | + 2381·2 | + 637·2 | + 966·84 | + 17827·4 |
| 10–11 | 170 8 7 | 64 12 0 | | − 18 | 64 11 12 | 2699·1 | + 2430·0 | + 1175·0 | + 3396·84 | + 19002·4 |
| 11–12 | 179 58 29 | 64 10 29 | | − 20 | 64 10 9 | 1767·5 | + 1581·9 | + 764·0 | + 4978·74 | + 19766·4 |
| 12–13 | 180 9 5 | 64 19 34 | | − 22 | 64 19 12 | 2744·3 | + 2473·2 | + 1189·2 | + 7451·9 | + 20755·6 |
| 13–14 | 110 37 49 | 354 57 23 | | − 24 | 354 56 59 | 1496·5 | − 131·74 | + 1490·7 | + 7320·16 | + 22346·3 |
| 14–15 | 179 40 15 | 354 37 38 | | − 26 | 354 37 12 | 3067·6 | − 287·62 | + 3054·1 | + 7032·54 | + 25390·4 |
| 15–16 | 170 16 40 | 344 54 18 | | − 28 | 344 53 50 | 2633·8 | − 686·24 | + 2542·6 | + 6346·30 | + 27843·2 |
| 16–17 | 179 55 56 | 344 50 14 | | − 30 | 344 49 44 | 3010·5 | − 787·75 | + 2905·6 | + 5558·55 | + 30748·8 |
| 17–18 | 180 21 55 | 345 12 9 | | − 32 | 345 11 37 | 2643·9 | − 675·68 | + 2556·1 | + 4972·89 | + 33304·9 |
| 18–19 | 179 57 36 | 345 9 45 | | − 34 | 345 9 11 | 1972·8 | − 437·17 | + 1552·6 | + 4377·4 | + 34857·5 |
| 19–20 | 178 49 39 | 343 59 24 | | − 36 | 343 58 48 | 1584·1 | − 505·51 | + 1507·0 | + 3940·23 | + 36744·5 |
| 20–21 | 179 49 23 | 343 48 47 | | − 38 | 343 48 9 | 2923·8 | −1094·5 | + 3708·0 | + 2845·73 | + 40312·5 |
| 21–22 | 182 28 56 | 346 17 43 | 346 17 4 | − 39 | 346 17 4 | 4565·4 | −1061·1 | + 4177·3 | + 1785·6 | + 44689·8 |

## Form H.
### Comparison of Traverse and Astronomical Amplitudes.

| Latitude station. | | Traverse amplitude. | Observed latitude. | Astronomical amplitude. | Remarks. |
|---|---|---|---|---|---|
| | | ° ′ ″ | ° ′ ″ | ° ′ ″ | |
| British | Padjai | 0 49 38·53 | 8 10 30·19 | 0 49 32·69 | Latitude mean of 2 pairs N. and S. stars p.e. = ± 0·37 |
| | Pillar | | 9 0 2·88 | | „ „ „ 3 „ = ± 0·194 |
| | Pakenju | 0 46 11·46 | 9 46 11·21 | 0 46 8·33 | „ „ „ 2 „ = ± 0·38 |
| | Gambaja | 0 45 27·15 | 10 31 42·10 | 0 45 30·89 | „ „ „ 2 „ = ± 0·101 |
| | | 2 21 11·14 | | 2 21 11·91 | Difference = 0·77″ |
| German | Dabamontu | 0 52 9·24 | 8 7 49·86 | 0 52 9·25 | Latitude mean of 5 pairs N. and S. stars p.e. = ± 0·73 |
| | Pillar | | 8 59 59·11 | | „ „ „ 7 „ = ± 0·31 |
| | Yendi | 0 26 42·82 | 9 26 39·40 | 0 26 40·29 | „ „ „ 1 „ = ? |
| | Gushiokho | 0 28 44·68 | 9 55 25·85 | 0 28 46·46 | „ „ „ 1 „ = ? |
| | Pion | 0 21 11·1 | 10 16 35·90 | 0 21 10·05 | „ „ „ 1 „ = ? |
| | | 2 8 47·7 | | 2 8 46·04 | Difference = 1·66″ |

Observations for latitude are made at each of the stations A, B, C, D, E, and F, and, wherever possible, telegraphic signals are used as a means of obtaining differences of longitude.

**Approximate Adjustment.**—For any two stations such as A and D, we have an amplitude in latitude from the traverse, and one from the observed latitudes at A and D. Take the mean of the amplitudes as the true amplitude, and take as the true values of the latitudes at A and D, values which differ by the same amount from the observed latitudes, in opposite directions. Treat the longitudes (if any have been determined telegraphically) in a similar way. For a more rigorous method of adjustment, see Chapter XVII.

**Recapitulation.**—To recapitulate, the principal points to be observed are :—

(1) The clearings should be as straight as possible;
(2)*The longer the legs of the traverse the better the result and the more rapid the work;
(3) Great care must be taken in centering the instrument;
(4) The shorter the leg the greater must be the care in centering, observing and bisecting;
(5) An azimuth should be observed at every 30 stations, if possible;
(6) Chain or tape each distance twice;
(7) Read at least two arcs in taking the included angle;
(8) Compute as you go along; the clearing generally delays the observing and an opportunity then presents itself for computing.

* This does not apply to subtense traverses.

## CHAPTER V.

### SUBTENSE METHODS AND TACHEOMETRY.

THE general principle of subtense work and tacheometry is the measurement of the angle subtended at the observer by a short measured length at a distance. In the simplest case suppose AB to be a bar placed horizontally, C the centre of the bar, O the position of the observer. Let AB be at right angles to OC. Let $AOB = a$, $AB = b$.

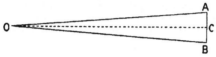

Fig. 23.

Then $\qquad OC = AC \cot AOC,$

$$= \frac{b}{2} \cot \frac{a}{2}.$$

Hence if the length of the bar AB is known, the measurement of the horizontal angle AOB is sufficient to determine the horizontal distance of the bar from the observer.

The angle AOB may be measured:—

    (a) On the arc of a theodolite;
    (b) By a movable micrometer wire in the diaphragm of a theodolite;
    (c) By fixed wires in the diaphragm of a theodolite, in which case AB must be a graduated bar, and the graduations intersected by the fixed wires must be read.

The distance AB may be:—

    (d) A bar with 2 fixed discs on it at a known distance apart;
    (e) The length between 2 signals, the distance between which must be measured each time they are set up;
    (f) A graduated bar, such as a level staff;
    (d) and (f) may be either vertical or horizontal.

Eliminating the combinations (c) (d) and (c) (e), which are impossible, and (a) (f), (b) (f), which are unsatisfactory, it is seen that there are eight practicable and useful methods of determining distances by measuring the angle subtended at the observer by a known short length. To this it may be added that, as regards (c), there are two distinct types of instrument which involve different computations.

*Nomenclature.*

**Tacheometry** (called also in American books Tachymetry or Tachyometry) is a system of "rapid measuring," as above described, and includes all the eight variations just mentioned. The system was first largely employed in Italy in 1820, but had been used in the eighteenth century in England. These various methods are clearly divisible into two main classes:—

    A. Those in which the angle at the instrument is a variable one and the distant base is constant or specially measured—those we shall call *subtense methods*;
    B. Those in which the angle at the instrument is constant and the base is variable i.e., measured on a graduated bar—these we shall call *tachrometer methods*.

A. **Subtense Methods.**—(i.) *With an ordinary Theodolite and a 10 or 20-foot Bar.*

Let a bar over 10 feet long be supported horizontally on a tripod or on trestles, the bar being at right angles to the line joining its centre to the observer's theodolite. Let there be two discs on the bar whose centres are exactly 10 feet apart. Intersect the left disc with the vertical wire of the theodolite; record the horizontal readings. Now move the upper plate with the slow-motion screw and intersect the right disc.

Leaving the upper plate clamped, move both plates with the slow-motion screw of the lower plate till the wire again intersects the left disc. The right disc is then again intersected by means of the slow-motion screw of the upper plate. This process should be repeated, say 10 times, and the final reading of the right disc is recorded. The observer then has the initial reading of the left disc and the final reading of the right disc; the difference between these two readings divided by 10 will give him a fair measure of the angle subtended by the bar.

The two readings mentioned are the only ones really required; but as a precautionary

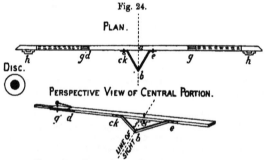

Fig. 25.* Subtense Bar used on the Survey of India.

measure, to give a check, it is a good plan to record also the reading of the right-hand end of the bar the first time it is taken.

The difference between the readings of the two ends of the bar will give a rough value for the subtended angle, sufficiently near to expose any error in counting the number of times the process has been repeated.

This method was introduced into the Indian Survey by Colonel Tanner, who used it largely in the Himalayas. Colonel G. Strahan also employed it exclusively for the traverses of the Nicobar Islands. The mean linear error in the latter case, which was very unfavourable, was about $\frac{1}{300}$.

Bars have been used in India of 2 feet, 10 feet and 20 feet. In the Himalayas the 20-foot bar was in general use; this had 12-inch circular discs, and under favourable conditions a ray 3 miles long could be measured with reasonable accuracy. A 10-foot bar with 8-inch discs will give good results up to $1\frac{1}{2}$ miles.

**Description of fig. 24.**—The bar is made of hard wood in three sections. The central section is square in cross-section, $1\frac{3}{8}'' \times 1\frac{3}{8}''$, with iron sockets 6 inches long, $g$, $g$, at each end, into which the outer portions of the bar fit, being pinned into place by the pins $g'$. The outer ends of the bar carry iron sockets, $h$, $h$, which have the recesses in them accurately machined out. Into these sockets the discs $i$ fit by means of carefully fitted hooks on their backs. The

* Extracted, by permission, from 'Hints to Travellers.'

discs are of wood, 10 inches in diameter, painted white with a black centre. Black cloth covers are also carried, to fit tightly over the discs, in case of working with a light back-ground.

In the centre of the bar is a brass socket plate, by means of which the bar can be attached to a tripod.

The sighting arrangement consists of a light iron frame hinged at $e$, $b$, and $k$. The pin of the hinge $b$ carries a point on the top, and a similar metal point is fixed at $a$, in the centre of the bar, and the other end is fixed by a thumb-screw, $c$, in such a position that the line joining $ba$ is at right angles to the line joining the discs. For travelling, the thumb-screw $c$ is unscrewed and the frame is closed up against the bar, in which position the thumb-screw screws into the hole $d$ in a metal plate affixed to the bar. The bar is fixed in position by an assistant looking along the sights $a$, $b$ and laying them on to the theodolite.

Fig. 26.*

Fig. 25 represents another form of rod and one more easily made, though not calculated to give such accurate results. A, A are two boards, 1 foot square, painted white, with a black cross on each. These are fastened on a bamboo, B, B, in such a manner that the centres of the crosses shall be a known distance apart.

When using the rod in a vertical position it will often be found convenient to fasten a stick to it, extending about 2 feet beyond one of the boards. This, when placed on the ground, takes the weight off the rod and helps the assistant to keep it steady.

(ii.) *With an Ordinary Theodolite and Two Signals.*

(See also p. 41).

The line joining the signals must be at right angles to the line joining the observer and the central point between them. The signals may be poles fixed vertically into the ground and stayed with rope or wire guys. It is desirable that the ground line should be visible from the observer, otherwise the poles must be very carefully plumbed. In fixing the poles the right angle may be determined by compass, or by tape, thus:

Fig. 27.

or, if available, by a theodolite. The space between the poles should subtend at the observer an angle of from 8 to 12 minutes. The reason being that if we admit the possibility of an error of 1", this will in a total measurement of 8' be equivalent to an error of about $\frac{1}{100}$, which should not be exceeded. The distance between the centres of the poles should be carefully measured with a tape; this distance will vary from 10 feet to 100 feet, according to the length of the ray. For long rays conspicuous marks must be fixed at the top of the poles to attract attention; in India and W. Africa wicker baskets, about 3 feet in

* Extracted, by permission, from 'Hints to Travellers.'

diameter, have been found useful for the purpose. The poles will be from 10 to 30 feet long. One of the longest rays thus determined was one of 11½ miles, measured on the Sierra Leone-Liberia Boundary Commission of 1903, the base (which on account of local conditions was somewhat small) subtended an angle of 5' 47" and was 101 feet long; 5 sets of 10 repetitions of the angle were observed. Experience shows that in such long rays it is desirable to use dark poles against a white back-ground.

*The Observations.*—As a rule, 3 sets of 10 repetitions should be taken for ordinary rays of 2 or 3 miles; for longer rays a larger number is necessary. Each new set should begin on a different part of the arc, *i.e.*, if the left signal reads about 0° on the first set, it should read about 35° on the next set, and so on. Vertical angles should be taken also, 2 on F.R., 2 on F.L. (Required for determination of heights.)

*The Computations.*—If $a$ is the mean observed angle,
$b$ the length of the base,
$d$ the required distance,

$$d = \frac{b}{2} \cot \frac{a}{2}.$$

(iii.) *With a Theodolite fitted with an Eye-piece Micrometer, and a Bar of Fixed Length.*

The eye-piece micrometer screw moves a wire in the diaphragm. The wire is made to intersect the discs or marks at each end of the bar, and the difference between the two readings of the micrometer head gives the subtended angle in terms of the divisions of the head. This process can be repeated many times with great rapidity and the mean, say, of 10 readings, will give a good value of the angle.

A table should have been previously prepared giving the distances corresponding to a series of intervals of the micrometer head. This table should be determined by *actual experiment*. The reason is that, owing to optical principles, the angles measured cannot be taken as those subtended at the vertical axis of the instrument, but at a point in front of this (see also p. 55). It is easy to show, if the focus has not been altered,* that if M is the micrometer reading of a staff of fixed length, A and B constants to be determined, $d$ the required distance *from the vertical axis of the instrument*, $d = \frac{A}{M} + B$. A few experiments at various distances will be sufficient to determine the constants A and B. A table should then be prepared.

This method is sometimes very valuable and was used in the traverse in Abyssinia, the horizontal angles being measured with the same theodolite, where the route followed was in many places so rough and blocked with troops and baggage animals, that any other system of measurement would have been impossible. The rapidity with which such a traverse can be carried on depends upon many considerations, but chiefly upon the length of the shots that can be taken. In the Abyssinian expedition of 1868, a traverse was carried with one of these instruments from Adigerat to Magdala without any break of continuity, the daily rate of progress averaging 5 miles, and being occasionally as much as 8 miles. After this run of some 300 miles nearly due north and south, the observed latitude differed from that given by the traverse by about ¼ of a mile.

The bar may be similar to those described on p. 52. For preference the bar should be horizontal. Rays a mile or two long are convenient.

The horizontal angles between the stations should be observed as in a chain or tape traverse.

The subtense methods just described are more generally useful to the topographer than the tacheometer methods, which will be next dealt with.

* The instrument must be used at solar focus.

**B. Tacheometer Methods.**—(i.) *An ordinary Theodolite, with Two Fixed Wires in the Diaphragm and a Graduated Bar.*

The theodolite is provided with two horizontal wires fixed in the diaphragm at equal distances from the axis of the telescope. The staff is held vertically, the divisions of the bar intersected by the fixed wires are read, and these are sufficient to determine the distance of the bar from the observer.

Now, an ordinary theodolite has a bi-convex object-glass and an eye-piece, and it is easy to show that the point at which the angle subtended by the fixed wires is constant is not on the vertical axis of the instrument, but at a distance $f + c$ in front of the vertical axis measured towards the bar, where

$f$ is the focal length of the telescope (at solar focus),
$c$ is the distance of the centre of the object-glass from the vertical axis.

$f$ can be measured by focussing the telescope on a distant object and measuring the distance from the centre of the object-glass to the diaphragm.

Hence, if $d$ is the required distance from vertical axis of theodolite to centre of bar, the bar being held vertically, the telescope being horizontal, if

$a$ is the distance intercepted on the bar by the wires,
$k$ a constant,
$$d = k \cdot a + (f + c).$$

$k$ should be determined by experiment by taking several readings of the staff at 100, 200 and 300 feet distant, measuring from a point $f + c$ in front of the vertical axis.

If the rod is held vertically, but on a different level from the telescope, so that the latter is inclined at an angle $a$ to the horizon when pointing at the centre of the rod, using the same notation as before,
$$d = k \cdot a \cdot \cos^2 a + (f + c) \cos a,$$

and the difference of height ($h$) between the axis of the theodolite and the mean reading of the staff is given by the expression
$$h = (f + c) \sin a + a \cdot k \frac{\sin 2a}{2}.$$

The rapid and effective use of tacheometer methods thus clearly involves the employment of tables. There are several such tables in existence.* It will be seen that the method is not so straightforward as might at first sight appear.

American experiments appear to show that the limiting length of the rays should be about 800 feet. In English practice this limit is usually about 600 feet, but in exceptional cases rays of 1,200 feet may be measured. On account of the shortness of the rays this method is more suitable for large scales than for small, and generally it is more applicable to engineering than to topographical surveys. Rods from 10 to 15 feet long should be used—they may be specially graduated or may be level staves.

(ii.)

By the introduction of a converging lens between the object-glass and the diaphragm of a theodolite, measurements can be read directly from the vertical axis without the necessity of considering the quantity $f + c$. The term "tacheometer"† is best confined to instruments which have this optical arrangement. If there are two fixed wires in the diaphragm of such an instrument and a graduated staff is observed, the distance from the vertical axis

---

* See "Theory and Practice of Surveying," S. B. Johnson. New York. Wiley and Sons.
† Messrs. Troughton and Simms make an instrument of this type.

to the staff varies directly as the number of graduations intercepted. If the central line of vision makes an angle $\alpha$ with the horizon, if $d$ is the required horizontal distance of the staff from the vertical axis, if $h$ is the height of the point at which the central line of vision strikes the staff, let $m$ be the number of divisions between the wires,

$$d = k \cdot m \cdot \cos^2 \alpha, \text{ when } k \text{ is a constant,}$$
$$h = d \tan \alpha.$$

It is desirable in this case also to use tables. The instrument is more generally suitable for large-scale surveys than for topographical work.

There is considerable literature on the subject of subtense methods and tacheometry. Amongst other books may be mentioned :—

"The Tacheometer," N. Kennedy, M.I.C.E.
"Hints to Travellers."
"Encyclopædia Britannica," vol. XXXIII.
"Theory and Practice of Surveying," S. B. Johnson. New York. Wiley and Sons.
"A Treatise on Surveying," Middleton and Chadwick, Spon, London.

There are many special instruments depending on the same principles, but they are not generally of much value in topographical surveying.

PLATE IV.

SERVICE PLANE-TABLE. 18" × 24".

## CHAPTER VI.
### PLANE-TABLING.

THE **plane-table** is merely a portable table. The table or board can be carried separately from the stand or legs; the stand is usually a folding tripod. For greater compactness, the stand is sometimes made of a collapsible pattern. The adjuncts to the plane-table are the sight-vane, called also the sight-rule or alidade, and the box or trough compass; the plane-table should also be provided with a waterproof cover.

A convenient size is the Service pattern, of which the board is 18" × 24" (see Plate IV).

The **sight-vane**\* should be about as long as the shortest side of the table, and the fittings should be of gunmetal or brass, not iron or steel. A common defect of sight-vanes is that the slit in the rear vane is too small. When working in hilly country the centres of the tops of the vanes should be joined by a string, tense when the vanes are upright, to enable rays to be taken on steep slopes. It is convenient, as in the Service pattern, for the sight-vane to be graduated with useful scales, notably 2 inches to 1 mile.

The **trough compass** should not be combined with the sight-vane. The 6-inch trough compass in mahogany box with sliding lid, which throws the compass off its pivot, is the best. Trough compasses should be stored horizontally. They can be adjusted for "dip" by tying string round the elevated end and dropping sealing-wax on to the string.

**Mounting a Plane table.**—With an ordinary plane-table which has a plain board (and all arrangements of clips should be eschewed), if time presses the only thing to do is to pin the paper on to the board with drawing-pins, and if the paper is linen-backed so much the better.

But if time allows, "mount" the plane-table as follows:—

Get some fine linen about 8 inches larger in each direction than the board, damp this slightly, and paste it on the underside of the board only. The paste should be made of the finest flour available; cornflour is the best. Now take a piece of drawing-paper the same size as the linen, damp it on the wrong side, paste the linen all over and smooth the paper down on to the linen; paste also the overlap of paper on the underside of the table. Put in a few pins temporarily on the underside: these can be removed in about 12 hours, when the table will be ready to work on.

When the work is finished and the field sheet is removed from the board, its edges should be bound with tape.

The reasons for mounting paper on linen are that it gives a good surface to work on and renders the paper less liable to be torn, and somewhat less liable to be affected by expansion caused by wet and damp.

SURVEYING AND SKETCHING WITH THE PLANE-TABLE.

The plane-table is by far the most useful instrument employed in filling in detail.

The following are **general maxims** which should always be observed:—

(1) See that the leg-screws and clamping-screws are taut before beginning work.

(2) Spread the legs widely, especially with collapsible patterns; the stiffness of the table largely depends on the spread of the legs. In soft soil the legs should be well pressed into the ground. The board should be horizontal.

(3) Always keep the pencil sharp; there should be no error in essential features greater than the thickness of a pencil line. With acute intersections, which must sometimes be used, fineness of line is absolutely necessary. Points required afterwards for

---

\* Telescopic sights are much used in America, and although they take away from the simplicity of the instrument, they undoubtedly increase the accuracy of the work.

interpolation should be fine dots surrounded by a circle. No good results can be expected with a blunt pencil.

(4) When it is required to draw a ray from a point, put the pencil upright on the table with the edge of the base just touching the point, and lay the sight-vane against the pencil. Or instead of a pencil some expert plane-tablers use the finger. *Pins should never be used.*

(5) When drawing a ray from a dot, set the point of the pencil into the dot first, so as to get the correct angle at which the pencil should be held when drawing it along the sight-vane. Without this precaution large errors creep in.

(6) Whenever the plane-table has to be set up or oriented by a previously drawn short ray, it is important that the ray, when it is first drawn, should be extended by marking short lengths of it at or near the edges of the board. These are called "repere" marks and give a good line along which to lay the sight-vane.

(7) When drawing a ray to a distinct object, mark its approximate position, as estimated, by a small circle on the ray. Write the name of the object along the ray; to assist identification it is sometimes useful to draw a small sketch of the object at the end of the ray.

The steps to be taken in making a plane-table survey are the following:—

1. **Plot the Graticule.**—The method of doing this is fully described in Chapter XI. Scales of latitude and longitude should be drawn on the board.

2. **Plot the Trigonometrical Points.**—The latitudes and longitudes of the trigonometrical points having been computed as described in Chapter III, the points are plotted with reference to the graticule lines and by means of the scales of latitude and longitude. At the outset, and before the graticule has had time to get distorted by expansion or contraction of the paper, it is perhaps somewhat more accurate to plot the points from the scale of inches used in constructing the graticule, by converting into inches the intervals in latitude and longitude from the graticule line.

3. **Test the Plotting** by going to a trigonometrical station; lay the sight-rule along the line joining this station and any distant visible station. Move the board until the distant station is intersected by the vertical wire of the sight-rule, then clamp the table. See if the other points are in their correct positions by aligning the sight-rule on them. Any point incorrectly plotted will easily be detected.

Now place the box compass on the table (away from the area to be surveyed), move the compass until the needle points to the centre graduation, and draw a pencil line along each side of the compass box. These lines will mark the magnetic north, and by their means the table can subsequently be set up approximately. Remove the compass.

4. **Fix Points by Intersection.**—At this station, and at every other point where the plane-table is accurately set up, rays should be drawn to all points which can be easily identified, or are likely to assist in the execution of the work. Along all such rays should be written a brief description; occasionally it is useful to draw a small sketch of the object at the end of the ray, and a small circle should be drawn on the ray in the supposed position of the object.

A position fixed by intersection of 3 rays may be relied on. If a point has been fixed by only 2 rays, it should be considered doubtful, and should not be used afterwards for interpolation.

5. **Fill in the Detail; mainly by Interpolation.**—Interpolation, or fixing by resection from three known points, sounds a very formidable process, but is in reality very simple. It is the principal means by which the bulk of the detail should be fixed. The method is as follows:—

Set up the plane-table at any point from which three previously fixed points can be seen. These should be trigonometrical points if possible. Orient the table roughly by the trough

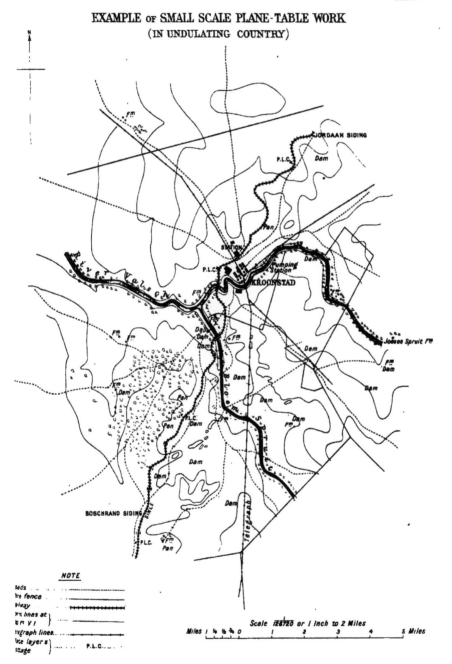

compass, and from the three fixed points draw back rays. If these three rays pass through a point, this point is the required position. If they do not pass through a point, the rays will form a small triangle called the "triangle of error." The true position is now to be determined by the following rules:—

    I. If the "triangle of error" is inside the triangle formed by the three fixed points, the position is inside the triangle of error; and if outside, outside.

    II*. In the latter case the position will be such that it is either to the left of all the rays when facing the fixed points, or to the right of them all. Of the six sectors formed by the rays, there are only two in which this condition can be fulfilled.

    III. Finally the exact position is determined by the condition that its distances from the rays must be proportional to the lengths of the rays.

In the attached example, for instance:—

By condition I the point must be outside the triangle of error;
    ,,   ,,   II   ,,     ,,     ,, in sector 6 or in sector 3;
    ,,   ,,   III ,,     ,,     ,, in sector 6, since the distances from it to the rays must be proportional to the lengths of the rays, and by estimation it will be where shown.

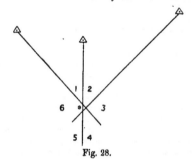

Fig. 28.

Having thus determined the position, place the sight-rule along the line joining this and the most distant of the points used, set the sight-rule on the point by shifting the plane-table; clamp and test on the two other points. If there is still an error (which should, however, be much smaller) go through the process again.

The best position is inside the triangle formed by the three fixed points, of which two are near and one is distant. Accuracy of *position* is ensured by two points being near, accuracy of *setting* is ensured by aligning on the distant point. And in general, fix from near points, set by a distant point.

It is very necessary to be able to interpolate by this method both rapidly and accurately. It is the essential foundation of all good topographical plane-tabling. The plane-tabler in open country need not pace or measure a yard after he has fixed his base. In addition to the speed which results from a free use of interpolation there is also a gain in accuracy, since there is no piling up of errors, and each fixing is dependent only on the main intersected or other well-fixed points.

*There is one case in which the method fails,* viz., that in which the observer's position and the three points lie on or near the circumference of a circle. The plane-tabler should be on the look-out for this, and not interpolate from points so situated.

                  * Rule II obviously includes Rule I.

It will have been noticed that the trough compass is only used to orient the plane-table roughly, the final fixing does not at all depend on the compass. It is important that this should be remembered.

**Resection from Two Points.**—It will, however, sometimes happen that only two fixed points or stations are visible; in this case the plane-table must be oriented by the compass, and rays must be drawn back from the two fixed stations intersecting in a point which must be taken as the required position. It is clear that such a fixing depends entirely on the compass, and may, therefore, be very inaccurate. In parts of S. Africa such a fixing would, on account of local magnetic attraction, be entirely worthless, and it is obvious that as a general rule compass fixings should be avoided.

This method should, if possible, only be used for putting in detail, and not for throwing out points on which future work will depend.

Whenever a fixing has been made the local detail should be drawn in; a combination of measurement and estimation should be used; the amount of estimation permissible will depend on the scale. Thus, on scales of 2 inches to 1 mile and smaller scales it is not necessary to pace the width of roads or size of houses; these are put in conventionally. On a scale of 2 inches to 1 mile a distance of 8 yards is less than $\frac{1}{180}$ inch, so that any distance may be estimated which is not likely to be 8 yards in error, and so on. When time presses, minute accuracy must be sacrificed.

**6. Traverse Detail which cannot be Fixed by Interpolation.**—It will sometimes happen even in open country that there are parts of the work, such as valleys and woods, in which plane-table fixings cannot be made; it is then necessary to have recourse to plane-table traversing. Plane-table traversing has to be much used in flat country unprovided with prominent land-marks, and was so used in the mapping carried out during the operations in China in 1901.

The method is as follows:—Set up the plane-table at some known point and orient it, draw a ray in the direction of some visible object (and the farther off this is the better), chain to this second point, set up the plane-table there, place the sight-rule along the ray previously drawn, and turn the table until the sight-rule is on the first point; clamp; measure the chained distance along the ray and mark the second point on the plane-table; draw a ray to some third point, chain to it, and go on as before. Continue the traverse until it is possible to make an accurate fixing by interpolation, and then adjust the traverse between this fixing and the starting-point. Errors are inevitably generated in this sort of traversing which should therefore be penciled in quite lightly, as it will have to be rubbed out when adjusted.

The adjustment is to be carried out as follows:—

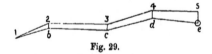

Fig. 29.

If 1, 2, 3, 4, 5 are the stations of a traverse, a fixing was made at 5, and the true position of this fixing was $e$; draw lines through 2, 3, 4 parallel to $5e$, and make $2b$, $3c$, $4d$ bear the same proportion to 1—2, 1—3 that $5e$ does to 15. Then the traverse as corrected will be 1, $b$, $c$, $d$, $e$.

Or, instead of orienting by back rays as above described, the plane-table may be oriented by compass alone. But it is not advisable to use this method where it can be avoided, as a little consideration will show; for a compass may easily be $\frac{1}{2}°$ out, which is equivalent to a lateral error of $\frac{1}{120}$; now, in aligning by a back ray, it is very unlikely that the sight-rule in a ray 1,200 yards long would be 10 yards off the object. However, where the rays are very short, or the traverse is long, or objects to line back on cannot be found, or where time is short, it may be necessary to traverse with a compass.

Plate V.(b)

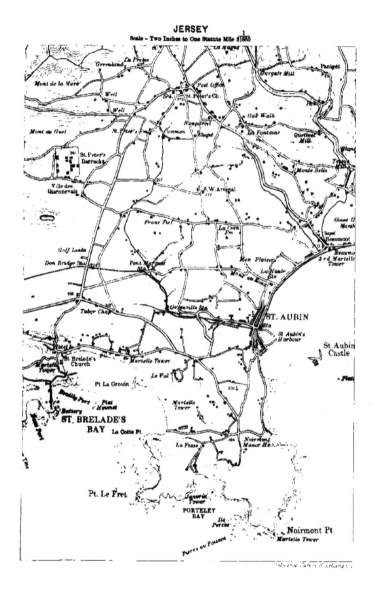

It is sometimes convenient to plot the traverse at the side of the board and subsequently to transfer it to its proper position when corrected.

**7. Contouring.**—In the preceding description of plane-tabling no mention has been made of contouring. As described in Chapter X, contours vary in accuracy according to the method adopted for the survey. They may be fixed by level, theodolite, water-level, or clinometer, or may be sketched in by eye. The method adopted depends on the scale of the map, the time and money available, and the purpose of the map.

Speaking generally, so far as the scale is concerned, contours would be determined as follows:—

| | | |
|---|---|---|
| 4 inches to 1 mile and larger ... ... | level, theodolite, or water-level, |
| 2 ,, 1 ,, ... ... | water-level or clinometer, |
| 1 inch to 1 ,, ... ... .. | ,, ,, |
| ½ ,, 1 ,, ... ... | clinometer or sketch contours, |
| ¼ ,, 1 ,, and smaller ... | sketch contours. |

Where contours are determined by level or theodolite, the detail should be surveyed before the contouring is commenced. Where the contours are approximate contours put in by clinometer or sketch contours or form-lines put in by eye, the contouring should be carried out during the survey of the detail.

To fix the ideas, we will now suppose that it is intended to contour a 1-inch field sheet with the aid of the Indian pattern clinometer.

**The Indian Clinometer** is 9 inches long and has two vanes which are upright when in use; there is a sight-hole in the rear vane, and the front vane is graduated in degrees on the left side of its central slit and in natural tangents on the right. The sight-hole is brought to a level with the zero of the front vane by a screw behind the rear vane, a level attached to the horizontal arm showing the position. The scale of natural tangents can be read by estimation to 3 places of decimals; the extreme range of elevation or depression is 23°.

Suppose now that the plane-tabler has fixed his position by means of resection from three known points of which the heights are known. He sets the clinometer on the table, levels it, and directs it on one of the known points, preferably the nearest. He looks through the sight-hole and notes against which reading on the scale of natural tangents the point appears. He now measures the distance from his fixing to the point, multiplies the distance in feet by the natural tangent obtained, and the result is the approximate difference of height in feet between the plane-table and the fixed point.

Thus, if the natural tangent is ·043 depression, the distance 2000 yards, difference of level = ·043 × 6000 feet = 258 feet; and if the height of the distant point was 401, and the plane-table board was 3 feet above ground, the height of the ground at the plane-table is clearly 258 + 401 − 3 = 656, and this height is marked against the fixing. It is not desirable to use this method for long rays on account of its comparative roughness, about 2 miles is a satisfactory limit. If values for the height can be got from more than one fixed point, it is advisable to take them and use the mean.

In place of actually computing the height a diagram may be used (such as Colonel Wabab's height indicator), from which the difference of height may be measured off.

Having now the height of the ground at the fixing, the next step is to determine the position of the nearest contour. Suppose that in the case quoted the fixing is on the top of a gentle rise, and that the ground slopes down uniformly in one direction at an angle of ½° as measured by the clinometer. Now a rule that all topographers should remember is that a slope of 1° is equivalent to a rise or fall of 1 in 57·3, for ½° this is 1 in 114·6, or 1 foot in 38·2 yards.

The height of the fixing is 656 feet, and if the contour intervals are 50 feet the nearest contour will be 650, 6 feet below the fixing, or a distance of 6 × 38·2 yards or 229 yards away. This point is marked on the table, and so are other surrounding points on the same contour.

We must now define the terms vertical interval and horizontal equivalent.

The **Vertical Interval** is the vertical distance between any two successive contours, and is usually given in feet.

The **Horizontal Equivalent** is the distance on plan or on the map between any two successive contours, and is usually given in yards.

The plane-tabler should prepare a scale of horizontal equivalents for the particular interval in use. Thus, suppose the scale to be 1 inch, and the vertical interval to be 50 feet, the scale would be prepared as follows:—

1° is equivalent to a rise of 1 foot in 19·1 yards,
1° ,, ,, ,, 50 feet ,, 955 ,,
2° ,, ,, ,, 50· ,, ,,· 478 ,,
&c., &c.

We have, therefore, this list of horizontal equivalents for a vertical interval of 50 feet.

| Degrees. | Yards. | Degrees. | Yards. | Degrees. | Yards. | Degrees. | Yards. |
|---|---|---|---|---|---|---|---|
| 1 | 955 | 6 | 159 | 11 | 87 | 16 | 60 |
| 2 | 478 | 7 | 136 | 12 | 80 | 17 | 56 |
| 3 | 318 | 8 | 120 | 13 | 74 | 18 | 53 |
| 4 | 239 | 9 | 106 | 14 | 68 | 19 | 50 |
| 5 | 191 | 10 | 96 | 15 | 64 | 20 | 48 |

If the slope is uniform, the positions of the other contours will be found by stepping off the horizontal equivalent for the particular slope in question, from the first fixed contour, along the ray down which the slope was taken.

But slopes, as a rule, are not uniform throughout, and this method should only be used for the distance up to which the slopes may be considered sensibly the same; the plane-table being then moved to the next change of slope and the same procedure carried out again there. In mountainous country it is best to contour the top of a ridge or hill first, then when the valley is reached the lower contours can be put in.

But if time presses, and the bottom of the valley is visible, the difference of level, and hence the number of contours, between the fixing and the bottom may be determined, and the contours spaced by eye, closer on the steep portions of the slope than elsewhere. This is rapid approximate contouring.

Generally in approximate contouring it is desirable to work in the first instance along ridges, spurs, and tops of hills, putting in the valley contours subsequently.

For a more elaborate method of contouring with the clinometer, see Chapter X.

# SPECIMEN OF SMALL SCALE PLANE-TABLE WORK
*(A portion of a survey in Somaliland during operations 1903-4)*

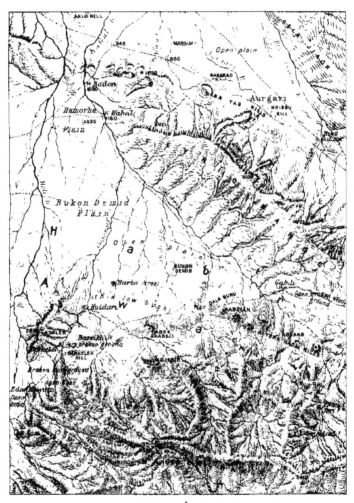

Scale 1/250,000

# FORM LINES
# FOR SMALL SCALE WORK

Scale $\frac{1}{126,720}$

*Facsimile of a portion of a plane-table field sheet.
Baluchistan Survey 1886*

## CHAPTER VII.

### SURVEYS UNDER ACTIVE SERVICE CONDITIONS.

**Rapid Triangulation.**—When officers are called on to undertake surveys, either on active service, or on the delimitation of a boundary, or when as explorers they map country with a view to their work being of use in subsequent military operations, the scene of their labours will generally lie in some uncivilised country and extend over a considerable area. On such occasions there is often opportunity for more accurate work than is obtainable by mere sketching.

The chief methods of securing accuracy in a rapidly executed map of an extended area are :—

   (I) Triangulation with the theodolite from a measured base;
   (II) A combination of triangulation with astronomical observations;
   (III) Astronomical observations.

Triangulation is sometimes regarded as only suitable for rigorously accurate surveys undertaken in peace time and carried out with deliberation. It has, however, been proved on many occasions that, even under the limitations of active service, it is possible, in a favourable country, to execute a rapid triangulation, which, though falling short of the accuracy to be attained in regular surveys, can still be of immense use in eliminating errors and in facilitating both the execution of the mapping and the subsequent compilation of different men's work, by providing fixed points on to which to hang the detail surveys.

The chief difficulties in this class of work are want of time, uncertainty as to when and in what direction the next move will be made, the possibility that ground once passed may never be revisited, and on some occasions the risk of going far from camp without a strong escort.

To meet these difficulties the procedure to be followed, though generally similar to that of regular triangulation, requires certain modifications in practice, of which the chief are :—

   (1) The use of smaller and more portable instruments:
   (2) Less refinement in measuring the base;
   (3) Less regard for symmetry in the triangles;
   (4) Disregard of the rule of always observing all three angles of main triangles;
   (5) The selection of natural marks to observe to, owing to the impossibility, as a rule, of making a preliminary reconnaissance and erecting artificial marks.

(1) It is obviously advisable to carry the lightest instrument that will give sufficiently accurate results.

Excellent small microscope theodolites are now made, i.e., theodolites in which the divisions on the limbs are read by micrometers instead of verniers; time is saved in reading, and the readings are more accurate.

With a 5-inch microscope theodolite, the micrometer-heads are graduated to 10 seconds and read to 1 second. These instruments are especially valuable when astronomical work has to be done. They will stand a good deal of rough usage. If this type of instrument is considered too heavy, a Vernier theodolite must be taken. The smallest that has been used is a 3-inch theodolite, reading to minutes, in two small boxes, with camera stand. Weight with stand and boxes 19½ lbs.

(2) The base would be measured as described in Chapter II.

In selecting a site for a base it is necessary to consider the facilities afforded by the natural features of the surrounding ground for the extension of the triangulation. A stretch of perfectly flat open ground, though admirably suited for the actual measurement, may be most unfavourably situated as regards extension; and, as a general rule, in surveys of the class now described, it is better to lay out the triangles symmetrically, and to run the risk of an error of a foot or so per mile, in the measurement of the base, than to have a more accurately measured base and badly-shaped triangles.

In the first case the error will be carried through the work consistently, and may be eliminated later by measuring a check base on more favourable ground; in the second case, it is probable that still larger and discordant errors will be introduced, which will cause endless trouble to the computer and, perhaps, throw out the detail surveyors.

Bases of this description are seldom more than one or two miles in length, and time rarely allows of the measurement being made more than twice.

As level ground is naturally selected for the base, and as high natural features are, if possible, chosen for the first stations fixed from it, it follows that the surveyor, when observing back from those stations to the ends of the base, will see them against a background of the surrounding plain or hills. The result is that any artificial marks erected on the ends of the base will be very hard to see unless they are flashing instruments, such as heliographs. Whenever, therefore, it is likely that a base may have to be measured, it is desirable that the surveyor should have a couple of heliographers with their instruments at his disposal. (N.B.—Camp looking-glasses can be made very useful.)

It may occasionally happen that there is no ground suitable for measuring a base at the starting point of a triangulation. In such a case, if it is convenient to start from a long base, running nearly north and south, and if there are suitable hills some 10 or more miles apart and nearly on the same meridian, an approximate value for the distance between their summits can be obtained by taking a careful series of latitude observations at each, and by observing the true bearing of one from the other. See Chapter XV.

Still, on account of instrumental errors and "local attraction," the resulting distance can be regarded as only approximate; and, although this method gets rid of the troublesome base measurement and extension of the triangulation from the base, the result obtained can never be regarded with confidence, and the earliest favourable opportunity should be taken of measuring a check base with a steel tape.

(3) As mentioned previously, the ideal shape for a triangle is the equilateral. Such triangles are, unfortunately, seldom obtainable in rapid work, and though it is worth while expending some labour to avoid having any angle of the main triangulation less than 20°, still the work must not be regarded as valueless if angles of 15°, or even 10°, occur occasionally. Distant intersected points can often be fixed within 200 or 300 feet by triangles whose apices do not exceed 4° or 5°.

(4) The more rapid the work the less chance there is of the surveyor being able to observe all the angles of his principal triangles. This need not, however, seriously impair the value of his work, if he makes a skilful use of his intersected points as described on p. 69.

The fact of all three angles not being invariably observed renders it far more necessary in rapid than in regular triangulation to fix each point by more than one triangle.

(5) Whenever time and circumstances permit, a preliminary reconnaissance should certainly be made for the purpose of selecting and marking all the stations and intersected points. Unfortunately in rapid work it is seldom possible to do so, and consequently natural objects must in the main be relied on. In these circumstances the measure of success with which rapid triangulation can be carried out must depend enormously upon the nature of the country. The greater part of Afghanistan, Persia, and Baluchistan, abounding as it does with lofty, well-marked peaks, which, owing to the clearness of the atmosphere, are visible from very great distances, offers an ideal field for this class of work. The forest-clad hills of Upper Burma and the Shan States, and of parts of tropical Africa, on the other hand, present very much greater, but not insuperable difficulties, and excellent work of this class has been

done in those regions, though with less rapidity than in the first-mentioned countries. When, however, the surveyor is called on to work in the forests of the Gold Coast, or on the rolling velt of parts of South Africa, rapid triangulation becomes an impossibility. Though at the first start, in a favourable country, it may be possible to make a small preliminary reconnaissance, and to erect marks on the first stations that are to be observed, still, as a rule, the triangulator will be compelled to utilise for his observing stations and intersected points any natural marks that offer themselves, such as tops of hills, trees, rocks, &c.; and it is in the selection of these, and in their identification from different points of view, that his skill and forethought will be most severely tried. *Always observe to and from the highest point of a hill only.*

**General Sketch of the Method of carrying out a Rapid Triangulation.**—Assuming that the surveyor with a military expedition has arrived, with his instruments, at the nearest point to the scene of operations to which he can proceed, and that he has a certain number of days in which to start his work before the force advances, he would probably proceed as follows :—

Having taken a general look round from the nearest suitable elevation, he would proceed to select a site for his base, and mark the two ends provisionally. He would then measure its distance approximately by pacing, or by any other rough method, and plot the two ends on a plane-table on any suitable scale, 1 or 2 inches to the mile for choice. From the two ends of his base he would intersect all the surrounding points that appear suitable for the first stations of his triangulation. If time permitted, he would extend his reconnaissance till he had embraced all the apparently suitable points in the required direction that he could fix approximately without moving his camp, and he would at the same time have cairns of stones or poles erected wherever the natural features were not sufficiently clearly marked for observing purposes. As a general rule, these cairns should be of the smallest size compatible with their being clearly visible through the telescope at the required distance. At very short distances a pole is better than a cairn; but if the latter is built, it need not be more than 2 feet high, and should have a sharp-pointed stone sticking up on its top.

It is now possible to decide where to place the first stations, and an inspection of the figures on the plane-table may show that an alteration is advisable in the length or position of the base, in order to make the triangles more symmetrical. The ends of the base are now finally chosen, and marked with wooden pegs covered with small cairns of stones or mounds of earth, so as to be easily recognisable.

The next step is to measure the base as accurately as possible under the circumstances, and to observe the angles from both ends of it. If time permits, these observations should be computed, and the lengths of the sides of the triangles fixing the first stations should be obtained. The true bearing of the base and the latitude of one end of it should also be determined astronomically.

Next, the projection should be drawn on the plane-table (p. 93). The latitudes and longitudes of the points should be computed and plotted; or, if time does not allow this, the points must be plotted by bearing and distance.*

On arriving at one of the near points that he has already observed, the cairn must be taken down, and the theodolite placed over the spot. In order to make sure of finding this spot when the stones are removed, it is well to lay a couple of sticks or lines of stones so that their prolongations cut the highest point of the cairn at right angles. When the cairn is removed, there will then be no difficulty in finding the exact point where the theodolite should stand. When the observing is concluded, similar referring marks should be placed, to ensure that, when the cairn is rebuilt, its top is exactly over the spot where the theodolite stood.

In addition to observing the ends of the base and the points observed from the base, fresh objects must be selected for observation in the direction in which the surveyor expects to move, with a view to their use as theodolite stations. All natural features, too, which are

\* The *triangulation chart* should not, as a rule, be on a larger scale than 4 miles = 1 inch (say $\frac{1}{250,000}$) or the triangulator will run out of it much too soon. The topography may be on any scale.

likely to be of service as intersected points must be observed, in order that, when fixed, they may facilitate the delineation of the topography.

In a regular survey, when the triangulator starts work he is provided with a plane-table, on which all the points he has to observe are plotted with sufficient exactness to render it impossible for him to mistake or omit one of them. In rapid work, on the other hand, owing to the absence of a previous reconnaissance, the triangulator has to construct the diagram of his triangulation on his plane-table as he goes along.

Once the triangulation gets started clear of the base-line and its surrounding stations—which, being necessarily rather close together, are comparatively easy to recognise—the work of identification becomes more difficult, and to effect it with success the triangulator must trust to a skilful use of his plane-table in the following manner :—

Having oriented his plane-table correctly, and having ascertained by calculation, intersection, or resection the spot on it corresponding to the station from which he is observing, the triangulator draws lines to each of the natural features that he has observed, and writes against each of them the name that he has given it in the angle-book. To this he adds an outline sketch of the hills as they appear to the naked eye and their estimated distances, together with any descriptive remarks that are likely to aid in their recognition from other points of view. When the telescope discloses any peculiarities of shape that are invisible to the unaided eye, a sketch of the hill-top as seen through the telescope should be made in the angle-book.

At the next observing station the same process would be repeated.

From these two initial stations (the line joining them being as nearly at right angles to the line of probable route as possible), several hills that appear likely to serve as the next observing stations, besides other objects to be used as intersected points will probably be fairly well fixed; but others will remain still undetermined, either on account of the acuteness of the two rays, or because they have been observed from only one of the two stations.

The force of circumstances, such as the direction of the next advance during military operations, or the position of water, the direction of the route, or the best site for a camp in other cases, will generally decide which of the points observed to is to be made the next observing station. When this point is a conspicuous hill, there will be no difficulty in recognising it; but the task is far from easy when, as often happens in mountainous country, on descending from the initial stations on to the low ground along which the line of march runs, the observer finds that most, if not all, of the points to which the observations have been taken have disappeared from view.

It is in such a case that the plane-table becomes of the greatest value. By its means the position of the observer on the line of march, with reference to all points previously fixed, whether visible or invisible, may always be ascertained by resection, provided two of these are visible and sufficiently well situated to admit of their being used for this purpose. From such resections rays should be drawn to other points, and by this means a new series of points is obtained, by the help of which the observer will be able to continue his advance without losing the means of finding his position.

At the same time, it will be necessary to ascend some sufficiently commanding hill near the line of march, the position of which can be resected, in order to take second rays to such points as have not yet been sufficiently well fixed, before the altered point of view changes their appearance beyond recognition.

Supposing, now, that a hill has been reached which is believed to be a point previously observed to, and from which it is now intended to observe, the accuracy of the surmise can be tested as follows :—

The plane-table having been set up, its position is resected from the fixed points that are most suitably situated. If the fixing thus obtained falls upon the previous intersection, or upon the single ray, should only one have been taken, there is little doubt that the hill has been rightly identified. Once all doubt as to this has been removed, all the other already intersected points can be easily recognised, if visible, by directing the ruler on them in turn,

For the identification of the actual spot observed (the highest point of the hill for choice), with the theodolite the observer must trust to his recollection, aided by such notes and sketches as have been previously made.

It will be seen from these observations how large a part of the difficulty in executing a rapid or reconnaissance triangulation consists in selecting suitable points, and in recognising them from different points of view; and how necessary practice with the plane-table is, in order to make a proper use of its aid.

This process will be much facilitated if circumstances allow of the surveyor visiting a hill some two to four miles on the right or left (with reference to direction of route) of his observing station, and of approximately the same height, from which he can get second plane table rays to his newly-observed points. As the point of view will be so similar, he should have little difficulty in recognising them; and though the intersections will be very acute if the points are distant, still the approximate positions they will give will be most useful in aiding him to recognise his points.

The same principle can be very advantageously applied to triangulation (see Fig 30).

Here the stations C and D have been fixed from A and B. On visiting D, the triangulator selects F and G as his next forward stations, and observes to them. There is

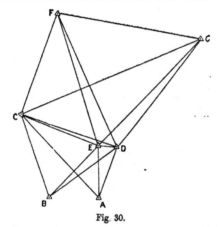

Fig. 30.

however, a considerable risk that when he visits C he may not be able to recognise these points with certainty. He therefore makes a station on E, a hill near D, which he should have observed from A and B, and from it he observes with the theodolite to F and G. If these hills are at all clearly marked, the triangles DEF and DEG, though very badly shaped, will suffice to enable him to fix the positions of F and G so nearly to the truth that, when they are plotted by their computed distances along the plane-table rays from D, he cannot possibly have any difficulty in recognising them from C.

In this way, too, a second triangle is secured to fix each of the points F and G. Should circumstances not allow of the theodolite being used at E, the intersection of plane-table rays drawn thence with those drawn from D will probably fix the positions of F and G sufficiently accurately for subsequent identification from C.

The length of sides permissible in the triangles depends upon the height of the stations and the extent of view from them, the nature of the objects observed, the clearness of the amosphere, and the maintenance of sufficient symmetry. In some countries, as far as view is concerned, triangles can be laid out of 30 and 40-mile sides, and intersected points as far distant as 100 miles may be fixed.

The difficulties connected with this class of work, however, generally make it advisable to keep the sides of the triangles between 10 and 20 miles in length.

**Theodolite Observations.**—Unless the points to be observed are extremely well defined, it is useless to multiply repetitions of the same angle. When observing to natural objects under Service conditions, one observation on each face is as much as can generally be obtained, but the triangulator should satisfy himself, before taking down his theodolite, that the observations on the two faces agree within small limits. If only one observation is taken to an object, no great reliance can be placed on it in view of the many possibilities of error in the observation, the reading of the verniers, and in the recording. When, however, two accordant values have been obtained on different faces, it is extremely unlikely that there is anything wrong. With regard to the selection of objects, it is a good rule to observe everything that is observable (i.e., every point that is sufficiently lofty to appear suitable for a station, and sharp enough to be intersected with accuracy), whether it lies in the probable direction of advance or not. It constantly happens that observations to points far away from the direction in which the triangulator expected to move turn out not only useful but indispensable.

In choosing intersected points, the great object is to fix such as will be useful for the topographers. As they do not always require to be fixed with the same accuracy as the stations, it is not necessary that they should be so sharply defined, but they should be conspicuous points, about which, when seen with the naked eye, there can be no mistake. Many a hill appears to be an admirable point to observe to until it is examined through a telescope, when it may turn out to have a flat, rounded top. Such a hill, unless previously marked with a pillar or pole, would make a bad station, but an excellent intersected point for plane-tabling purposes.

Generally speaking, the highest point of a hill should always be observed; as a rule, there is little chance of making a mistake as to which of a group is the highest, but when there are several close together of apparently equal height, they should all be observed. Occasionally it happens that the highest point, though suitable for the theodolite to stand on, and easily marked by a cairn when visited, offers no point, when first seen, that can be intersected with any accuracy. In such a case, if a little below the top there is a conspicuous rock or bush, it should be observed to, and the angle can be corrected afterwards to the actual position of the theodolite.

As each point is observed for the first time, its outline, as seen in the theodolite, is carefully sketched in the angle-book. If possible it is entered under its real name, but as this is seldom obtainable for a distant hill, it is usual to give it some descriptive name, such as black rock, jagged peak, &c., or else to give it a distinguishing letter or number. Whatever designation is given should be adhered to throughout the angle-book and computations. If, eventually, a reliable name is obtained, all the entries can be changed together.

**Computations.**—The method of computing rapid triangulation is similar to that already described, except that the computer must allow himself more latitude in dealing with any excess or defect on 180°, when the three angles of a triangle are observed, as it is seldom the case that all three are equally trustworthy, owing to the varying nature of the marks observed. If the computer is familiar with the method of calculating the latitudes and longitudes of the triangulated points, he will probably find it best to make it his practice to compute them, in spite of the additional labour involved. Plotting by latitude and longitude is far more accurate than building up triangles by distances; and the advantage for compilation of having all the topographical sketches properly located as regards latitude and longitude is alone sufficient to repay the labour of calculation. The distance between points not directly connected can be readily calculated, see Appendix III. This is particularly useful for interpolations. The advantage is, perhaps, most apparent when it is the triangulator's business to provide fixed points for several sketchers, more particularly if they are working some distance apart.

If his only means of communication with his assistants is by letter or heliograph, it is no easy matter to indicate to one of them how some newly fixed point should be plotted by distances, supposing, as may well happen, that the stations from which it has been fixed lie off his board. On the other hand, if the latitude and longitude have been computed and sent to the plane-tabler, he can have no difficulty whatever in plotting them correctly on his projected plane-table sheet.

It has been mentioned previously (p. 64) that in rapid work it may be sometimes impossible to observe all three angles of the triangles, and that by a skilful use of intersected

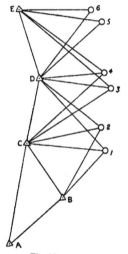

Fig. 31.

points triangulation may still be carried on without appreciable loss of accuracy. Instances of this are given in figs. 31, 32.

In these it is assumed that the observer finds it impossible, from the nature of the country or for other reasons, to visit any points better situated than ABCDE, and that AB is the last side that he has been able to fix in the ordinary way by observing all three angles.

A glance at the figures shows that BCDE are too nearly in a line for it to be possible to calculate with accuracy the triangles they form. If there are no well-marked points on either side, rapid triangulation of a continuous nature becomes impossible; but, if there are well-marked points on one flank, such as 1, 3, 5, the difficulty can be surmounted. If such points can be found in pairs, as 1, 2; 3, 4; 5, 6, the situation becomes much improved, and the triangulation would be computed as follows :—

Compute the side CB of the triangle ACB.

Then with BC as base, compute the triangles BC1 and BC2, the angles at 1 and 2 being the supplements of the angles at B and C.

With the base 1C compute the value of the side CD. Repeat the same operation with the base 2C.

Take the mean value of CD as base, and compute D3 and D4. From these two bases obtain DE, and from DE calculate E5, E6, and so on.

The most favourable case for this sort of triangulation is when points can be obtained on both sides of the line of stations, as in Fig. 32. Here the method of computing would be exactly similar, two values being obtained for each of the sides CD and DE.

In spite of all the surveyor's foresight and energy cases may arise, even in a favourable country, where continuous triangulation becomes impossible. For instance, after fixing one or more peaks on a range of hills, over which his route passes, he may be obliged to cross to the other side without having an opportunity of visiting any of these points. In such a case, if circumstances allow, the best course is to start a fresh triangulation on the further side and

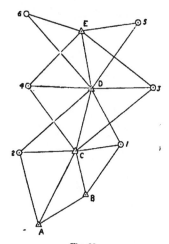

Fig. 32.

from it to intersect the peak or peaks already included in the previous triangulation. If this is done, the values of latitude and longitude of the new triangulation can be computed from the old one, and all the points fixed by both can be plotted correctly relatively to one another.

In this case also the methods of trigonometrical interpolation may be of great assistance (see p. 32).

If the march is very hurried indeed, even this may be impossible, and if the route lies east and west the relative longitudes of the different points along the line can be ascertained only by astronomical observations, or by a traverse. Which of these two methods was the more reliable would depend upon the circumstances under which they were carried out.

If, however, the route runs approximately north and south, a combination of observed latitudes with the azimuths of conspicuous points and the measurement of a few base lines will enable the surveyor to maintain a good value of longitude as long as the country is of a suitable character. For, where the latitudes of two points and the azimuth of the line joining them are known, the difference between their longitudes can easily be computed (see Chapter XV.)

Plate VIII.

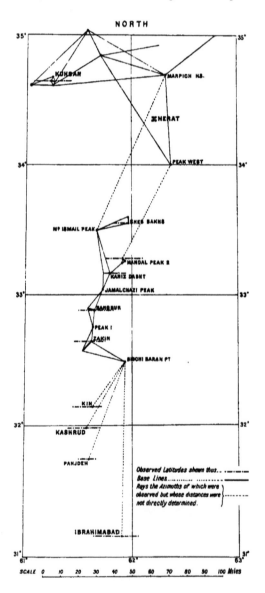

For purely topographical purposes the same result could be obtained by plotting the nitial point A and the azimuth from it to the second point, B. Then the point B will be at the spot where the azimuth cuts the parallel drawn through the observed latitude of B. A good instance of both this method and that above described, of starting a fresh triangulation when it is impossible to carry on the old one, is to be found in the march of the Afghan Boundary Commission to Herat in 1884, when it was considered of great importance to carry the true longitude from India up to the Afghan frontier (see Plate VIII.).

By various methods the longitude had been carried with fair accuracy as far as Ibrahimabad. At this point the route turned north, and a lofty hill, Bibchi Baran, became visible on the horizon in the direction of the line of march.

At Ibrahimabad, a latitude was observed and an astronomical azimuth taken to the Bibchi Baran peak, some 95 miles distant, almost due north. The column then marched through Panjdeh, Kashrud, and Kin, at which places latitudes were observed and azimuths taken to the same Bibchi Baran hill; these were afterwards used to fix the positions of these three places. At Zakin a 7-mile base was measured with a perambulator along the line of march, and observations were taken to Bibchi Baran and other peaks from each end; the true azimuth of the base line and latitude of Zakin being also observed.

From these data the distance from Zakin to Bibchi Baran peak, the difference of longitude between the two places, and the latitude of the latter, were computed. Then from the latitudes of Ibrahimabad and Bibchi Baran peak, and the observed azimuth of the ray Ibrahimabad to Bibchi Baran, the true longitude of Bibchi Baran peak, and thence that of Zakin were obtained. From Zakin onwards to Sangbur, Karez Dasht, and Mandal, short bases were daily measured on arrival in camp, and triangulation executed, by means of which each day points were fixed on ahead, which were observed back to from the next camp.

At night azimuths and latitudes were observed, the latter serving merely as checks on the triangulation. By these means the latitudes and longitudes were computed on from Zakin to Mandal. At Mandal an azimuth was observed to "Peak West," a hill some 60 miles to the N.N.E. The march was then continued through various places to Kuhsan, bases being measured almost daily and triangulation executed as before. It was found afterwards to be unnecessary to utilise these latter bases for the purpose of bringing up the longitude, but they served their purpose as aids to the topography at the time. At Kuhsan a base was carefully measured, and a comprehensive system of triangulation commenced, based on an observed latitude. This triangulation, which was afterwards extended to Mashhad in one direction and to Bamian in the other, included, among other points fixed, that called "Peak West," which had been observed from Mandal. The latitude of this "Peak West" and its difference of longitude from the Kuhsan base were then computed. With the latitudes of "Peak West" and Mandal, and the observed azimuth of the line Mandal to "Peak West," the difference of longitude between these two points was now calculated, and the true longitude thus brought up from Mandal through "Peak West" to Kuhsan.

A check was also obtained by an azimuth taken from Marpich Hill Station—a station of observation of the Kuhsan main triangulation—to Mahomad Ismail Peak, a point that had been fixed from the Karez Dasht base. The result on the whole was most satisfactory, the work standing rigorously any check applied to it by observations taken to various alternative points.

While on the march it was impossible to compute all the observations; in fact only those that were of immediate use to the topographers, such as the latitudes of the camps and the distances of some of the prominent peaks, were computed daily. Still, even with this limitation, the work could not have been carried through unless there had been a very strong party employed upon it, and it was found that the trigonometrical and astronomical observations, together with a plane-table and perambulator traverse, on the scale of 1 inch to 4 miles, provided full employment for three officers and three native surveyors.

As a measure of the rapidity with which this method can be conducted in a very favourable country, such as this was, it may be noticed that the whole distance from Ibrahimabad to Kuhsan, some 310 miles, was covered in 19 days including halts, during which the connection in longitude was kept unbroken. In the diagram of the triangulation, Plate VIII., many of the daily local triangulations are omitted, as it was found unnecessary to use them when the triangulation was subsequently extended from Kuhsan. As regards the accuracy, it was estimated that the error in the longitude of Kuhsan relatively to Ibrahimabad, computed as described above, probably did not exceed 5 seconds of arc, that is less than 200 yards. If there had been no opportunity of starting an extensive triangulation from Kuhsan and working back, and the daily work of the march had been alone available, it is probable that the error in the longitude of Kuhsan would not have exceeded half a mile.

**Interpolations.** (See also p. 30).—In rapid work interpolations play a more important part than they do in regular surveys. By means of this method useful stations can sometimes be made in the middle of a valley where there is no marked spot to which it would be possible to observe.

In interpolating a station it is always advisable and sometimes indispensable to observe the azimuth of one of the points from which the station is to be fixed.

For instance, supposing that the surveyor finds, that, owing to the configuration of the country and the exigencies of the occasion, his triangulation ends in an intersected point A

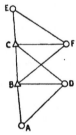

Fig. 33.

and a hill station B situated on a range of hills running parallel to the direction in which he proposes to travel, and that from B he has observed to C, another point on the same range. The surveyor having crossed the range ABC is encamped at D on flat ground.

Now, if a third fixed point were visible in a suitable position, it would be possible for him to interpolate the position of D as described on p. 32; but as there are only two fixed points visible he must observe at D the azimuth of either A or B as well as the angle between them, as described on p. 33.

In the same way the position of the next camp, F, can be interpolated, and if the observer visits C, and observes E from it, the position of E can be fixed, and so on.

In this way it would be possible to carry a triangulation along a single ridge of hills without ever observing from it to any points on a flank, and though the work would not be of very great accuracy, still, if the triangles were fairly well conditioned, the points on the hills were well marked, and the azimuth observed with accuracy, no great error could creep in.

In all rapid triangulation, where it is always an object to observe from as few stations as possible, the important thing to remember is to keep a continuous series of azimuths. This is attained by ensuring that from each station at least one back station is observed, from which it has itself been observed. When it is impossible to do this, as in the instance just described, the azimuth must be observed astronomically.

Of course, the use of azimuths involves the calculation of the latitudes and longitudes of all the stations and intersected points of the triangulation, a procedure which has been advocated above on other grounds.

Perhaps the most successful application of the method of interpolation ever utilized in a new country was made when triangulation was carried across the "snowy range" of the Himalayas into the Pamirs, for the Boundary Commission of 1895. The interpolation was made from peaks of one of the secondary series of the Himalayan triangulation.

**When Triangulation is impossible,** owing to the nature of the country or the extreme rapidity of the march, if north to south or diagonal rays can be obtained the next best thing is to provide a framework of points by means of latitudes and azimuths (see Chapter XV.).

If, on an expedition, even this is impossible, positions must be fixed by rough compass traverses controlled by astronomical observations. This is the explorer's method of mapping, and is necessarily of a low standard of accuracy. If the march is generally north to south, observed latitudes will serve to check the distances traversed; if east to west, they will check the bearings observed; if the march is in a diagonal direction, the observed latitudes will serve to check both bearings and distances. For the methods of determining latitudes see Chapter XIV.

## CHAPTER VIII.

### LEVELLING.

**Levelling** is the art of determining the relative heights of various points referred to a hypothetical surface which cuts the direction of gravity everywhere at right angles. This surface is consequently slightly irregular where the direction of gravity is influenced by local and abnormal attractions.

Mean sea-level, at some selected station on a coast, is the usual datum to which such heights are referred, though other arbitrary levels may be assumed if necessary.

There are three principal methods by which relative heights may be determined.

(i.) The first and most accurate method is by spirit levelling.

(ii.) The second, in order of accuracy, is by trigonometrical or angular measurement. This method has the advantage that it enables the relative heights of inaccessible points to be determined.

(iii.) The third method is by the comparison of the barometric or atmospheric pressure at the required points.

The last two are indirect methods of levelling, and are described elsewhere in this book. The first is the only method which is, technically speaking, included in the term "levelling."

By levelling is meant the direct determination of the differences of height of various points from the readings of the heights at which graduated staves, held vertically over these points, are cut by the horizontal plane which passes through the eye of the observer.

Direct levelling may be subdivided into :—

(a) Geodetic, or precise levelling.
(b) Ordinary, or engineering levelling.

These two classes of levelling are similar in general principles, but differ chiefly in the details of their execution, particularly as regards the degree of accuracy to which personal and instrumental errors are eliminated.

**Geodetic or Precise Levelling** has for its object the determination of the heights above mean sea-level of a number of fixed points, known as "bench-marks," scattered over the country. The main object is to afford reliable points of reference for mapping and engineering purposes, consequently every precaution should be taken to make these heights as accurate as possible. Geodetic lines of levels are run so as to cross and recross each other, forming a network over the country. The crossing points, where there are bench marks common to two or more lines, afford a means of checking the accuracy of the work. This class of levelling is of a special character and of a high order of accuracy.

We are here only concerned with **Ordinary** or **Engineering levelling**, such as is required for general engineering purposes, as the construction and laying out of irrigation works, roads, railways, canals, buildings, &c. ; also for the accurate contouring of maps.

#### INSTRUMENTS.

The instrument generally employed is known as a **level**, though any theodolite, which carries a spirit level attached to its telescope, may be used for this purpose, and is, at times, more expeditious though less accurate than the regular level.

The essential part of all levels is a telescope provided with a long spirit level and containing a diaphragm carrying a horizontal and (usually) two vertical wires.

PLATE IX.

Cooke's Reversible Level.

The telescope is capable of being revolved in azimuth about a vertical axis, and is secured to the usual tripod stand. It is supplied with three foot screws for the purpose of making the axis, round which the telescope revolves, truly vertical.

Some old pattern levels are mounted on four instead of three foot screws; this is a mechanically unsound method of support.

A level is described by the focal length* of its telescope, and sometimes, in addition, by the aperture of the object-glass.

Fourteen inches focal length is the size in most common use. Ten-inch levels are used for rough surveys, and 16 to 26-inch levels for very exact work.

**Patterns of Levels.**—There are several kinds of levels, of which the Dumpy, Cooke's Reversible and the " Y " pattern are the ones most frequently used. The two latter are more readily adjusted than the Dumpy, and are generally used for ordinary levelling operations.

It will be sufficient here to describe Cooke's reversible level, which is the pattern most generally used in the Service, and to explain the adjustments of all three above-named patterns.

### COOKE'S REVERSIBLE LEVEL.

An illustration of this instrument is given on Plate IX.

Its construction is shown below, where fig. 34 is the telescope (in longitudinal elevation). This telescope has two equal cylindrical flanges, F and F', turned on it concentrically with the tube.

The cross-wires, consisting of one horizontal and two subsidiary vertical wires, are lines ruled upon parallel glass, carried by a perforated block in the eye end which is adjustable vertically to correct for collimation by the two antagonising diaphragm screws $d$ and $d'$.

The eye-piece is focussed by pushing it in or out. The object-glass is focussed by turning the milled head M.

The two flanges F and F' fit closely inside two corresponding collars, S and S', *vide* figs. 35 and 36.

When the cover of the object-glass and the screw Sc are removed, the telescope can be introduced indifferently from either end of the socket SS' (fig. 35) and pushed home until the stop flange St (fig. 34) comes in contact with the socket end S or S'.

Great care should be used in reversing the instrument, not to allow any grit to enter the sockets.

One end S of the socket is furnished with a threaded bolt which passes through a hole in the base plate PP' (fig. 36), and is adjustable vertically by means of the two lock-nuts N and N'.

**1. Adjustment for Parallax.**—To make the focus of the eye-piece and the image formed by the object-glass coincide on the plane of the cross-wires.

First adjust the eye-piece by carefully pushing it in or out in its tube, until the cross-wires are seen as distinctly as possible, after which there should be no necessity to adjust the eye-piece again, then focus the telescope on some well-defined object at the average distance at which the levelling staff may be expected to be held (say 200 feet).

If, when the eye is moved gently up and down behind the eye-piece, the image of the object does not move in the slightest degree with respect to the horizontal cross-wire, then the telescope is correctly focussed, but if the image seems to follow the movements of the eye, that is, to tend to move above the cross-wire when the eye is raised and to move below it when

---

* The focal length of a level is usually measured from the front shoulder of the object-glass to the cross-wires when the telescope is focussed for parallel rays. It would be sufficiently accurate to focus on an object 2 miles distant. This is near enough for most purposes, though not theoretically correct, the true focal length being measured from the cross-wires to the optical centre of the object-glass, and the latter point is a varying distance (according to size) behind the front shoulder.

the eye is lowered, then the image is not focussed exactly upon the cross-wire, but lies on the side towards the object-glass, which must consequently be made to approach the cross-wires by means of the milled head M (*vide* fig. 36). The reverse is the case when the image tends to move below the cross-wire when the eye is raised and *vice versa*, that is, the error of parallax must be corrected by moving the object-glass farther away from the cross-wires.

The above adjustment for parallax is the same for all levels and theodolites, and must be carried out before any correct observations can be taken. This adjustment should precede all others, including even the levelling of an instrument for ordinary use.

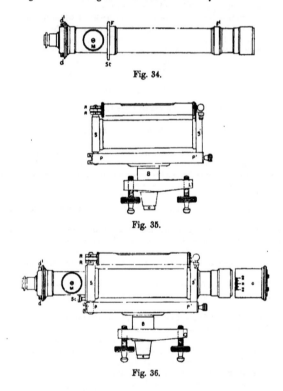

Fig. 34.

Fig. 35.

Fig. 36.

Generally, when once set for the day, there is no occasion for altering the focus of the eye-piece, but the focus of the object-glass will obviously require alteration as the distance of the object varies.

**2. Adjustment for Collimation.**—To make the line of collimation\* coincident with the axis of rotation of the telescope in its sockets.

---

\* The *line of collimation* is the line joining the optical centre of the object glass and the intersection of the cross-hairs.

Having set the telescope over a foot screw, direct it on an object as in the last adjustment and bisect this object with the horizontal cross-wire; then clamp the instrument, withdraw the fixing screw Sc (*vide* fig. 36) and twist or rotate the telescope half way round in its socket, so that the screw hole, through which Sc passed, is brought to the top, and the cross-wire is again horizontal but reversed. If the object is not still bisected, the error of collimation is corrected half by means of the foot screw lying under the telescope, and half by the antagonising diaphragm screws $d$ and $d'$.

**3. Adjustment for making the "Horizontal" Axis of the Reversible Socket exactly at Right Angles to the "Vertical" Axis.**—The terms "horizontal" and "vertical" are here used in an explanatory sense, as the two axes are not made truly horizontal and vertical until the final adjustment is carried out. Referring to fig. 36, this adjustment is to make the "axis of figure" of the reversible socket SS' exactly at right angles to the "vertical" axis B.

First remove the cover of the object-glass and the screw Sc (*vide* fig. 36), then set the telescope over a foot screw, direct on and bisect an object, as in the last adjustment. Withdraw the telescope carefully from its socket SS'; then turn the socket round, end for end, and carefully insert the telescope again, keeping the screw hole in the stop flange St, through which the screw Sc passes, vertically below the centre of the tube when the telescope is pushed home. The socket SS' will now be reversed with reference to the telescope. Again direct on the object, and if it be not now bisected by the horizontal cross-wire, correct half the error by the foot screw below the telescope and half by the lock nuts N, N.

This operation of reversing the socket on the telescope and the correction of the error, half by the foot screw and half by the lock nuts, should be repeated until the bisection of the image remains undisturbed by the reversal of the socket. When the adjustment is complete, the telescope ought to lie in its socket with its stop flange St against the end S of the collar, so that it may be secured by the screw Sc.

Care should be taken that no dust is allowed to settle on the flanges of the telescope during its insertion in the socket.

**4. Adjustment of Vertical Axis of Rotation.**—That is, to make the vertical axis of rotation truly vertical, and to make the telescope level at right angles to it.

This is the same adjustment which is gone through for levelling a theodolite on first setting up.

Place the telescope parallel to two foot screws, and with them alone bring the bubble approximately to the centre of its run, noting the reading at one end of the bubble (say the end near the capstan-headed screws).

Turn the telescope through 180 degrees until it is again parallel to the same two foot screws, and again note the reading of the same end of the bubble. By means of the two foot screws, bring this end of the bubble to the mean of the two readings. Now turn the telescope until it is over the third foot screw, and by it alone bring the same end of the bubble to this mean reading.

Repeat until one end of the bubble reads the same in any position, when the upright axis will be truly vertical. The bubble can then be brought to the centre of its run by means of the two lock nuts R, R at one end of the level (*vide* fig. 36).

It is obvious, that if the third adjustment is carried out and both axes of the instrument are thereby made exactly at right angles to each other, then the last adjustment necessarily makes the horizontal axis of the reversible socket truly horizontal.

All the above adjustments should be repeated until the required degree of accuracy is attained.

Slight errors in adjustment may, in many cases, be eliminated as explained later in the procedure for levelling.

## THE "Y" PATTERN LEVEL.

An illustration of this instrument is given in Plate X.
The following are the necessary adjustments :—

1. Parallax ;
2. Collimation ;
3. Telescope level ;
4. Vertical axis of rotation.

The first and last adjustments are the same as those already described for the Cooke's pattern level.

**Adjustment for Collimation.**—Having set the telescope over a foot screw, turn the telescope, by a rotary motion in the Y's, until one pair of diaphragm screws is vertical, and direct the intersection of the cross-wires on some distant and well-defined object. Now turn the telescope half round in the Y's, and if the horizontal wire has moved vertically from the object, bring it back half by the foot screw and half by the vertical diaphragm screws.

**Adjustment of Telescope Level.**—To make the telescope level parallel to the axis of revolution of the telescope in the Y's, and therefore parallel to the line of collimation, bring the bubble of the telescope level to the centre of its run by means of two foot screws ; reverse the telescope end for end in the Y's, and if the bubble does not still remain in the centre, adjust by means of the screws, giving vertical motion at the other end of the telescope level. Repeat until perfect.

## THE DUMPY LEVEL.

This is the usual pattern of non-reversible level. An illustration is given in Plate XI.
The adjustments of the Dumpy level are for :—

1. Parallax;
2. Vertical axis of rotation ;
3. Collimation.

The first two adjustments are the same as the first and last adjustments of the Cooke's reversible level already described, but, owing to the Dumpy level not being reversible, it is impossible to correct for collimation in the same way.

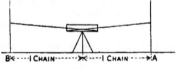

Fig. 37.

Therefore, instead of this, the line of collimation (*i.e.*, the line joining the optical centre of the object-glass and the intersection of the cross-wires) is made at right angles to the vertical axis of rotation.

The method of carrying out this **adjustment for collimation** of a Dumpy level is as follows :—

Select a level piece of ground, set up the instrument in the centre, and carry out the first

PLATE X.

Y LEVEL.

PLATE XI.

Dumpy Level.

two adjustments explained above. Then drive two pickets, A and B, into the ground about a chain away, but exactly equidistant from the instrument and on either side of it (see fig. 37). Place a levelling staff on A (if the lower picket), and note the reading. Then place the staff on the picket B, and drive it in until the reading of the staff is the same as it was at A. The pickets A and B will now be on the same level, notwithstanding any collimation error there may be.

Now set up the level beyond either picket and in the same line, as at C (see fig. 38), making the distance CB some simple fraction of the distance AB, say = ¼ AB, or ½ chain.

Level the instrument and take readings of the staff held at A and B in turn. If the readings are not the same, make them so by using the diaphragm screws only. Thus if $x$ is the

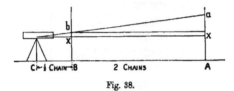

Fig. 38.

required reading on the staff in fig. 38, to which the cross-wires should be lowered (or raised), and if $a$ is the reading on A, and $b$ the reading on B, then by similar triangles $b - x = \frac{1}{4}(a - b)$ = amount by which the readings on B must be diminished (or increased) to complete adjustment.*

### LEVELLING STAVES.

There is a great variety of these, but all English patterns are divided into feet, tenths and hundredths of a foot, and are usually from 10 to 14 feet long. Some staves are jointed or telescopic for convenience in use and in transport, but these patterns are liable to introduce inaccuracy in the work, and for this reason a staff in one single length is to be preferred, like the Ordnance Survey pattern, which is 10 feet long. A full-sized section of this staff is shown in fig. 39. The method of reading the numbers is shown by the figures against the dotted lines. This staff is graduated on both sides; on one side the readings start at zero, on the other they start at 0·3 of a foot, or any other arbitrary number. This obviates all bias in taking readings on both sides of the staff, as is done in most accurate work. The difference between the two readings should be a constant, if correctly observed.

When one side of the staff only is read, care must be taken that the same side is always held to face the observer.

To obtain a correct reading on the staff, it is evident that it must be held in a vertical position. This is usually done by the staff-holder swaying the staff slowly backwards and forwards on both sides of the vertical, when the lowest observed reading will correspond to the vertical position of the staff. The staff-holder, standing behind the staff, can usually see whether it is in the same vertical plane as the instrument, if not, the observer at the instrument can always direct him by the vertical cross-lines in his level.

Another means of ensuring the perpendicularity of the staff is to attach a pendulum to the side, so that the point of the pendulum fits loosely in a ring below the attachment, which

---

* The flaw in this method is the necessity of having to alter the focus between readings of the staff at A and B, in fig. 38, on account of B being so much nearer the instrument than A. There is a method (Gauss') which obviates this, but it requires two levels or theodolites to carry it out.

is so adjusted that the staff is perpendicular in all directions when the pendulum-point hangs in the centre of the ring.

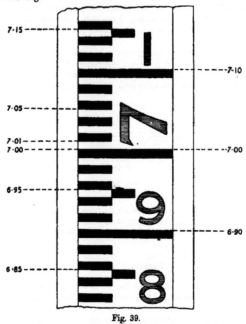

Fig. 39.

## CURVATURE.

By means of the level a line can be traced tangential to the surface of the earth at the point where the instrument is set up. Owing to the sphericity of the earth it is evident

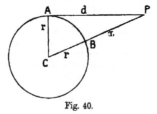

Fig. 40.

that the further this tangential line is produced the further it moves from the centre of the earth.

In fig. 40, representing a vertical section through the centre of the earth, let A be the

observer's position, P that of the distant object, situated in the plane tangential to the surface of the earth at A. Let C be the centre of the earth. Join CP, cutting the circumference at B. Call BP, $x$, and AP, $d$, and AC and BC, each representing the radius of the earth $r$.
Then
$$(r + x)^2 = r^2 + d^2 \quad \text{or} \quad (2r + x)\,x = d^2.$$

But the diameter of the earth, $2r$, being so great in comparison with the quantity, $x$, at all distances to which a levelling operation usually extends, $2r$ may be taken for $2r + x$ without sensible error.
Then
$$2rx = d^2 \quad \text{and} \quad x = \frac{d^2}{2r}.$$

The mean diameter of the earth may be taken as 7,916 miles, therefore if $d = 1$ mile, the excess $x$ of the apparent over the true level $= \frac{1}{7.916}$ of a mile, or a little over 8 inches; at 2 miles it is four times that quantity, or 32 inches, and so on, increasing in proportion to the square of the distance. Hence the simple formula is deduced that the *correction for curvature in feet is two-thirds of the square of the distance in miles.* That is to say, if a man were swimming in a perfectly smooth sea, with his eyes on a level with the surface of the water, he would be just able, as regards curvature, to see the top of an object floating in the sea a mile off, if it protruded 8 inches above the surface.

### CURVATURE AND REFRACTION.

The distance at which objects are visible is, however, augmented by the effects of refraction, which makes objects, as a rule, appear higher than they really are.

Taking the normal amount of refraction in the middle of the day, it would increase the distance at which an object of a certain height could be seen, taking curvature only into consideration, by about $\frac{1}{15}$.

This gives the approximate formula for both curvature and refraction, that the height in feet of the distant object equals $\frac{4}{7}$ of the square of the number of miles that it is distant from the observer, or $h = \frac{4}{7} d^2$.

The effect of refraction, however, varies so much with the place, season, and time of day, that it is necessary to adopt some method of observing that will eliminate it. This is attained by placing the instrument at an equal distance from any two points to which observations are taken. If this is done, the errors in the readings of both staves will be equal in amount even if the telescope is not perfectly collimated, and consequently the differential reading will be correct.

This method also obviates the necessity of changing the focus of the telescope and so avoids the possibility of introducing an instrumental error caused by the focussing tube not moving truly parallel to the visual axis of the instrument on a change of focus.

### PROCEDURE WHEN LEVELLING.

Suppose the spirit-level to be set up and adjusted at A (fig. 41), directed on the levelling staff at 1, and the figure noted which is cut by the horizontal wire. This figure would be the difference of level between 1 and A, minus the height of the horizontal cross-wire of the instrument above the ground at A (which might be measured and deducted). Suppose the staff was now held at 5 and the telescope directed on it, the difference between the readings at 1 and 5 would evidently be the difference of level between these two points, and is independent of the height of the instrument above the ground which is common to both; assuming that points 1 and 5 are equidistant from the instrument at A, they would be equally affected by curvature,* refraction, and instrumental errors, which might therefore be ignored.

The levelling staff might also be held at intermediate points 2, 3, 4, &c., and the readings noted. It would then only be necessary to measure the horizontal distance between these points in order to obtain the data for drawing a section of the ground between them, as shown in fig. 41.

It is evident that, unless the intermediate points 2, 3, and 4 are the same distance from the instrument at A as 1 and 5, the accuracy of the determination of their heights will be slightly affected by curvature, refraction, and instrumental errors.

These errors will not be cumulative if entries of level readings are made as shown in the form of field book described on the next page, in which the levels of successive sets of readings depend only on the heights of the forward and back stations.

If it were required to continue the levelling beyond 5, the spirit-level would be carried forward and set up in such a position B and at such a height that it might intersect some part

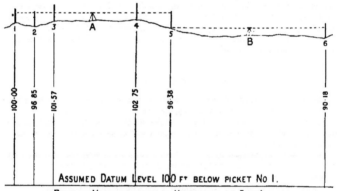

Fig. 41.

of the staff at 5, care being taken not to alter the position of the staff meanwhile, but merely to turn the required face towards the level.

As regards the length of sights, it is advisable not to exceed 100 yards for accurate work with a 14-inch level. For relatively unimportant work, the length is only limited by the power of the telescope used.

In the operation represented in fig. 41, where the levelling commenced at 1 and proceeded forward to the right, the reading of the staff at 1 would be called the "back reading" and the reading of the staff at 5 the "forward reading." The readings at 2, 3, and 4 are called "intermediate readings."

On setting up the level at B, the point 5 becomes the back station and point 6 the

* This is only strictly correct when the readings on the staff are taken at the same height above homogeneous ground, as refraction varies with the altitude of the ray and is also influenced by the nature and temperature of the soil.

FORM J.

| Number of Picket. | Horizontal Distance in Feet. | Back Reading. | Intermediate Reading. | Forward Reading. | Height of Instrument. | Reduced Level. | Remarks. |
|---|---|---|---|---|---|---|---|
| 1 | 0 | { 5·98<br>6·26 | | | 105·98 | 100·00 | Assumed |
| 2 | 63 | | { 9·13<br>9·43 | | | 96·85 | |
| 3 | 126 | | { 4·41<br>4·71 | | | 101·57 | |
| 4 | 399 | | { 3·23<br>3·53 | | | 102·75 | |
| 5 | 504 | { 0·61<br>0·91 | | { 9·60<br>9·90 | 96·99 | 96·38 | |
| 6 | 1008 | | | { 6·81<br>7·11 | | 90·18 | |
| &c. | &c. | | | | | &c. | |
| | Totals | 13·78 | | 33·42<br>− 13·78 | | | |
| | For double readings divide by ... | | | 2 ) 19·64 | | | |
| | Total amount of fall | | | = 9·82 | | | |

Heights from above Column of Reduced Levels—

Picket 1 .... . 100·00
Picket 6 . 90·18

Total fall . 9·82

forward station. Similarly each forward station becomes a back station when the instrument is moved forward. The stations are called "back" and "forward" with reference to their relation to each other on the line of levels, and not as regards the instrument which might be to the side and before or behind both.

For ordinary levelling operations the form of field book on p. 83 will be found generally convenient; the entries refer to fig. 41 :—

In this case, two readings have been taken at every picket, one on each side of an Ordnance Survey pattern staff. This insures greater accuracy, as the difference of two readings at any picket must be a constant if the readings are correctly taken and if the staff-holder is careful not to knock the picket or press it into the ground while swaying the staff on it.

For rough work it is sufficient to take single readings on a staff, but when it is graduated on both sides, great care must be taken to ensure that the same side is always made to face the instrument.

In the above field-book form the "reduced level" of picket No. 1 is assumed as 100 feet, and the heights of all the points along the line of levels are referred to a hypothetical datum plane passing 100 feet below picket No. 1.

It is convenient that the datum level should be lower than any part of the section, so that all measurements may be taken above it.

The horizontal distance of each picket from the starting point is entered in column 2 of the field-book form, and is measured by chain or tape, the deduction for the degree of slope being made by holding the chain horizontally when measuring up or down hill.

The entries in columns 3, 4, and 5 are the readings of the levelling staff at the "back," "intermediate," and "forward" stations respectively.

The "height of the instrument" is entered in column 6, and is obtained by adding the "back reading" to the "reduced level" of the back station or starting point, thus the height of the instrument at A = 100·00 + 5·98 = 105·98 and at B = 96·38 + 0·61 = 96·99.

The "reduced level" of an intermediate or of a forward station is entered in column 7, and is found by adding the "back reading" to the "reduced level" of the station on which the back reading is taken (shown in column 6 as "Height of instrument") and subtracting the intermediate or forward reading from this. Thus the "reduced level" of picket No. 4 would be 105·98 − 3·23 = 102·75 feet, and of picket No. 5, 105·98 − 9·60 = 96·38 feet above datum level.

When it is only required to find the difference of level between two terminal points, it is only necessary to record the back and forward readings at the various stations, and the difference of their sums will give the required difference of level; if the sum of the back readings is greater than the sum of the forward readings, then the difference of level is a rise, and *vice versâ*. If the levelling operation is carried back to the starting point, it is clear that the sum of the back readings will be equal to the sum of the forward readings, if the work is accurately performed.

This method of finding the difference of height between two terminal points is usually called "check levelling," and is useful for checking a series of levels taken in greater detail or in "closing" a detailed series on to the starting point or on to the nearest known level, such as a bench mark of the Ordnance Survey.

In extensive levelling operations it is usual to leave permanent marks, called "bench marks," on objects such as milestones and walls, and to record their exact heights for future reference.

In Ordnance Survey maps the levels of various bench marks are figured with their height in feet and decimals of feet above mean sea-level at Liverpool, which is the datum to which all altitudes in this survey are referred.

The Ordnance Survey bench-marks are of this pattern, ⏀, cut about half an inch deep into the vertical face of brick or stone structures, the exact point of reference being the centre of the V-shaped cut in the horizontal line.

## Levelling with the Theodolite.

Sections of ground may be taken with the theodolite more expeditiously, though less accurately, than with the regular level. Thus :—

Let the theodolite be set up as shown in the figure and its telescope directed parallel to the general slope of the hill. Observe this vertical angle with all precautions for accuracy proper to the operation.

Keeping the telescope clamped in this position, note the successive readings of a levelling staff held by an assistant at the various points which will best define the irregularities of the

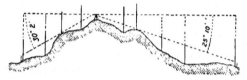

Fig. 42.

section; the height of the theodolite and the horizontal distance between these points being measured, a section can be drawn thus :—

Draw the horizontal line passing through the telescope. Protract the observed angle of inclination (as shown in the figure). On the horizontal line set off the measured distances to scale. Draw vertical lines from these points cutting the hypotenuse. Beneath each point thus found set off the staff reading, being the level of the ground where the foot of the staff stood for each observation.

Should the staff be so distant that the markings are illegible, a system of signals can be agreed upon, so that the assistant can move a visible mark along the face of the staff until signalled correct; the assistant can then note the reading.

## CHAPTER IX

### HEIGHTS.

THE datum to which all heights should be referred is mean sea-level. For the determination of M.S.L. see p. 30.

Heights may be determined by any one of the methods in the following list; the methods are given in order of their accuracy:—

    (i.) By levelling.
    (ii.) Trigonometrically (by vertical angles with a theodolite).
    (iii.) By clinometer (for short distances).
    (iv.) By barometer (mercurial or aneroid).
    (v.) By hypsometer (boiling-point thermometer).

As regards accuracy (i.), a line of good (not geodetic) levels should not produce an error of 0·1 foot per mile or 1 foot in 100 miles, or 2 feet in 400 miles. (The error may be assumed to increase as the square root of the distance.)

(ii.) In principal and secondary triangulation it is found that for any given ray of 30 miles or so an error of, say, 1 or 2 feet may be expected, hence we might expect an accumulated error of about 4 to 8 feet in 500 miles. In topographical triangulation the accuracy of the heights will largely depend on the nature of the signals used. In rapid triangulation under favourable conditions, the error might be 10 feet in 200 miles.

(iii.) Clinometer heights are only used for short rays up to 2 miles, and depend on neighbouring trigonometrical heights; each clinometer height will be correct within 2 or 3 feet.

(iv.) Barometer heights should be used in dependence on level or trigonometrical heights; judiciously used, heights may be determined with an aneroid barometer within 10 or 20 feet. "Absolute" determinations with a barometer will give large errors similar to those resulting from the use of the hypsometer.

(v.) Hypsometer heights depend on the actual air pressure at the moment; the hypsometer is necessarily a rough instrument; an error of $\frac{1}{10}°$ Fahrenheit corresponds roughly to an error in height of 55 feet. When every precaution has been taken, we may assume that a hypsometer height taken inland in a new country is liable to be at least 100 feet in error.

    For determinations of level heights, see Chapter VIII.
             "     "     trigonometrical heights, see Chapter III.
             "     "     clinometer heights, see Chapter VI.

**Barometer Heights.**—Barometer heights are frequently of great use in places where no trigonometrical station can be fixed, such as in forests, valleys and deep ravines. Barometer heights should always, where possible, be based on level or trigonometrical heights. Thus suppose a trigonometrical station to be on the top of a 2,000-foot hill; suppose, as often occurs, that the adjoining valleys are thickly wooded, it is nevertheless very desirable to fix some heights in the valley. This is best done by reading the barometer and thermometer on the top of the hill, then reading them at various points in the valley, and then, if possible, reading them again on the top. Aneroid barometers are most convenient for the purpose.

**Use of the Aneroid Barometer.**—Barometers should not be used near the limit of their range; thus a barometer graduated to 5,000 feet should not be used much above 4,000 feet.

Barometers should be tapped gently with the nail or a pencil before reading.

Barometers take some little time to record the correct reading if a change of altitude has been rapidly made, they should therefore be given a rest of a few minutes.

They should always be read in the same position, either when horizontal, or vertical with the needle on a level with the eye, which should be always in the same relative position to the needle to avoid the effect of parallax.

The thermometer (in the shade) should be read at the same time as the barometer.

If possible get a Kew certificate for each barometer in use, or compare with a mercurial barometer to ascertain index error.

For determining differences of height with an aneroid barometer, see Tables V. and VI.

**Use of Tables.—**

If $\beta$ be the reading of barometer at lower station in inches,

$\beta'$ „ „ „ upper „ „

$t$ „ „ Fahrenheit thermometer at lower station,

$t'$ „ „ „ „ upper „

A the number in Table V. corresponding to $t$ and $t'$,

C „ „ VI. „ latitude,

make $\quad\quad\quad\quad\quad D = \log \beta - \log \beta'$,

then, $\quad\quad\quad \log D + A + C = \log$ difference of altitude in feet.

*Example.*—In lat. 51° 31' the aneroid barometer at the lower station read 29·81 inches at the upper station 25·55. The readings of the thermometers in the shade were 63° and 50°·5 F.

What was the difference of altitude?

$$\text{Log } 29\cdot81 = 1\cdot47436$$
$$\text{Log } 25\cdot55 = 1\cdot40739$$
$$D = 0\cdot06697$$

$$\text{Log } D = \bar{2}\cdot82588$$
$$A = 4\cdot80339$$
$$C = \bar{1}\cdot99973$$

$$\text{Log } h = 3\cdot62900$$

$$h = \quad 4256 \text{ the required difference of height (in feet).}$$

When time does not allow of this computation, a good aneroid will give fair indications of differential heights taken within a short interval of time, if the scale of heights given in feet is used.

**Hypsometer Heights.—**The hypsometer, or boiling-point apparatus, consists of a thermometer graduated from 180° to 215°, a spirit lamp, boiler, and a telescopic tube which screws on to the top of the boiler, and in which the thermometer, supported by a rubber washer, is placed when in use. The boiler should be about a quarter full, and the telescopic tube so adjusted that the bulb of the thermometer is just clear of the water. When the water boils, the steam ascends through the tube and causes the mercury to rise. When the mercury becomes stationary, read the thermometer, and at the same time take the air temperature with an ordinary thermometer.

To compute the difference in altitude between two stations by boiling-point observations, take from Table VII. the numbers opposite to the boiling-points at the two stations. The difference between these numbers is the difference in height when the mean temperature of intermediate air is 32° F. When the temperature differs from 32°, the difference between the numbers obtained from Table VII. must be multiplied by the multiplier given in Table VIII. for the given temperature.

The third column of Table VII. gives a ready means of comparing the readings of the aneroid with those of the hypsometer.

When observations are only taken at the upper station, and its absolute altitude has to be determined, the mean barometric reading for the time of the year reduced to the sea-level in the locality must be taken as the reading of the lower station. The best approximations to the true values will be found in the isobaric charts given in "Hints to Travellers."

Whenever the observations at the two stations under comparison are not simultaneous, or when the mean barometer is taken in place of the lower station, the correction for the diurnal variation must not be omitted. In the tropics there is a regular diurnal oscillation amounting to one-tenth of an inch approximately, the maximum daily readings occuring at about 9 a.m. and 9 p.m., and the minimum readings about 3 a.m. and 3 p.m. In other respects the barometer is very steady.

Determinations of height by the hypsometer can only be looked upon as very rough approximations. The determination is affected by atmospheric changes, by errors in graduation of the thermometer, by errors in reading, by the purity of the water used, and by other factors.

As an illustration of the discrepancies to be expected in the determination of the height of an inland station by this means, the following list of hypsometric values of the height of Lake Tanganyika is given:—

| | |
|---|---|
| Baumann | 2,886 feet |
| Stanley | 2,755 ,, |
| Hore | 2,749 ,, |
| Cameron | 2,709 ,, |
| Stairs | 2,690 ,, |
| Wissmann | 2,670 ,, |
| Popelin | 2,663 ,, |
| Reichard | 2,558 ,, |

the mean value being 2,713 and the range 328 feet. It is not known what the true value is.

## CHAPTER X.

### CONTOURING AND HILL SKETCHING.

**A Contour** is the representation on a map of an imaginary line on the ground joining all immediately adjacent points which are the same height above mean sea-level. For small areas a contour may be considered to be the outline of the intersection of the surface of the ground with a horizontal plane.

Contours vary in accuracy from deliberately surveyed contours determined by instrumental levelling, to sketch-contours drawn in by eye; such sketch contours are called **form-lines**.

In determining the accuracy necessary for the contouring of a map, the principal factor is the scale of the map. Suppose, for instance, that it is required to show the hill features in a plane-table survey on a scale of $\frac{1}{4}$ inch to 1 mile, it would be waste of time to attempt any instrumental contouring. Hill features on such a scale must be generalized, and it is sufficient to show them by form-lines controlled by scattered heights on the hills and in the valleys. On the other hand, a 1-inch map, such as the Ordnance Survey 1-inch map, can show contours with great exactness, and they may be fixed by instrumental levelling, or by using the water-level, or may be approximate contours determined by clinometer.

**Methods of Contouring.**—The principal methods of contouring, using the term in its widest sense, are:—

    (1) With theodolite or level;
    (2) With water-level;
    (3) With clinometer;
    (4) With aneroid barometer;
    (5) By eye (sketch contouring).

(1) **A small Theodolite** (used as a level) and a contouring staff, 8 feet long, provided with a sliding vane, are required, and pickets, as in the operation of levelling.

The contourer begins by choosing a bench mark or known height as near the required altitude as possible, and levelling from it up or down till he has established a point on the contour and marked it by a picket driven in the ground at the correct height. The instrument is then set up nearly on the same level as the picket, and at a distance of about 10 chains, beyond which a correction would be required for the curvature of the earth; the staff is held on the picket, and the vane raised or lowered till it is intersected by the cross-hairs of the telescope; it is then clamped, and the staff with the vane fixed at this height is then moved along the contour and set up at every change in the form of the ground, the assistant who holds it being directed into position by a signal from the contourer at the instrument; a mark is made at each station, and when a point is reached at about 10 chains from the instrument (or nearly the same distance as the first picket) another picket is driven, and carefully adjusted to the correct height. The instrument is now moved, and set up again in a similar position to the second picket as it was to the first; the vane of the staff is re-adjusted on the new picket, and the process of marking out the contour continued.

Where the vertical interval is small, more than one contour may often be marked out without shifting the position of the instrument if the staff is long enough. Too much should not, however, be attempted at one time.

In marking out it often happens that "isolations," or small detached contours on the level of the main contour, are passed, and care should be taken that they do not escape observation; examination of the horizon in all directions with the telescope will point out ground where isolations will probably be found. Should this be close to the main contour the isolation may be marked from one of the ordinary positions of the instrument, but when it is distant a point on it must be found by levelling it in the ordinary way.

When a sufficient length of the contour is marked the contourer plots it on the map, and fixes the points marked on the ground either by traverse or by chaining as in ordinary chain survey, or on a plane-table by intersection and resection; the latter method being more suitable for the smaller scales.

(2) **The Water-Level** is an instrument of great simplicity which never requires any adjustment. As it has no telescope, it is impossible to take long sights with it, and it is, of course, not susceptible of minute accuracy. On the other hand, no gross errors can creep into work with this instrument. It is easy to improvise a water-level.

Plate XII. shows the pattern of water-level in use on the Ordnance Survey. The distance between the bottles is 1 foot, the length of tripod when closed is 3 feet 1 inch. The tripod has one leg which can be lengthened for setting up on a hill side. A prismatic compass can be fitted to a projection between the bottles. The bottles are partially filled with coloured water through a small opening at the top of the tube frame; a cap fits over the opening. The bottles are connected above and below with hollow tubes. Weight of level 4 lbs. 2 ozs.

To use the instrument it is best to look diagonally across the two bottles, say from the right edge of the near bottle to the left edge of the far bottle, and get the bottles in such a position that the distant object comes between the two edges.

The level is to be placed at the starting point of the contour which has been determined previously by levelling or by careful clinometer work. The sight vane on a staff is to be adjusted to the height of the water in the bottles, and the staff man sent on with it some 400 or 500 feet, or to any change in direction of the contour; the level man aligns him right or left until the vane is on a level with the top of the coloured water in the bottles. A picket may now be driven in by the staff man, and the position of the picket is determined either—

On a plane-table by forward bearings and resection;
By the prismatic compass and chaining, or pacing.

The level man now moves up to the picket and proceeds as before. In contouring round a hill an excellent check is obtained by closing on the starting point.

(3) **Deliberate Method of Contouring with the Indian Clinometer and Plane-Table.**—To carry out this method two men are required, viz., one plane-tabler and one flagman, and it is necessary that three fixed points should generally be in sight, and that the plane-tabler should always set up his table at the same height above the ground.

A convenient way of doing this is for the plane-tabler to always set up his table to the level of some point on his person, say, for instance, one of the buttons of his coat, but if this is not considered sufficiently accurate he can carry a plumb-line of fixed length. The flagman carries a small disc or board, and notes the height of the table, so as to hold up his disc always at that height.

The plane-tabler sets up his table at station 0, the starting point of the contour, and interpolates his position. He then levels his clinometer and sends on the flagman approximately along the contour line, and when the latter has gone sufficiently far ahead, he aligns him by means of his clinometer until he is on the same level as himself.

The plane-tabler takes a ray on the flagman and moves up to him. He sets up his plane-table at that point and again interpolates his position, while the flagman goes on ahead as before, and so the work proceeds. This will be found a very convenient method on large scales.

**Rapid Method of Approximate Contouring with Indian Clinometer and Plane-Table.**—This is described in Chapter VI.

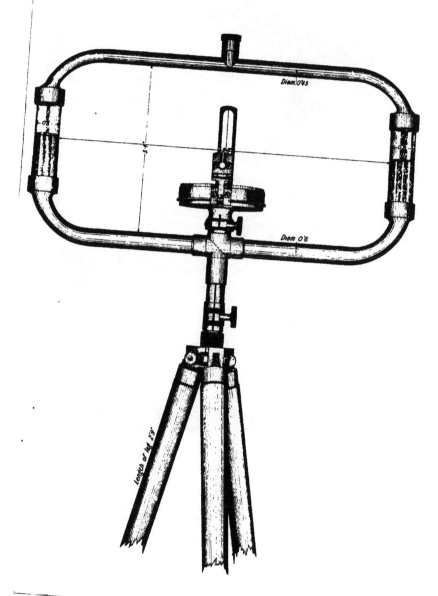

## ORDNANCE SURVEY
### Scale.- 1 Inch to 1 Mile

SHOWING THE COMBINATION OF CONTOURS AND VERTICAL HACHURES

(4) **Contouring with Aneroid Barometer.**—Where the vertical interval of the contours is large, say at least 25 feet, approximate contours can be put in by means of the aneroid barometer. This is a particularly convenient method of approximate contouring for scales of ¼ inch to 1 mile, and a vertical interval of 100 feet, or 1 inch and an interval of 50 feet. In variable weather this method cannot be used.

The barometer should be carried loose in the pocket.

Before reading it should be lightly tapped; in reading it should be held with the face vertical; the eye should always be in the same position relative to the needle to avoid the effect of parallax.

A 2½-inch barometer is a convenient size. The scale of feet should be used. The barometer should not be used near its extreme range; thus a barometer graduated to 5000 feet should not be used much above 4000 feet.

In using the barometer to determine the approximate positions of contours, it is necessary to work between points of which the heights have been previously fixed. Thus supposing the height of the top of a hill is known and also a height in the valley below; at the lower station set the instrument to the proper height; move up along any convenient spur, stopping where the barometer reading shows the required contour levels, fix these points on the plane-table, sketch in the neighbouring features, and close on the top of the hill. If the reading differs from the known height of the top of the hill, a proportionate correction must be made to all the contours drawn in. It is desirable that the distance between the fixed heights should not, as a rule, exceed a few miles, and it is clearly necessary that the elapsed time between starting and closing should be as short as possible.

Aneroid barometers do not at once take up the reading of a new height if a change of altitude has been rapidly made; a few minutes rest is necessary.

(5) **Sketch Contouring, or Hill Sketching.**—As has been explained, on a small scale such as ¼ inch to 1 mile, it would be waste of time and money to determine contours instrumentally.

The hill features are, therefore, sketched in by eye by means of **form-lines**. Form-lines serve to show the shape of the ground rather than its exact height at any point, and on small scales it is obvious that the hill forms must be generalized. In fact, hill sketching on small scales is not a mechanical science with exact rules, but partakes somewhat of the nature of an art; the hill sketcher must, of course, show the main features with accuracy, but it is impossible to show all the minor features.

The summits of the hills, the ridges, spurs, valley lines and streams will be accurately fixed, and the sketching will be controlled by a few trigonometrical (and in some cases barometrical) heights.

In drawing the form-lines it is desirable that the sketcher should observe a rule as to the number of lines to fit in for a given difference of level. A very convenient rule is that the interval (in feet) between two adjacent form-lines = 50 × number of miles to the inch. Thus on the ¼-inch scale the vertical interval between successive form-lines would be 200 feet. Supposing, for instance, that the sketcher was at a point which he knew, or judged, to be 1000 feet above the valley, he should draw five form-lines below the one representing the crest of the hill.

In hill sketching it is important to get on to the tops of hills, ridges, and spurs. The positions of these tops, ridges, and prominent points are fixed by plane-tabling in the usual way, and the sketching of the hill features is carried on at the same time as the fixing of the detail. The form-lines need not be continuous. Plates VI. and VII. are examples of hill sketching with form-lines.

## CHAPTER XI.

### MAP PROJECTIONS AND GRATICULES.

**A map projection** is the system on which the terrestrial meridians and parallels are represented on paper.

*A graticule* is the net-work of lines representing meridians and parallels on paper; the term is usually applied to the net-work of meridians and parallels plotted for topographical work in the field.

The term *projection*, though sanctioned by long usage, is an unfortunate one. It gives the beginner the idea that map projections are arrived at in the same way as the projections treated of in solid geometry. Now, the great majority of useful map projections are not obtained in any geometrical way. A map projection is to be treated as the representation on a plane, by any law, of the terrestrial meridians and parallels; in fact, any definite system of drawing the meridians and parallels is a map projection.

Before dealing with the general subject of map projections, a subject which every topographer should have some acquaintance with, we will describe the projections used in the field.

It is obvious that since the surface of the earth is curved, no portion can be represented on a plane with entire truthfulness. It is evident, also, that the smaller the portion to be represented, the more nearly will it be represented truthfully. If, for instance, we represent

Fig. 43.*

on paper a portion of the earth's surface extending 30′ from north to south and 30′ from east to west, and draw meridians and parallels as straight lines, then the portion ABCD of the earth's surface is represented by the trapezium *abcd*, in which *ab, bc, ad, cd* are the rectified lengths of AB, BC, AD, CD.

Now it can be shown that if AB, BC, CD, AD do not exceed 30′, then the surface ABCD can be represented by the rectilinear figure *abcd* on scales of $\frac{1}{4}$ inch to a mile and under without causing the topographer to make sensible errors. If the parallels were drawn on paper as portions of circular arcs cutting the meridians at right angles, the result would be still more accurate. But in the field it is inconvenient to plot circular arcs, and the topographer must, as a rule, be content to plot his graticules with straight lines.

A graticule should be drawn on *every small-scale map*, which, as a rule, *should be bounded by meridians and parallels*. The object of the graticule is to identify positions, to enable new points to be plotted, to enable maps to be compared with each other, and to facilitate joining maps of the same scale together.

It is not very important what the intervals are. The following would be suitable:—

On a scale of $\frac{1}{4}$ inch to a mile the meridians and parallels should be drawn for every $\frac{1}{2}$ degree.
On a scale of $\frac{1}{2}$ inch to a mile the meridians and parallels should be drawn for every $\frac{1}{4}$ degree.
On a scale of 1 inch to a mile the meridians and parallels should be drawn for every $\frac{1}{8}$ degree.

* The convergence of the meridians is much exaggerated in this figure.

**Example of Plotting a Graticule.**—It is required to plot the graticule of a map on a scale of ¼ inch to 1 mile, between 10° and 11° N. lat. and 25° and 26° E. long.

From Table X., by simple proportion, take the values in *miles* of the 30′ lengths of meridians and parallels. Divide these by 4 to get the lengths in inches.

The diagonals of the squares are obtained as shown at the foot of Table X.

Thus
$$BE = \sqrt{(8·59^2 + 8·51 \times 8·50)}$$
$$= 12·09 \text{ inches.}$$

Now, having all the necessary data, take the plane-table or map and draw the straight line AB (fig. 44) for the central meridian.

Measure AC, CB the lengths of 30′ of latitude, then with centre B and radius BE

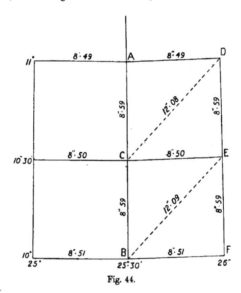

Fig. 44.

(diagonal) intersect a circle with centre C and radius CE (30′ of longitude). Similarly for the other trapeziums.

The result will be a graticule in which the meridians converge, and the parallels are concave to the pole.

**Second Example of Plotting a Plane-Table Graticule.**—Suppose it is required to plot a graticule on a scale of ¼ inch to 1 mile. Limits 30° and 31° S., 25° and 26° E.

Proceed as follows (fig. 45):—

Draw a straight line ABC down the middle of the paper to represent the central meridian 25° 30′ E

Look out in Table X. the length of 30′ of latitude, from 30° to 30° 30′. This is 34·440 miles, or on this scale 8·61 inches; mark this length CB. From 30° 30′ to 31° is found in the same way to be 8·61 inches; mark this BA.

Next look out the length of 30′ longitude on parallels 30° and 30° 30′, viz., 7·50 and 7·46 inches.

Then the length of the diagonal of the northern quadrilaterals

$$= \sqrt{\{(8·61)^2 + 7·50 \times 7·46\}} = 11·41 \text{ inches.}$$

In a similar way it is found that the diagonal of the southern quadrilateral is 11·37 inches.

### Plane-Table Graticule.

Dimensions in inches for a scale of 4 miles to 1 inch.

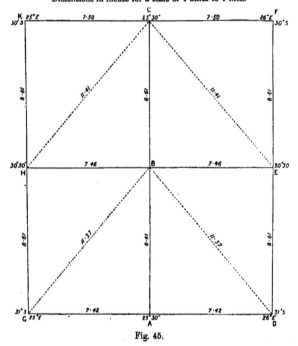

Fig. 45.

Now with centre C describe a circle of 11·41 inches, intersect a circle with centre B and radius 7·46 inches, thus fixing E, and so on for the other points F, D, G, H, K.

This method can be used for graticules on all topographical scales, and for any intervals between the meridians or parallels.

But, to save labour in plotting graticules on the scale of 2 miles to 1 inch (a very useful topographical scale), Table IX. has been prepared and from this the lengths of 15′ spaces on meridians and parallels and the lengths of the diagonals can be taken directly. This table is not to be used for other scales or other intervals.

The field projection above described is to be considered as a rectilinear approximation to a conical projection. It is all that a topographer absolutely requires in the field, but if it is possible, it is better (though not actually essential) to plot the parallels as portions of circular arcs. Labour-saving tables for the purpose are "The Construction and use of Marginal Lines for Maps," J. O'Farrell, 1862; a War Office pamphlet "On the Projection of Maps," by Major L. Darwin, R.E., and "Tables for a Polyconic Projection of Maps," published by the United States Coast and Geodetic Survey.

We now proceed to the general discussion of the subject of map projections. The remainder of this chapter need not be read by the beginner.

GENERAL DISCUSSION OF THE SUBJECT OF MAP PROJECTIONS.

The term map projection is sometimes misunderstood to mean a projection by means of straight lines from the earth's surface on to a plane, cylindrical, or conical surface. It is actually a term of much wider meaning. Any representation on a plane, by any law, of the network of terrestrial meridians and parallels, is a map projection, and the great majority of useful map projections are not arrived at in any obvious geometrical way.

It is necessary to bear in mind that, in general, a map projection is not a simple geometrical projection, but the result of any definite system of drawing the earth's meridians and parallels.

There is a common idea that because there can be no perfect plane projection, all projections are equally bad or good, an idea encouraged by the customary geometrical way of looking at the question. Now, the choice of a projection for a map is in general by no means a matter of indifference. Projections may indeed be considered as graphic representations of certain properties of the curved surface; and if we know the special purpose of the map, we also know the special property or quality which it is most important to represent truthfully.

Useful theories of map projection are based on this fact, that although it is impossible to represent on a plane surface any portion of a spherical or spheroidal surface with entire truthfulness, yet we may select some quality or property of the curved surface and represent that accurately.

For instance, there is a whole group of projections in which areas are correctly represented, another group in which there is no local distortion, another group which preserves the azimuths from a central point, one projection which preserves distances and azimuths from a central point, one projection in which circles are still represented by circles, whilst in most projections we can represent at least one line correctly in addition to other properties.

There is also a class of projection in which no general property is accurately represented, but all properties are represented fairly well; a sort of golden mean.

FIELD PROJECTIONS.

All map making is divided into two parts—work in the field and work indoors. The former includes surveys, explorations, and reconnaissances, the latter is a matter of drawing, reducing, or compiling.

*Field Projections.*

First: Large-scale cadastral surveys, such as $\frac{1}{2500}$ maps of the Ordnance Survey, or the 16-inch to the mile maps of the Survey of India, are always based on a triangulation plotted by rectangular co-ordinates; latitude and longitude are not usually marked, if they were, the resulting projection would be that known as Cassini's. Each sheet represents such a very small part of the earth's surface, and the boundaries are so near the central meridian, that the matter need not be gone into further.

Secondly: Rapid explorations usually result in maps of a rough and ready order, most interesting, but not most accurate. The intrepid explorer draws a thin red line across a continent, and there is little on each side of it. If his march is east and west, he should use a simple conical projection; if north and south, a polyconical; and if in a diagonal direction, a so-called secant conical projection.

Thirdly: What projection should be used in the field by the topographer who works on scales varying from $\frac{1}{10000}$ to $\frac{1}{100000}$? As the difficulties evidently increase as the scale decreases, we can deal with the smaller scale.

When the plane-tabler looks along the sight-rule, any points which are bisected by the central vertical thread are considered to be on the same straight line. That is, great circles are represented by straight lines. Hence, considering this condition only, the plane-table is on the gnomonic projection.

But the plane-tabler also expects that the scale shall be the same all over the map.

If we only consider one point, and set up the plane-table at this point, all rays from this point are to represent great circles, and the scales of all the rays are to be the same. This gives the equidistant-zenithal projection.

If the plane-table is set up at other points, we get conflicting conditions, which, in theory, are best reconciled by Airy's projection, but in practice are approximately satisfied by some form of conical projection, provided that the area of the plane-table does not exceed about 1 square degree.

In the field it is inconvenient to plot such flat arcs as are required for the parallels; it is therefore customary to substitute for them rectilinear inscribed polygons, or, which comes to much the same thing, to plot, instead of a small portion of the curved parallel, a straight line of the same length.

The resulting projection may be known as the field rectilinear projection; it should not be used where the extent of latitude is more than 2 degrees.

There is a case on record in which a survey party commenced work on Mercator's projection, which is, of course, entirely unsuited for land surveys.

### LIST OF PROJECTIONS.

We can now deal with the more general and interesting question of the projections used to represent much larger portions of the earth's surface, or even the whole of it.

Projections have been termed:—

a. **Orthomorphic** (or *conform*, or *conformable*).—In these the scale is the same in all directions round a point in its immediate neighbourhood. It follows that very small areas are similar in outline to the originals on the earth's surface. It also follows that the meridians and parallels intersect at right angles, though the fact that they intersect at right angles does not show that a projection is orthomorphic. Mercator's is a projection which belongs to this group, so does the stereographic.

b. **Equal-Area** (or *equivalent*, or *surface-true*).—In these, equal areas on the map represent equal areas in reality, *i.e.*, areas are correctly represented, though distances and angles are not. Bonne's modified conical projection belongs to this group.

c. **Perspective**, a name which explains itself. The eye may be imagined to be inside, outside, or on the surface of the earth. The gnomonic projection is perspective.

d. **Zenithal**.—In these the azimuths round a central point to all other points are correctly represented, and equal distances from this point represent equal distances in reality. Perspective projections are zenithal, though all zenithal projections are not perspective. The equidistant projection belongs to this group.

e. **Conical**, which can be imagined as drawn on the surface of a cone (not necessarily *projected* on to the cone).

f. **Cylindrical**, which can be imagined as drawn on the surface of a cylinder. Mercator's is cylindrical.

*g.* **Conventional.**—Those which are projected by arbitrary rules for convenience of drawing. The globular projection is conventional, so are polyconics.

These terms do not strictly represent classes of projections, as they are not all mutually exclusive, nor do they comprise all possible projections, and to treat them as classes would be to fall into his error who divided the human race into Frenchmen, red-haired people, and cannibals.

Thus we may have an orthomorphic cylindrical projection (Mercator's), or an equal-area cylindrical projection (Lambert's), or an equal-area zenithal projection (Lorgna's), and so on. (But a projection cannot be be both orthomorphic and equal-area, which would imply a perfect representation.) The seven terms given are, in fact, just descriptive terms.

The following is a list of the thirty principal projections; a description of each is given on pp. 100 to 111.

**Orthomorphic :—**
1. Mercator's (cylindrical orthomorphic) ;
2. Transverse Mercator's (Lambert's first) ;
3. Conical orthomorphic (Lambert's second) ;
4. Circular orthomorphic (Lagrange's). Also the stereographic (17).

**Equal-Area :—**
5. Cylindrical equal-area (Lambert's third) ;
6. Transverse equal-area (Lambert's fourth) ;
7. Modified conical equal-area (Bonne's)* ;
8. Sinusoidal equal-area (Sanson-Flamsteed) ;
9. Polar equal-area (Werner's) ;
10. Simple conical equal-area (Lambert's fifth) ;
11. Conical equal-area (Albers') ;
12. Zenithal equal-area (Lambert's sixth—called Lorgna's) ;
13. Rectilinear equal-area (Collignon's) ;
14. Homalographic equal-area (Mollweide's).

**Perspective :—**
15. Orthographic ;
16. Central or gnomonic ;
17. Stereographic ;
18. Minimum error (Clarke's) ;
19. James's ;
20. La Hire's.

**Zenithal :—**
21. Equidistant zenithal ;
22. Projection by balance of errors (Airy's) ;
    Also the zenithal equal-area (12),
    And perspective projections (15–20).

**Conical :—**
23. Simple conical ;
24. Conical, with rectified meridians and two standard parallels, sometimes wrongly called "secant conical" ;
25. Simple polyconical ;
26. Rectangular polyconical ;
    Also Nos. 3, 7, 10, and 11.

* Sometimes, though wrongly, called "Flamsteed's Modified."

**Cylindrical:—**
    27. Simple cylindrical ;
        Also Nos. 1, 2, 5, 6, and 29.

**Conventional:—**
    28. Globular ;
    29. Projection by rectangular co-ordinates (Cassini's) ;
    30. Field rectilinear ;
        Also Nos. 25 and 26.

This is not a complete list of all the projections which have been in use, there are a number of others (especially perspective and conventional projections) which have been proposed or used, but they are not now of any importance, and their interest is only historical.

Those interested in the subject will find a very complete account, both technical and historical, in an admirable work, entitled "Traité des Projections des Cartes Géographiques," A. Germain, Paris. A. Bertrand, Editeur, Libraire de la Société de Géographie.

## The Choice of Projections.

The choice of a projection falls under one of the following cases :—
    (1) *For field work.* This has been discussed.
    (2) *For a single sheet map ;*
    (3) *For a series, which is to be capable of being accurately fitted together ;*
    (4) *For a series, without the last condition.*

(2) *Single-sheet maps :—*

This second case includes any map, though it may be made up of several parts, which is meant to be used as a unit. The Cape Colony $\frac{1}{500000}$ map is an example. This case also includes all atlas maps. All possible projections are available, though all are not equally suitable for the special purpose in view. For instance, statistical maps should be on equal-area projections, great circle charts on gnomonic projections, and so on.

A single-sheet map of Africa may be taken as an example, and the following is a preliminary selection :—

        Equal-area ... ... ... Sanson-Flamsteed or
                                            Zenithal Meridian (Lambert's).
        Zenithal ... ... ... Balance of Errors (Airy's) or
                                            Equidistant meridian.
        Conical ... ... ... Simple polyconical or
                                            Rectangular polyconical.
        Orthomorphic ... ... Mercator's or
                                            Stereographic meridian.
        Perspective ... ... ... Minimum error (Clarke's) or
                                            Orthographic meridian.

Of these, Mercator's is out of the question for any land purpose.

If the map is a political or statistical map, one of the equal-area projections should be taken, and of these the zenithal is the best, and Sanson-Flamsteed the easiest to draw.

If the map is for general purposes, we may reject equal-area and orthomorphic projections as being extreme cases and the polyconic projections as conventional, and finally choose from

                Balance of errors (Airy's) ... ... ... No. 22
                Minimum error perspective (Clarke's) ... No. 18
                Zenithal equidistant ... ... ... ... No. 21

These are all good projections for the purpose.

*(3) A One-Projection Series :—*

The third case is that of a series of maps which are to be capable of being accurately fitted together, *i.e.*, which are to be on one projection. This case, which is practically very convenient, can only be used for a comparatively small part of the earth's surface and on medium scales (*i.e.*, smaller than topographical and not so small as atlas scales).

We may take as a typical country for this treatment, Africa south of the Zambesi, and as a typical scale $\frac{1}{1,000,000}$. It is not absolutely necessary, but it is convenient, to bound the sheets by meridians and parallels. All this points to the use of a conical projection; and one of the best for the purpose is that form of conical projection in which the meridians are their true lengths and in which the errors of scale along the extreme parallels and parallel of maximum error are made equal. In the case of S. Africa, the errors on the parallels mentioned are less than $\frac{1}{2}$ per cent., whilst the scale along the meridians is correct throughout.

*(4) A Topographical Series :—*

The fourth case is that of a series of maps, of which each sheet is independent, the sheets are not meant to fit accurately together, and the accuracy of each sheet is the principal consideration, *i.e.*, the case of ordinary topographical maps.

Each sheet will be constructed on its own central meridian, and the projection (as for the field) will be a conical projection. But on account of the smallness of the area dealt with, simple or rectangular polyconics give sensibly the same results. The present War Office system is the rectangular polyconic, p. 110, and does very well for topographical scales.

*Note.*—When the country is of small extent in either latitude or longitude, topographical maps may be projected on one projection, as in Case 3, above. Ordnance Survey one-inch maps of England, for example, are a series on Cassini's projection, and the extreme error due to the projection does not amount to $\frac{1}{8000}$, *i.e.*, less considerably than may be expected from the distortion of the paper.

$\frac{1}{1,000,000}$ *Map :—*

The question of the projection to be chosen for the universal $\frac{1}{1,000,000}$ map is interesting, and Africa may be taken as an example.

First, it is clear that a general projection for the whole of Africa is out of the question on such a scale.

Next, we *may* construct each sheet independently, and this may sometimes be the best system; but some difficulties would be found in fitting the sheets together, since each will represent some 24 square degrees.

Perhaps the best system is a compromise; special areas, such as Egypt, Nigeria, S. Africa, &c., should be treated as units, each on the special projection best suited for its particular case. This course would have the balance of local advantage, which is the chief thing to consider, and the projection can be so chosen that the resulting error is negligible. The only objection to this occurs when it is required to fit on sheets of an adjoining block. This can be obviated by arranging that the sheets shall well overlap the boundary.

PROJECTIONS USED FOR SOME IMPORTANT MAPS.

*Conical Orthomorphic* (Lambert's second); Generally known as Gauss's. (No. 3.)
    Used for a map of Russia published by the Geographical Society of St. Petersburg in 1862.

*Modified Conical Equal-area*, known as Bonne's. (No. 7.)
    Adopted in 1803 by the Depôt de la Guerre for the map of France.
    Ordnance Survey 1-inch, $\frac{1}{2}$-inch, $\frac{1}{4}$-inch of Scotland and Ireland.

Ordnance Survey 10-mile maps of Scotland and Ireland.
Indian Atlas 4 miles to 1 inch.
Cape Colony $\frac{1}{500,000}$.

*Orthographic.* (No. 15.)
French map of Africa $\frac{1}{2,000,000}$, 1895, where its defects are very marked.

*Central or Gnomonic.* (No. 16.)
Some charts have been published on this projection by the U. S. Hydrographic Department.

*Zenithal Projection by Balance of Errors.* (Airy's.) (No. 22.)
Ordnance Survey 10-mile map of England.

*Conical with rectified Meridians and two Standard Parallels.* (No. 24.)
Ordnance Survey $\frac{1}{1,000,000}$, 1904.
Survey of India, India and Adjacent Countries, $\frac{1}{1,000,000}$, 1904.

*Simple Polyconic.* (No. 25.) U. S. Coast Survey maps.

*Rectangular Polyconic.* (No. 26.) Intelligence Division, War Office maps.

*Projection by Rectangular Co-ordinates* or *Cassini's.* (No. 29.) .
Ordnance Survey 1-inch, $\frac{1}{2}$-inch, and $\frac{1}{4}$-inch of England.
    ,,   ,,    $\frac{1}{6}$ ,, of The United Kingdom.
    ,,   ,,    $\frac{1}{2,500}$ ,,   ,,

## Brief Description of the Thirty Principal Projections.

### ORTHOMORPHIC PROJECTIONS.

1. **Mercator's,**\* or cylindrical orthomorphic.

In this projection the meridians and parallels are straight lines at right angles to one another.

The meridians are spaced at their equatorial distances.

The distance of any parallel from the equator is:—

$$\frac{a}{M} \left\{ \log \tan\left(45° + \frac{l}{2}\right) - e^2 \sin l - \frac{e^4 \sin^3 l}{3} \ldots \right\},$$

where $a$ and $b$ are the semi-axes,

$$e = \frac{\sqrt{a^2 - b^2}}{a},$$

M the modulus of common logarithms, $l$ is the latitude.

The scale at any point is the same in all directions, and small outlines are represented without distortion. A straight line represents a loxodrome or line of constant true bearing; the projection is therefore universally used at sea; it is of no value for land surveys. A loxodrome continuously pursued describes an endless spiral round the pole.

2. **Transverse Mercator's**, or transverse cylindrical orthomorphic (Lambert's first).†

Mercator's projection may be considered as developed on a cylinder touching the earth along the equator; imagine this cylinder turned round so as to touch some chosen meridian. The lines which formerly represented meridians now represent great circles at right angles to the chosen meridian, and the lines which represented parallels now represent small circles parallel to that meridian. It is not necessary to consider this further.

3. **Conical Orthomorphic Projection** (Lambert's second), generally called Gauss's.

---

\* Mercator, born A.D. 1512, died 1594
† Lambert, German mathematician, born 1728, died 1777.

In this projection the parallels are represented by concentric circles and the meridians by their radii. Two selected parallels are of their proper lengths, but the lengths of the other parallels and of the meridians are determined by the following law, treating the earth as a sphere :—

If $z_1, z_2$ are the co-latitudes of the selected parallels

$$\text{let } \lambda = \frac{\log \sin z_1 - \log \sin z_2}{\log \tan \frac{z_1}{2} - \log \tan \frac{z_2}{2}},$$

then the angle two meridians make with each other

$$= \text{true difference of longitude} \times \lambda ;$$

then if $r$ is the required radius of a given parallel (of co-latitude $z$), $k$ a constant which defines the scale,

$$r = k \left( \tan \frac{z}{2} \right)^{\lambda}.$$

This is a remarkable and useful projection.* Small outlines are correctly represented, and the scale is the same in all directions round any point. The scale along two parallels is correct, and is consequently correct in their neighbourhood in every direction. Between the selected parallels the scale is very slightly too small, outside them it is too large. It has been used for a map of Russia and is very suitable for large areas having a great extent in longitude.

**4. Circular Orthomorphic Projection** (Lagrange's).

In this both meridians and parallels are represented by circular arcs. More interesting to the mathematician than useful to the map-maker.

The stereographic projection is orthomorphic. See No. 17.

### EQUAL-AREA PROJECTIONS.

**5. Cylindrical Equal-area Projection** (Lambert's third).

If the earth is considered a sphere, this is the simplest of all projections. Imagine a cylinder to circumscribe the earth, touching it at the equator, and from every point on the earth let a straight line be drawn at right angles to the earth's axis and produced backwards to intersect the cylinder. When the cylinder is unrolled, the meridians and parallels are clearly straight lines at right angles to one another, and the distance of any parallel from the equator is equal to $a \sin$ lat. when $a$ is the radius.

It follows, from the elementary geometry of sphere and cylinder, that the projection is equal-area.

This projection is only useful for maps which do not extend more than a few degrees from the equator.

**6. Transverse Cylindrical Equal-area Projection** (Lambert's fourth).

If the cylinder touches the earth along a meridian and points are projected as above, then, treating the earth as a sphere, if $x, y$ are the co-ordinates of a point $l$, $a$ its latitude and longitude,

$$x = \sin a \cos l,$$
$$\tan y = \tan l \sec a.$$

The meridians and parallels can be drawn by giving $a$ and $l$ respectively constant values.

This projection can be used for countries having a great extent in latitude and but little in longitude. It is not, however, the best for the purpose.

---

* This last expression may be taken as giving the whole group of orthomorphic conical projections of which this is a particular case. Mercator's is another.

**7. Modified Conical Equal-area Projection** (Bonne's), sometimes called the modified conical or Bonne's projection, and known in France as "projection du dépôt de la guerre." This, with its two modifications (8 and 9), is an important projection.

Imagine a cone, having the same axis as the earth, to touch it along any selected parallel; imagine the cone spread out and a radius of this parallel taken as the central meridian. Along this central meridian, on each side of the parallel, measure the true rectified distances to the other parallels. Through these points describe circles having the same centre as the chosen parallel, and along each parallel measure the true distances to the successive meridians.

The projection is clearly equal-area, since all of the small elements between any two parallels bounded by two infinitely close meridians are equal to each other, and to the corresponding elements on the earth's surface.

The true spheroidal shape of the earth has been taken account of.

If $l_1$ is the latitude of the selected parallel, and $l$ of any other parallel, the radius of the latter is $r \operatorname{cosec} l_1$ + the rectified arc of meridian between $l$ and $l_1$. Where $r = \dfrac{180}{\pi} \times$ length of 1° of chosen parallel. This can be found from geodetic tables.

The projection has the following properties :—

(1) It represents areas correctly.
(2) Parallels are their true lengths.
(3) The meridians cross the chosen parallel at right angles.

The meridians are not simple curves, though so easily constructed.

The projection has been much used in France, and the $\frac{1}{800,000}$ map of Cape Colony published by the Surveyor-General is also on this projection, also the Ordnance Survey 1-inch maps of Scotland and Ireland.

The less its extent in longitude the better. It is frequently used in ordinary atlases for countries for which it is not suited—Asia, for example.

**8. Sinusoidal Equal-area Projection** (Sanson's).—Sometimes, though wrongly, called Flamsteed's, or Sanson-Flamsteed.

If Bonne's cone (No. 7) touches the equator, it becomes a cylinder, and each parallel becomes a straight line at right angles to the central meridian, cutting it at the true distance from the equator. Along each parallel distances are measured from the central meridian equal to the portion of the arc of the parallel represented.

Assuming the earth a sphere, each meridian becomes a sine curve, hence the name of the projection. But it is clearly equally easy to plot the true spheroidal lengths. The general remarks on the last projection apply to this. It is a good equal-area projection for the whole of Africa.

**9. Polar Equal-area Projection** (Werner's).

If Bonne's cone touches the pole, the cone becomes a plane. The parallels are then incomplete concentric circular arcs with the pole as centre; the central meridian is still a straight line, which is cut by the parallels at true distances. The meridians as before pass through points on the parallels at true distances from the central meridian measured along the arc.

This projection, though interesting, is of little use. It is not possible to use it for Polar regions, and is clearly inferior to the two previous projections for other parts of the earth's surface.

**10. Simple Conical Equal-area Projection** (Lambert's fifth).

In this projection the parallels are concentric circles of which the pole is the centre, the meridians are radii. The radius of each parallel is

$$2a \sec \frac{z_1}{2} \sin \frac{z}{2},$$

when $a$ is the earth's radius, $z$ the co-latitude, and $z_1$ the co-latitude of a selected parallel, which is then correctly represented. The angle any two meridians make with each other is $\alpha \cos^2 \frac{z_1}{2}$, when $\alpha$ is the difference of longitude.

The cylindrical equal-area projection is *not* a particular case of this, because here we have the condition that the pole is the centre of the circular parallels.

The properties of this projection are :—

(1) It represents areas correctly ;
(2) The meridians cut the parallels at right angles ;
(3) The selected parallel is correctly represented ;
(4) It is easy to draw.

The objection to it is that it misrepresents distances both along the parallels and along the meridians. It cannot be used near the poles, nor for countries having a great extent in latitude.

If in the above we put $z_1 = o$, the radius of each parallel is $2a \sin \frac{z}{2}$ ; the meridians are then inclined at their true angles to each other. (See No. 12.)

**11. Conical Equal-area Projection** (Albers').

In this projection any two selected parallels are correctly represented. The projection consists of concentric circles with radii for the meridians, but the pole is not now the centre.

If $l_1, l_2$ are the chosen latitudes ; $r_1, r_2$ the radii on paper ; put $k = \dfrac{1}{\sin \frac{l_1 + l_2}{2} \cos \frac{l_1 - l_2}{2}}$,

$a$ being the earth's radius.

Then, $r_1 = ka \cos l_1$,

$r_2 = ka \cos l_2$,

the inclination of two meridians $= \dfrac{\text{angular difference of longitude}}{k}$, and if $r$ is the radius of any other parallel of latitude $l$, $r^2 = 2a^2k (\sin l_1 - \sin l) + r_1^2$.

The pole is now represented by the arc of a circle. Two parallels are correctly represented, and the properties are otherwise the same as those of the preceding projection. Except for very large areas it does not differ materially from No. 24 ; where representation of areas is important it should be used in preference. It should not be used near the pole. It would make a good projection for Central Asia.

**12. Zenithal Equal-area Projection** (Lambert's sixth). Sometimes, though wrongly, called Lorgna's.

This is a particular case of 10. Meridians are represented by straight lines at their true angles to each other. The parallels are circles of which the meridians are radii. If $z$ is the co-latitude of any point, and $a$ the earth's radius, or, better, the radius of curvature at the pole, the radius of any parallel is

$$2a \sin \frac{z}{2}.$$

As described it may be called the polar equal-area projection.

The same projection can, however, be applied to any part of the earth's surface, if we imagine great circles to radiate from some selected point, in the same way as the meridians radiate from the pole. The meridians and parallels can be drawn by a transformation of spherical co-ordinates. In the special case where the central point is on the equator we have (assuming the earth spherical)—

If $\theta$ is the angular distance of any point from the centre, $l, \alpha$, the latitude and longitude of this point,

$$\cos \theta = \cos \alpha \cos l.$$

If A is the azimuth of this point at the centre, $\tan A = \sin \alpha \cot l$, A is then truly drawn, and the distance $\theta$ is represented by

$$2a \sin \frac{\theta}{2}.$$

The properties of this projection (which is on the spherical assumption) are—

    (1) It is zenithal, i.e., azimuths are correct at the central point and equal distances are represented by equal distances ;
    (2) It is equal-area ;
    (3) There is but little deformation for 30° from equator or central meridian.

It is a very good projection for a single-sheet equal-area map of Africa.

**13. Rectilinear Equal-area Projection** (Collignon's).

The meridians are straight lines which radiate from the pole, and the parallels are parallel straight lines cutting the central meridian at right angles.

Interesting but of little use.

**14. Homalographic Equal-area** (Mollweide's). Sometimes called Babinet's.

The parallels are parallel straight lines and the meridians are ellipses, the central meridian being a straight line at right angles to the equator, which is equally divided.

The distance of any parallel from the equator is $\sqrt{2} \sin \delta$ where $\pi \sin \text{lat} = 2\delta + \sin 2\delta$.

The whole sphere can be represented in an ellipse of which the major axis is twice the minor.

### PERSPECTIVE PROJECTIONS.

Generally speaking these projections are of no value for land surveys.

The principal consideration is the position of the point of sight. The projection plane is always supposed to be at right angles to the line joining this and the earth's centre; the distance of this plane from the centre only affects the scale.

**15. Orthographic Projection.**—In this the point of sight is supposed to be infinitely distant, and the perspective rays consequently at right angles to the projection plane. If the plane of projection is the equator, the result is the polar orthographic; if a meridian, a meridian orthographic, and if any other plane, a horizontal orthographic.

In the latter case, which is the most general, it is clear that the central meridian is a straight line, and all other meridians and parallels are ellipses ; obviously only a hemisphere can be represented. From the nature of the case it is a pictorial projection, and from its simplicity it is a good deal used in popular works. It is not the best perspective projection for large areas, and for practical purposes is of little use. The laws of its construction are very obvious.

It has been used for the French $\frac{1}{2,000,000}$ map of Africa. The distortion on the extreme sheets, e.g., St. Louis, is very marked. It is a projection to be avoided.

**16. Central or Gnomonic Projection.**—In this projection the eye is imagined to be at the centre of the earth. As before, we may have polar, meridian, or horizontal, central projections. In the latter case, which is the most general, the meridians are straight lines which meet at the pole; the parallels are conic sections. The term gnomonic was applied in ancient days to this projection because the projection of the meridians is evidently a similar problem to the graduation of a sun dial. If $\theta$ is the angle any meridian makes on paper with the central meridian,

    $\lambda$ the inclination of the projection plane to the earth's axis,
    $\alpha$ the difference in longitude of the meridians,

$$\tan \theta = \sin \lambda \tan \alpha ;$$

Plate XIV.

Part of the
## ATLANTIC OCEAN
on a
### MERIDIAN GNOMONIC PROJECTION

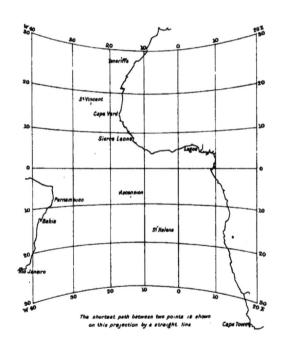

The shortest path between two points is shown on this projection by a straight line

Plate XV.

# AFRICA
## on
## CLARKE'S PERSPECTIVE PROJECTION

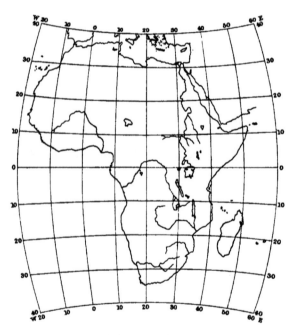

The distance of the point of sight
from the centre of the earth is 1·625 × radius

and if $l$ is the geocentric latitude of any other point, the distance from the equator along the meridian is

$$m \frac{\sec \alpha \sin l}{\sin (l + x)},$$ where $\tan x = \cot \lambda \cos \alpha$, and $m$ is a constant defining the scale.

The equator is, of course, a straight line at right angles to the central meridian.

The special property of the projection is that all great circles on the sphere (*i.e.* shortest lines) are represented by straight lines. There is no projection which accurately possesses this property for the spheroid, but one which does so very nearly is that which results from the intersection of terrestrial normals with a plane. In this case the normal sections between two points, either on the same parallel or the same meridian, are strictly represented by straight lines.

The central projection is useful at sea, for star charts and for the study of direct routes in great countries.

An example is here given of part of the Atlantic Ocean on a meridian gnomonic projection. (Plate XIV.)

**17. Stereographic Projection.**—In this the eye is imagined to be on the surface of the sphere, which is supposed to be hollow, and the surface represented is that of the opposite portion of the sphere. The projection is orthomorphic, and has the remarkable property (on the spherical assumption) that circles are represented by circles. It has, of course, three aspects, polar, meridian, and horizontal. In the latter case, which is the most general, the central meridian is a straight line, the others are circles intersecting at the pole. The parallels are circles, but are not concentric. The graphic construction is simple. As it is only used for general maps of the hemisphere on very small scales, and not much for these, there is no need to consider the compression of the earth

**18. Minimum Error Perspective Projection** (Clarke's).[*]—In this projection (due to Colonel Clarke) the point of sight is outside the sphere, and the distant surface is that to be represented. The position of the point of sight is chosen, so that at each point the sum of the squares of the errors of scale in two directions at right angles to each other shall be a minimum over the whole surface. In other words, the total misrepresentation is a minimum. The surface to be represented is that cut off by a plane parallel to the projection plane, and the boundary is therefore a circle.

If $u$ is the spherical arc between any point and the adopted centre, $h$ the distance of the point of sight from the centre, $k$ a constant defining the scale, then the distance at this point from the centre (on the map) is

$$\frac{k \sin u}{h + \cos u}.$$

The following table shows the value of $h$ for several values of $\beta$ (the spherical radius of the map):—

| $\beta$. | $h$. |
|---|---|
| 40 | 1·625 |
| 54 | 1·61 |
| 90 | 1·47 |
| 108 | 1·40 |
| 113½ | 1·367 |

**19. Sir Henry James's Projection.**—This is a particular case of Clarke's projection in which the spherical radius of the area to be represented is $113\frac{1}{2}°$; in this case $h = 1·367$.

[*] Colonel A. R. Clarke, C.B., R.E., F.R.S.

(As originally described by Sir H. James, $h$ was taken as 1·5, which does not give the minimum error.) The total area represented is about two-thirds of the earth's surface.

**20. La Hire's Projection.**—A perspective projection in which the point of view is outside the sphere at a distance from the centre equal to $\left(1 + \dfrac{1}{\sqrt{2}}\right) \times$ radius. The middle point of the radius of the map of a hemisphere, therefore, represents the middle point of the opposite quadrant. It is used for maps of a hemisphere, but has no great merit. It is sometimes confused with the globular projection, which is a conventional one. (No. 28.)

### ZENITHAL PROJECTIONS.

**21. Equidistant Zenithal Projection.**—In this projection, any point on the sphere being taken as the centre of the map, great circles through this point are represented by straight lines of the true length to scale, and intersect each other at the true angles. There may thus be polar, meridian and horizontal equidistant zenithal projections.

The position of the intersection of any meridian with any parallel is given (on the spherical assumption) by the solution of a simple spherical triangle.

Thus, if $z_1$ is the co-latitude of the centre of the map (longitude 0),

    $z$  ,,    ,,    any other point.
    $a$  ,,  longitude of  ,,    ,,
    $A$  ,,  required azimuth.
    $c$  ,,    ,,  spherical distance.

Let $\tan \theta = \tan z_1 \cos a$, then $\cos c = \cos z_1 \sec \theta \cos (z - \theta)$ and $\sin A = \sin z \sin a \operatorname{cosec} c$.

The projection is neither orthomorphic nor equal-area; the scale in the direction of the centre is always correct, and at right angles is less than 5 per cent. in error at 30° from the centre. It would make a good projection for a single-sheet map of Africa, and is especially useful for polar maps.[*]

**22. Zenithal Projection by Balance of Errors** (Airy's).—At any point of a zenithal projection we may consider two scales, one along the radiating straight line which represents a great circle, and one at right angles to this.

If these two scales are equal at each point, though they differ from point to point, the result is an orthomorphic projection—the stereographic. (No. 17.)

If the product of the two scales is constant, the result is the zenithal equal-area projection—Lambert's sixth. (No. 12.)

But we may deal with them in a different way, and make the sum of the squares of the errors of the two scales at each point, a minimum over the whole surface to be represented. The result is Airy's projection, which we are now dealing with; in this the total misrepresentation is a minimum.

If $\beta$ is the spherical arc from proposed centre to proposed circumference, $u$ the spherical arc of any point from the centre, then the distance of this point from the centre on the map is

$$2 \cot \frac{u}{2} \log_e \sec \frac{u}{2} + C \tan \frac{u}{2},$$

where

$$C = 2 \cot^2 \frac{\beta}{2} \log_e \sec \frac{\beta}{2}.$$

All this is on the assumption that the earth is spherical.

---

[*] The earliest known instance of the use of the Polar Equidistant Zenithal Projection occurs in a treatise by Glareanus, written about A.D. 1510.

This projection is, so to speak, the golden mean between an orthomorphic and an equal-area projection. It has been used for the Ordnance Survey 10-mile map of England.

The zenithal equal-area projection, Lambert's (No. 12), and perspective projections (15 to 19) have been discussed.

CONICAL PROJECTIONS.

These are really developments on a conical surface, not projections on to the surface.

23. **Simple Conical Projection.**—Imagine a cone to touch the earth along any chosen parallel, the cone and the earth having the same axis. Unroll the cone, then the radius of the chosen parallel is $\nu$ cot lat., where $\nu$ is the normal terminated by the minor axis. Along this parallel measure the true lengths of the degrees of longitude (for the parallel in question), and join the apex of the cone and these points by straight lines—these are the meridians. The other parallels are circular arcs, concentric with the chosen parallel, and distant from it the true lengths of the rectified arc of the meridian.

This projection is unsuited for a country having a great extent in latitude, but it may stretch any extent in longitude. When the chosen parallel is the equator, the projection becomes the *equidistant cylindrical projection* (27), a projection of no value. When the chosen parallel is the pole, the projection becomes the *polar equidistant zenithal projection*, a very valuable projection already described (No. 21).

*Note.*—It is possible, though it has never been done, to construct transverse conical projections, in which the cone touches some parallel to a meridian. In the same way it is possible to have transverse secant conical projections.

Again, these cones may be diagonal, the pole of the small circles being neither at the earth's pole nor on the equator.

24. **Conical Projection, with 2 Standard Parallels.**\*—This is one of the most useful of all projections. (It is sometimes wrongly called " secant conical.")

Two selected parallels are represented by concentric circular arcs of their true lengths; the meridians are their radii. The degrees along the meridians are represented by their proper lengths, and the other parallels are circular arcs through points so determined, concentric with the chosen parallels.

This construction evidently fixes the position of the centre of the parallels, for if $p_1 p_2$ are the true lengths of the chosen parallels, $m$ the length of the intervening meridian (all of which can be taken from geodetic tables), then $h = \dfrac{p_2 - p_1}{m} \times \dfrac{180}{\pi}$, where $h$ = ratio between the inclination of two meridians on paper, and their true difference of longitude.

The parallels may be chosen arbitrarily from a consideration of the country to be mapped; but it is better to adopt one of the following methods :—

24a. The absolute errors for a given difference of longitude along the centre and extreme parallels may be made equal. This is Euler's projection. For the purpose of choosing the projection it will be sufficient to treat the earth as a sphere, and in the construction to plot the true spheroidal lengths of the meridian arcs;

Or (24b), the errors of *scale* on these parallels may be made the same as follows :—if $c$, $c^1$ are the co-latitudes of the extreme parallels, $z$ the distance in degrees of the centre of the parallels from the pole, $h$ the ratio between the inclination of two meridians and their difference in longitude,

$$\frac{h(z + c)}{\sin c} - 1 = \frac{h(z + c^1)}{\sin c^1} - 1 = 1 - \frac{h\left(z + \dfrac{c + c^1}{2}\right)}{\sin \dfrac{c + c^1}{2}}$$

from which $h$ and $z$ are easily found.

\* The earliest known instances of the use of one of the cases of this projection occurs in maps of the two hemispheres by Ruysch, dated A.D. 1508.

Applying this to South Africa, south of the Zambesi, taking the extreme parallels as 18° and 34° S.; $h = .439$, $z = 52°.7$, and the error of scale on the extreme and mean parallels is 0·0049, i.e., less than ½ per cent., or less than may be expected from ordinary contraction of paper.

Or (24c), as suggested by Colonel Clarke, the mean length of all the parallels may be made correct, still keeping the meridians their true length.

This is equivalent to equating the total areas of the zones, and must be combined with another condition, for example, that the errors of scale on the extreme parallels shall be equal. (This is not the same as Albers' projection, No. 11, in which the meridians are not kept their true length.) We then have

$$z = \frac{c^1 \sin c - c \sin c^1}{\sin c^1 - \sin c}, \quad h = \frac{\cos c - \cos c^1}{(c^1 - c)\left(z + \frac{(c + c^1)}{2}\right)}$$

in the case of South Africa, as above,

$$h = .440, \quad z = 52°.7,$$

i.e., practically the same result as 24b.

It is easy to imagine other variations; for instance, we may consider the maximum error instead of that of the mean parallel (24d). This projection was used for the $\frac{1}{1,000,000}$ Ordnance Map of the British Isles. There is no difficulty in working out the projection for the spheroid.

As an example of the choice of a conical projection we may take the case of South Africa, south of the Zambesi, the limits being 15° S. lat. to 35° S. lat.

To save labour, in the preliminary calculations the earth will be treated as a sphere.

Let $c$ be the co-latitude of any parallel which is represented by $cm$, let P correspond to the pole, let $OP = z$ and the angle $com = hw$, where $w$ is the true difference of longitudes of the meridians $oc$, $om$; $Pc$, $pm$ are their true lengths.

Then the error of scale along any parallel is $1 - \frac{h(z + c)}{\sin c}$. This is a maximum when $\tan c - c = z$.

Now, let it be a condition that the scale errors on the extreme parallels shall be equal, then the largest error will be a minimum when these two errors are equated to that on the parallel of maximum error.

Then

$$\frac{h\left(z + 75 \times \frac{\pi}{180}\right)}{\sin 75} - 1 = \frac{h\left(z + 55 \times \frac{\pi}{180}\right)}{\sin 55} - 1 = 1 - \frac{h(z + c)}{\sin c},$$

where

$$\tan c - c = z.$$

From these equations we get

$$z = 56°·62$$
$$c = 64°·66$$
$$h = ·424,$$

and the maximum error of scale is less than 0·7 per cent. It occurs on the parallels of 15°, 25° 20′, and 35° S. The errorless parallels, to the nearest degree, are those of 18° and 32° S.

To plot the projection of which the elements have been thus determined, we shall use the true spheroidal lengths, making the parallels of 18° and 32° S. errorless; this will slightly alter the value of $h$ as follows:—

The true length of the meridian between 18° and 32° S. is 963·70 miles.
The true length of 1° on the parallel of 18° S. is 65·808 miles.
,, ,, ,, ,, ,, 32° S. is 58·720 ,,

whence $h = ·421$, which does not differ materially from that found above.

Plate XVI.

## SOUTH AFRICA
on
A CONICAL PROJECTION WITH RECTIFIED MERIDIANS
AND TWO STANDARD PARALLELS.

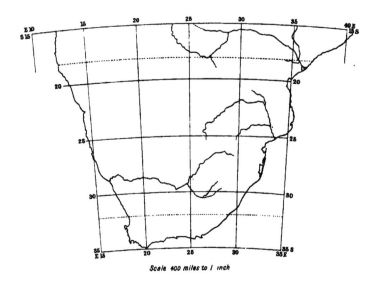

Scale 400 miles to 1 inch

The meridians and two parallels (18° and 32° S.) are true to scale

The radius for parallel 32° is now easily found to be 7983·7 miles.
The radius for parallel 35° is this—the length of the meridian between 32° and 35°, 7983·7 - 206·7 = 7777 miles.
The length on the map of 1° of longitude on 35° S. is now

$$7777 \times \cdot 421 \times \frac{\pi}{180} = 57 \cdot 143.$$

The true spheroidal length is 56·728, an error of about 0·7 per cent.
The radii of the other parallels are obtained in the same way.
Next to calculate the co-ordinates for plotting. Take the parallel of 35° S.
The radius of this parallel is 7777·0 miles.
Take the meridian, of 25° E. as the axis of $y$, and the intersection of this meridian with the parallel of 35° S. as the origin.
To determine the co-ordinates of the point 35° S., 30° E., difference of longitude = 5°.
Angle between meridians on map = $hw$ = ·421 × 5 = 2° 6′ 18″.
Then

$$x = 7777 \times \sin 2° 6' 18'' = 285 \cdot 65 \text{ miles},$$
$$y = 7777 (1 - \cos 2° 6' 18'') = 5 \cdot 25 \text{ miles},$$

and so on.
The following table gives the results :—

Conical projection for S. Africa.
Co-ordinates in miles.
Origin, point 35° S., 25° E.

| Latitudes S. | | Longitudes E. | | | |
|---|---|---|---|---|---|
| | | 25° | 30° | 35° | |
| 35° | $x$ | 0 | 285·7 | 570·9 | The parallels of 18° and 32° S. are errorless, as also are all the meridians. |
| | $y$ | 0 | −5·3 | −21·0 | |
| 30° | $x$ | 0 | 296·3 | 596·2 | |
| | $y$ | 344·5 | 339·0 | 322·6 | |
| 25° | $x$ | 0 | 310·9 | 621·5 | See Plate XVI. |
| | $y$ | 688·7 | 683·0 | 665·9 | |
| 20° | $x$ | 0 | 323·6 | 646·7 | The projection is symmetrical, and the co-ordinates are to be also used for points W. of 25° E. |
| | $y$ | 1032·7 | 1026·8 | 1008·9 | |
| 15° | $x$ | 0 | 336·2 | 672·0 | |
| | $y$ | 1376·5 | 1370·3 | 1351·8 | |

**25. Simple Polyconical Projection.**—In this projection, which is a conventional one, any parallel is represented by a circular arc of radius $\nu \cot$ lat., where $\nu$ is the normal terminated by the minor axis. The central meridian is a straight line through the centres of the parallels, which are their true distances apart measured along this meridian. The other meridians are curves which cut the parallels at the true distances, measuring along the parallels. Thus the meridians do not cut the parallels at right angles.

The projection is best suited for countries having a great extent in latitude and little in longitude. The meridian curves are sometimes for convenience replaced by the rectilinear inscribed polygons.

The great usefulness of this projection and of the Rectangular Polyconical Projection consists in the fact that it is easy to construct universal tables, since each particular parallel has always the same radius ($\nu$ cot lat.).

Single sheets, not exceeding in latitude four degrees, when plotted on either system, do not materially differ in any of the principal qualities from sheets plotted on specially devised conical projections, such as 3, 11, or 24 (*b*), (*c*) or (*d*). But this only applies to small areas and independently plotted sheets. The best tables to use are those published by the United States Coast and Geodetic Survey—"Tables for a Polyconic Projection of Maps," 1900.

Polyconical projections should not be used for large areas, and are especially unsuited for maps having a great extent in longitude.

**26. Rectangular polyconical projection**, sometimes called the **rectangular tangential**. The parallels are formed as in the last projection, but the meridians in this case are curves which cut the parallels at right angles. The lengths along the parallels are consequently not true.

This projection is described in the War Office pamphlet "On the Projection of Maps," by Major. L. Darwin, R.E., 1890. A geometrical construction (due to the late Mr. O'Farrell, of the Ordnance Survey) determines the intersection of the meridians and parallels. (See War Office pamphlet which contains the necessary tables, or "Encyclopædia Britannica.")

For small areas it does not differ appreciably from the simple polyconical. It is best suited for countries having a great extent in latitude and not much in longitude. It will also do well for a topographical series, each sheet being plotted independently.

The conical orthomorphic, Lambert's (No. 3),
The modified conical equal-area, Bonne's (No. 7),
The simple conical equal-area, Lambert's (No. 10),
The conical equal-area, Albers' (No. 11), have been described.

### Cylindrical Projections.

**27. Simple Cylindrical Projection.**—The parallels and meridians are straight lines at right angles to each other. The lengths along the meridians are correct, and those along some selected parallel. A projection of no value.

Mercator's (1),
Transverse Mercator's (2),
Cylindrical equal-area Lambert's (5),
Transverse ditto (6),
have been described.

### Conventional Projections.

It is rather difficult to draw the line between regular and conventional projections, but the distinguishing mark of the latter is that they are devised for simplicity of drawing and not for special geometrical or analytical properties. The globular projection is certainly conventional, and it is probably fair to call polyconical projections conventional.

Ancient conventional projections are often more arbitrary than these; though sometimes of historical interest, they are of no practical importance. An account of them will be found in "Germain."

**28. Globular Projection.**—A conventional projection of a hemisphere, in which the equator and the central meridian are two straight lines at right angles, their intersection is the centre of the circular boundary. The meridians cut the equator at equal distances and are arcs of circles passing through these points and the poles. The parallels are arcs of circles which cut the central and extreme meridians at equal distances.

This name is sometimes incorrectly given to La Hire's perspective projection (20), and also equally incorrectly to the polar equidistant zenithal projection (21).

It differs, however, but little from La Hire's. It is used in popular atlases, but has no claim to be treated seriously.

29. **Projection by Rectangular Co-ordinates.**—Sometimes called Cassini's or Cassini-Soldner's. Any point on the earth is fixed if we know the length of a perpendicular from it on to some fixed meridian, and the distance of the foot of this perpendicular from some fixed point on the meridian. On this projection, these spheroidal co-ordinates are plotted as ordinary plane rectangular co-ordinates.

The perpendicular spoken of is really a plane section of the surface through the given point at right angles to the chosen meridian, and may be briefly called a great circle. Such a great circle clearly diverges from the parallel. The divergence of latitude in seconds is, approximately, $\frac{x^2}{2\rho\nu} \tan \lambda \operatorname{cosec} 1''$ where $x$ is its length, $\rho$ that of the radius of curvature to the meridian, $\nu$ the normal terminated by the minor axis, $\lambda$ the latitude of the foot of the perpendicular.

The difference of longitude from the meridian in seconds is $\frac{x}{\nu} \sec \lambda \operatorname{cosec} 1''$.

The resulting error consists principally of an exaggeration of scale north and south, and is, approximately, equal to sec $x$; it is practically independent of the extent in latitude.

This projection should not be used when the map extends more than, say, 200 miles from the central meridian. It is on this projection that Ordnance Survey 1-inch maps of England are drawn, and also the 6-inch Ordnance Survey maps of the United Kingdom; and for these comparatively small areas the projection is a very suitable one.

30. **Graticules and Field Rectilinear Projections.**—These have been described (see p. 93).

## CHAPTER XII.

### THE REPRODUCTION OF MAPS IN THE FIELD.

It is important that the Intelligence Division of an army in the field should possess the means of reproducing maps and sketches. The most suitable methods where large numbers of copies are required are lithography and zincography. In the former, the printing surface, as its name implies, is stone, in the latter, zinc.

Every officer employed with a survey and mapping section should possess some knowledge of these methods, of the equipment and personnel which would be available, and of the time which would be required for the work, though it may not be essential for him to be an expert lithographer.

In the South African War of 1899-1902, the Survey and Mapping Sections supplied themselves with stones and lithographic presses, which worked admirably at Orange River, Bloemfontein, and Pretoria.

The experience gained in this war has resulted in a new equipment. In the heavier equipment, stone is carried in preference to zinc.

The former is, of course, much heavier than zinc, and is also fragile, but this disadvantage is compensated for by the greater ease with which it is worked, especially under service conditions.

It is considered that any photographic process, except sun-printing, is unsuitable for use in the field.

The recently approved Field Lithographic Equipment consists of a portable press, and the necessary stores as well as a tent. Four double-faced stones are carried, and the whole equipment, with the exception of the tent, is carried in six boxes, details of which are given in the following table :—

DETAILS OF FIELD LITHO. EQUIPMENT (No. 1).

| Number of Box. | Contents. | Measurements. | Weight. |
|---|---|---|---|
| | | | lbs. ozs. |
| 1 | Lithographic press | 4' 3½" × 2' 2" × 1' | 401  8 |
| 2 | Two double-faced stones | 2' 2¾" × 1' 9¾" × 10" | 226  0 |
| 3 | Ditto | Same as 2 | 226  0 |
| 4 | Paper | 2' 0" × 1' 8" × 1' 2" | 197 14 |
| 5 | Various stores | 2' 2" × 1' 10" × 1' 3" | 112 12½ |
| 6 | Ditto | 2' 1" × 2' 1" × 1' 3" | 129 14 |
| Total weight of equipment | | | 1,294  0½ |

This equipment, which can be reduced on emergencies, is intended to be carried in any ordinary cart or wagon, and will reproduce maps up to 18" × 23".

A light field litho. press weighing about 190 lbs., capable of reproducing maps up to double foolscap size, i.e., 13" × 16", has also been designed and approved.

PLATE XVII.

FIELD LITHOGRAPHIC PRESS (FOR CART TRANSPORT).

Zinc plates will be used with this press, which is intended for mule transport.

**The personnel required** for the lithographic equipment is two lithographic draughtsmen and two lithographic printers.

The draughtsmen are specially trained in the art of drawing reversed on stone or zinc, but have nothing to do with the actual printing of the plans, which is done by the printer.

**Rates of Work.**—A skilful draughtsman should be able to trace, in transfer ink, a map of enclosed country, with a fair amount of lettering and detail, at a scale of 1 inch to 1 mile, at the rate of 10 minutes per square inch. If required to draw direct on to the stone or plate, double this length of time would be necessary.

A printer should be able to print about 40 copies of the same map in an hour, under ordinary circumstances.

**Lithography and zincography** are methods of reproducing maps or plans in greasy ink, in large numbers, by mechanical means from stone or zinc. Both methods depend on the fact that stone and zinc are to a certain extent porous, and capable of absorbing both water and grease. Where a grease spot exists, water is repelled, and *vice versa*.

If a drawing in greasy ink, or transfer as it is called, be laid face down on a clean stone, and subjected to pressure, the grease leaves the paper, and is absorbed by or takes on to the stone or zinc. If the stone or zinc plate be then damped and a roller charged with a greasy ink be passed to and fro over its surface, the ink will only adhere on the lines of the drawing. Should a piece of clean paper be placed on the stone or plate and passed through the press, it will lift some of the ink, and a reproduction of the original picture will be obtained. By damping and rolling up alternately, a very large number of copies can be produced.

### Preparation of Transfers.

The first step in the reproduction of a map, unless it is to be drawn direct on to the stone or plate, is the preparation of a transfer, that is to say, a copy of the plan or map in greasy ink.

**The ink** used for this purpose is known as litho. writing ink, and is applied with ordinary pen, ruling pen, or brush. It is obtained in sticks and prepared for use by warming the bottom of a saucer, rubbing the ink on, adding water, and stirring up. Its chief constituent is some form of soap, *i.e.*, a substance of a greasy nature which has been acted upon by an alkali, and which can be subsequently rendered greasy by the application of an acid.

The transfer would usually be traced on a specially prepared paper known as **tracing transfer paper**.

Where an alteration in scale from that of the original is required, the enlargement or reduction would first be drawn on ordinary paper in pencil, and then traced in litho. writing ink on tracing transfer paper.

Drawings in litho. writing ink should not be blotted or rubbed.

### Preparation of Stone.

**The Stone** used in lithography is a kind of lime stone, and comes from Bavaria.

It is ground with a levigator and silver sand, passed through a 40-mesh sieve. This gives a level surface, and removes old work.

Polish with pumice stone and water.
Polish with snake stone and water.
Wash with cold water.

TRANSFERRING LITHO. WRITING INK TRANSFER, MADE ON ORDINARY PAPER, TO STONE.

The stone is placed in front of a fire to warm.

The transfer, which should be freshly prepared, is placed face downwards on a clean piece of paper, and damped from the back with a sponge wetted with dilute nitric acid (1 in 80), about 4 times, at intervals of 2 or 3 minutes. The object of the acid is to convert the ink from the soluble to the insoluble form; it also renders the paper more pliable. The acid must not be strong, or it will rot the paper. If the ink were soluble it would wash off the stone. The impression would probably roll up, but it would not be so strong, and the lines possibly rotten, i.e., broken.

The warmed stone is then placed in the press, and the transfer face down on it. Grease takes more readily on a warm than on a cold stone.

The transfer is pulled through the press about half a dozen times with light pressure, transfer damped, scraper or stone reversed in case pressure is not equally distributed over the surface of the stone, and pulled through another six times.

These operations are repeated till transfer is damp throughout. Transfer is removed, composition washed off with cold water, and the stone gummed in.

Light pressure has to be used, as the composition on the paper is soft and might "smash."

The transfer is then carefully removed from the stone, which is then gummed in with a gum solution of the consistency of thick treacle. The stone is then allowed to dry and cool. The gum fills up the pores of the stone where clear of ink; when once dried cannot be removed with water. It prevents the lines from spreading, and lessens the chance of scum; that is to say, it prevents the printing ink from the roller taking all over the stone. The gum is then washed off from the surface of the stone with a sponge, and the ink is strengthened by rolling up and damping alternately. If a hot stone is rolled up the ink may spread in spite of the gum, and as it dries quickly scum might appear. A stone should always be damp while being rolled up.

The stone is then gummed again, cleaned up where necessary with the point of a knife or piece of snake stone, and etched.

The etching solution is dilute nitric acid (about 1 to 80), such as will give a little roughness on the teeth, but no effervescence on the stone. It is applied with a sponge followed up with a second sponge with water. The stone is then cleaned with a damping cloth, gummed again to replace any gum destroyed by the acid, and allowed to dry.

The object of etching is:—

1. To prevent the work from spreading.
2. To prevent scum appearing on the white portions.
3. To leave the work in slight relief, and to prevent fine work from filling up with ink.

PRINTING.

**The ink** used for printing is composed of burnt linseed oil, lamp black and copal varnish (as a drier).

It is bought ready ground, but for use is thinned down with lithographic varnish.

Chalk printing ink is a superior kind of litho. printing ink used for fine work.

The ink is applied with a roller. This is made of calf skin, flesh side outwards, over two layers of flannel fastened on to a wooden cylinder.

When using black ink the roller need not be washed, but it should be scraped about three times a day.

**The stone** is washed to remove surface gum, and if the ink has had time to get hard (8 or 9 days with average ink) it is removed with turps. The stone is then damped with a cloth, and rolled up with the roller charged with ink. For fine work the ink should be thick.

The paper is put down to "**lays**," *i.e.*, marks or lines drawn as a guide on the stone with printers' lead, a mixture of lead and antimony. These lays do not wash out, and, not being easy, do not take the printing ink. The stone should be damp when the lays are drawn.

Only paper of inferior quality requires damping to soften it, and allow it to press better into the work.

The right side (*i.e.*, smooth side) should be used.

The paper having been laid on the inked stone, is pulled through press once, with moderate pressure, and removed.

The stone is damped and rolled up again after every impression is taken.

If work is stopped, although only for a short time, the stone should always be gummed in.

### Drawing on Stone Direct.

To draw on stone direct needs special training, as the drawing has to be reversed.

The outline at least would be usually laid down from a tracing by means of red chalk paper in a similar manner to that employed for duplicating bills, &c., with carbon paper. The wrong side of the tracing would, however, be uppermost. This outline is called an offset or faint.

The stone should be clean and free from scratches.

The outline is inked in with litho. writing ink; all, except the actual portion being inked in, should be covered over to prevent the chalk dust from being rubbed off by the hand or arm.

When the drawing has been completed, the stone is flooded with a mixture of gum and dilute nitric acid (1 in 60).

Solution washed off; stone gummed and dried. Then oiled, washed with turps, damped, rolled up, cleaned, etched, washed, gummed and dried. It is then ready for printing from.

### Alterations on Stone.

Wash off the gum with water. Take out any part to be removed as lightly as possible with snake stone or knife, and wash with water. Blot. Make the correction in litho. writing ink. Allow ink to dry. Gum. Wash off gum. Roll up. Etch. Gum in again and dry.

If an addition has to be made on a part where nothing has yet been drawn, the process is:—wash off gum with water. Wash with citric acid (1 in 80) to eat gum out of stone, wash with water, and draw in new work as before.

### Printing in more than one Colour.

When a map has to be printed in more than one colour, a **keystone** has first to be prepared for printing, and on it is laid down all the details of the complete map.

Two lines or **lays** are drawn across the plate, at right angles to each other, to mark where two adjacent sides of the paper are to rest.

These lays are marked as follows:—

Where the lays will come, wash off gum with water. Apply citric acid (1 in 80); dry with blotting paper. Draw lays with litho. writing ink in the manner just described for additions on stone.

In printing the offset the keystone is allowed to dry after being rolled up, so as to lessen the chance of the paper stretching or contracting. For the same reason glaze board or enamelled paper should be used. If these are not procurable use a bit of stout paper which has just been pulled through the press.

The **offset** is then dusted over with red chalk powder and the surplus shaken off.

The offset is then laid on a stone prepared in the usual way to receive a transfer, and pulled through the press. This leaves the chalk on the stone. The portions required in one colour are then drawn in in litho. writing ink, as described in drawing on stone.

The marks of the lays are cut with a knife and filled in with common ink.

A colour stone is prepared for each separate colour, a fresh offset being required for each stone.

When this is done the keystone is prepared for work by removing all the detail which is not required in black, in the manner described for alteration.

Before printing from the colour stones the black ink must first be washed off with turpentine, and the stone rolled up with the coloured ink.

**The roller** used is similar to the black litho. roller except the smooth side of the skin is outwards, as this makes it easier to clean.

The lightest colour is printed first, and the print is placed accurately on the next stone, and so on until all the colours have been printed.

**Register.**—Great care is necessary that the paper is layed accurately to the lays on each stone, or the colours will overlap.

## ZINCOGRAPHY.

**The preparation and putting down of transfers** on to zinc is done in the same way as on to stone.

The plates are usually obtained with a smooth surface (termed planished). Before use, they are grained with fine sand, the plate being kept wet and the graining done with a muller.

**Printing.**—On removal of the transfer the plate is etched with a solution of gum, galls and phosphoric acid. The solution is allowed to act for $\frac{1}{2}$ to $1\frac{1}{2}$ minutes, producing a fine layer of phosphate of zinc. It is then washed off and a gum solution, to which $\frac{1}{2}$ oz. of etching solution to 1 pint has been added. This solution is slightly thinner than that used for stone. The plate is fanned dry, damped, sprinkled with oil followed by turpentine, and rubbed with a piece of flannel. It is then rubbed over with a turpentine cloth and rolled up with ordinary litho. ink, which takes all over the plate owing to the presence of the turpentine. Water is then sprinkled on the plate, which is rubbed with a cloth. This causes the ink to leave the plate except on the lines. The plate is again rolled up, gummed in and fanned dry. Gum washed off and plate rolled up. It is then ready to take the first impression. Plates should always be gummed in and fanned dry whenever it is necessary to stop work, even for a short time.

**The number of impressions** which can be obtained from a zinc plate depends on a variety of circumstances, but usually from 500 to 1,000 should be obtained. From stone the number would run into thousands.

## SUN PRINTING.

When the lithographic equipment is not available for the multiplication of maps in the field, recourse must be had to sun printing.

There are several papers now on the market which are suitable for this purpose, and although this method is not nearly so satisfactory or expeditious as lithography, there is a very great advantage in the gain in the lightness of the equipment required.

The most suitable paper for this purpose is **Bemrose No. 5, Brown Process**, to be had from Bemrose and Sons, 4, Snow Hill, London. It is made in three qualities, viz.:—

> E. 101 thick.
> E. 100 thin tough.
> E. 102 linen.

**The Prints** produced by this method are either white on a deep chocolate ground, or chocolate on a white ground. The former would be produced by printing from a tracing, the latter by printing from a negative print made on the thin paper (E. 100).

A tracing on tracing cloth or paper should be made from the original. It should be drawn with Indian ink, made as thick and opaque as possible. Creases and folds should be avoided. It is possible to print direct from an original, unless made on very thick paper, but the printing is of course slower than with a tracing.

The tracing is placed in a printing frame, face to the glass, and the sensitized paper laid over it, prepared side next to the tracing, the felt pad and back of frame next in succession, care being taken that the paper is in perfect contact with the tracing.

The frame should be placed in the sun if possible for $\frac{1}{4}$ to 8 minutes. The print should be examined at intervals and removed from the frame as soon as the detail is completely visible.

The print is next placed in water for 3 to 5 minutes, and then in a bath of hyposulphite of soda, 1 oz. to $3\frac{1}{4}$ pints of water, for $\frac{1}{2}$ to 1 minute, or, if preferred, the hyposulphite can be applied with a brush or sponge.

The print is then well washed and dried.

The paper is supplied in rolls of 11 yards, width 26, 30 and 40 inches; sufficient developer is supplied with each roll.

For quick work two men would be required, one to expose the prints, and the other to develop and wash them.

The following apparatus would be required (cost about £3):—

> 1 printing frame, 26″ × 21″;
> 2 zinc baths, 28″ × 21″;
> 1 developing brush;
> 1 tin case for sensitive paper, 26″ long;
> 1 bucket or can for water;
> 1 pot for developing solution;
> 1 roll of ordinary tracing paper.

The price of the paper per roll of 11 yards is as under:—

> Thick ... ... ... 7s. 0d., 30″ wide; 9s., 40′ wide.
> Thin .. . ... 5s. 3d. ,, 7s. ,,
> Linen ... . .. ... ... 21s. ,,

## CHAPTER XIII.

### BRITISH AND FOREIGN GOVERNMENT MAPS AND SURVEYS.

**National Surveys.**—A good topographical map is such a vital military necessity, that the initiative in the matter of national surveys has, in nearly every country, been taken by the army. This is the case in the United Kingdom, India, France, Germany, Italy, Spain, Austria, Russia and other countries.

**Mapping v. Sketching.**—The necessity for a topographical map having been realised, the method of its formation would become the next point for consideration. It would occur to any officer accustomed to sketching (as are the officers of most armies) that a workable map might be formed by combining military sketches. As an instance of the effect of such a system we may quote the state of the maps of India before the year 1802. These maps were compiled from sketches and were subsequently tested by the results of the accurate survey. In the short distance across India in the latitude of Mysore, the maps so compiled were found to be 40 miles in error.

As another instance may be quoted the compilations made in Burma after the annexation in 1885. A vast amount of sketching was carried out by officers during the operations, these sketches were compiled; the result, though better than nothing at all, was highly unsatisfactory.

Akin to these were the compilations made from farm surveys during the South African War of 1899-1902. These were reported by many generals to be worse than useless. They gave, however, an abundance of names and served to show the location of farms and the lines of railways. The hill features shown on them were quite unreliable, which will not surprise those who have seen the original farm surveys from which they were compiled.

Another instance of the futility of combining sketches to make maps is the present condition of the maps of most British possessions in Tropical Africa. Indeed all experience demonstrates the fact that *no system of sketching, no combination of sketches, and no compilation of previously unconnected material, whether sketches, isolated surveys, or farm plans, will result in a trustworthy topographical map.*

**The Ordnance Survey.**—In the United Kingdom, the first step towards the production of a topographical map was taken in 1747, two years after the rebellion in the Highlands. General Watson, D.Q.M.G. to the Duke of Cumberland's forces, conceived the idea of making a map of the Highlands, this was subsequently extended to the rest of Scotland. Various schemes were now advocated for the accurate mapping of the whole kingdom, but it was not until 1783 that the Ordnance Survey may be said to have come into existence by the measurement by General Roy of a base line on Hounslow Heath; and it is to be noticed that the whole work was undertaken by soldiers, the working party being furnished by the 12th Regiment; the reason given by Roy being, that, "not only was this obviously the most frugal method," but, also, because "soldiers would be more attentive to orders than country labourers."

The subsequent history of the Ordnance Survey is in brief:—Until 1824 work was almost entirely confined to the 1-inch scale. At that date the 6-inch map of Ireland was commenced; in 1840 the same scale was commenced for Great Britain. The map of Great Britain on a scale of $\frac{1}{2500}$, approximately 25 inches to the mile, was commenced in 1854. This is periodically revised, so that no sheet is more than 15 years out of date. The second revision is now in progress in England and Scotland. The same scale was commenced in Ireland in 1895, about half Ireland is completed. The interest of the 25-inch map for a soldier lies in the fact that

Plate XVIII.

## ORDNANCE SURVEY
Scale.- 1 Inch to 1 Mile *(Coloured)*

SHOWING THE COMBINATION OF CONTOURS AND VERTICAL HACHURES

it is reduced to form the 6-inch, and on this latter, the 1-inch, and smaller scales, are based and revised.

The following maps are formed by reduction from the 1 inch :—$\frac{1}{2}$ inch, $\frac{1}{4}$ inch, $\frac{1}{10}$ inch and $\frac{1}{1,000,000}$.

These maps, from 1 inch to $\frac{1}{1,000,000}$, form the most interesting feature of the Ordnance Survey at the present day.

They should always be bought in open book form, mounted on linen.

The 1-inch Ordnance map is published in 4 forms :—

1. Detail and contours black, no hachures.
2. Detail, contours and hachures, black.
3. Detail and contours black, hachures brown.
4. In six colours.

**The Survey of India.**—Like the Ordnance Survey, the Survey of India owes its origin to a soldier, Major Lambton, 33rd Regiment. After the fall of Tippoo, Major Lambton submitted a scheme for the Trigonometrical Survey of the Peninsula. Actual work was commenced in 1802.

Major Lambton, who died in 1823, was succeeded by Sir George Everest; "the whole conception of the trigonometrical survey as it now exists was the creation of Everest's brain," and no man has ever left a greater monument. The first attempt at a topographical survey was made by Col. Colin Mackenzie, who surveyed some 40,000 square miles in the Deccan between 1790 and 1809. Everest was succeeded by Sir Andrew Waugh in 1843, and from this date a great extension of topographical surveys took place. The names of Capt. Robinson and Capt. Montgomerie are specially to be remembered as topographers.

It is to Capt. Robinson that we owe the present accurate method of using the plane-table.

It is impossible even to mention the vast numbers of topographical maps of the Survey of India. They cover an area of about 2,000,000 square miles, and vary in scale from $\frac{1}{8}$ inch to 4 inches to the mile. Generally speaking, a new province is surveyed first on the $\frac{1}{4}$ inch scale, and, subsequently, if required, on the 1 inch. The system of surveying on service has been brought to a high degree of excellence. Perhaps the most interesting of recent surveys is the $\frac{1}{2}$ inch work carried out during the operations in China. A preliminary edition of this was printed in the field. The special merit of the Indian topography lies in its frontier work; no country can turn out such rapid and accurate maps under such difficult conditions.

**South Africa.**—In South Africa the only topographical map which existed on the outbreak of war in 1899 was a 1-inch map of a small area north of Ladysmith. This map had been rapidly made some years before; both the time and means allowed for its construction were inadequate.

**Foreign Maps.**\*—The principal topographical maps of other countries are :—

*Austria*, $\frac{1}{75,000}$. Hills are shown by contours and vertical hachures, the later are coarse and heavy. Not a good example of a topographical map.

*Belgium*, $\frac{1}{40,000}$, in colours; hill features shown by contours. $\frac{1}{160,000}$, military map of Belgium. The hill features are badly defined by feeble contours in brown, which are difficult to follow.

*France*, $\frac{1}{80,000}$. "Carte de l'État-Major," the staff map of France, engraved on copper; hills shown by vertical hachures. Similar to, but inferior to, the black edition of the Ordnance Survey 1 inch (Plate XXIII).

The coloured $\frac{1}{50,000}$ map of Algeria and Tunis, prepared by the Service Géographique de l'Armée. Hill features shown by contours at 10 metre intervals, and by shading with an oblique light. These are excellent maps (Plate XX).

\* *See* Pamphlet by Col. Sir John Farquharson, K.C.B., on Foreign Topographical Maps.

The $\frac{1}{100,000}$ "Carte Vicinale." Hill features shown only by stump, difficult to read except in hilly country.

*Germany*, $\frac{1}{100,000}$, engraved on copper, printed in black, somewhat inferior to the Ordnance Survey 1 inch.

$\frac{1}{25,000}$, "Positions-Karte," the maps of Prussia, Baden, Saxony and Wurtemberg on this scale are good clear maps.

*Italy*, $\frac{1}{100,000}$ contours at 50 metres, with vertical hachures (Plate XXV).

*Russia*, $\frac{1}{126,000}$, or "3 verst" map, engraved on copper, printed in black. Hills shown by vertical hachures when there are any. $\frac{1}{126,000}$ or 3 verst maps of the Balkan Peninsula. Hills shown by contours in brown, and the contours are somewhat close; otherwise a good clear map.

*Spain*, $\frac{1}{50,000}$, in colours, contours at 20 metre intervals.

*Switzerland*, $\frac{1}{25,000}$, and $\frac{1}{50,000}$, contours brown, detail black; beautiful maps.

*United States*, $\frac{1}{62,500}$, $\frac{1}{125,000}$, and $\frac{1}{250,000}$, according to population of country. In colour; contours in brown; water, blue. Very excellent maps (Plate XXI).

**Analysis of Scales of Foreign Maps.**—We can now analyse the scales used. Where two maps of a country are of equal importance, they are both put down in the accompanying list:—

SCALES.

| | | |
|---|---|---|
| The United Kingdom | 1 by | 63,360 |
| ,, | ,, | 126,720 |
| India | ,, | 63,360 |
| ,, | ,, | 253,440 |
| Austria | ,, | 75,000 |
| Belgium | ,, | 40,000 |
| ,, | ,, | 160,000 |
| France | ,, | 80,000 |
| ,, | ,, | 50,000 |
| Germany | ,, | 100,000 |
| ,, | ,, | 25,000 |
| Italy | ,, | 100,000 |
| Russia | ,, | 126,000 |
| Spain | ,, | 50,000 |
| Switzerland | ,, | 25,000 |
| ,, | ,, | 50,000 |
| United States | ,, | 125,000 |
| Mean, about 1 by | | 90,000 |

This list of the scales of the principal topographical maps of the world, most of which are military maps, is sufficient to show how important it is that officers should accustom themselves to the use of small scales. The same opinion would have been arrived at on the merits of the case, by considering those instances in which maps are used in war, and the purposes which they have to serve. Even in the case of a small country like England it is simply inconceivable that an army could make any general use of 6-inch maps—the 6 inch map of England is over 15,000 square feet in area.

Plate XIX.

## ORDNANCE SURVEY
Scale. - 4 Mile to 1 Inch *(Coloured)*

**Uses of Maps in War.**—The uses to which maps are put in war are :—

(1) For strategical purposes ;
(2) For finding the way ;
(3) To give the Commander of a force topographical information concerning roads, villages, woods, hills, streams, bridges, ferries, railways, and water supplies, &c. ;
(4) To enable a general outpost, offensive or defensive position to be taken up ;
(5) And generally to convey information which cannot be given at the moment by the eye alone.

It is clear that, as a general rule, large scales are not required.

**Analysis of methods of showing hill features.**—It is worth noting that most government maps represent hill features by a combination of contours and shade, contours and vertical hachures, or by contours alone. On very small scales the contours become generalized form-lines which may be replaced by vertical hachures. As regards the intervals of the contours, the following list gives the number found by multiplying the contour interval in feet by the scale in inches to the mile.*

| | | | | | |
|---|---|---|---|---|---|
| France, | $\frac{1}{50,000}$ | ... | ... | ... | 41 |
| ,, | $\frac{1}{80,000}$ | ... | ... | ... | 52 |
| Prussia, | $\frac{1}{25,000}$ | ... | ... | ... | 41 |
| Baden, | | ... | ... | ... | |
| Bavaria, | $\frac{1}{25,000}$ | ... | ... | ... | 81 |
| Wurtemberg, | | ... | ... | ... | |
| Italy, | $\frac{1}{50,000}$ | ... | ... | ... | 20 |
| ,, | $\frac{1}{100,000}$ | ... | ... | ... | 102 |
| Spain, | $\frac{1}{50,000}$ | ... | ... | ... | 81 |
| U.S.A., | $\frac{1}{62,500}$ | ... | ... | ... | 50 |
| ,, | ,, | ... | ... | ... | 20 |
| O.S. ¼-inch | | ... | ... | ... | 50 |
| ,, Colonial | | ... | ... | ... | 50 |
| ,, 1 inch | | ... | ... | ... | 100 |
| | Mean | ... | ... | ... | 57 |

We may, therefore, say that the weight of authority is in favour of a contour interval of about 50 feet for a 1-inch scale, and for other scales in proportion, thus, 100 feet for a ½-inch map, &c.

These same results might have been arrived at on the merits of the case. All practical topographers know that in ordinary country the contours on the Ordnance Survey unshaded 1-inch map are not sufficiently close to show the ground. It should be added that the rule of 50 feet for a 1-inch scale is a tacit rule on which the form-lines of the small scale Survey of India maps are made ; e.g., on the transfrontier ¼-inch maps the form-lines may be considered to be 200 feet apart.

These considerations have led to the decision to adopt the same rule for military sketching, both on it own merits, and because sketches will necessarily be used in conjunction with Government maps. The rule may be stated thus :—The contour interval in feet is found by dividing 50 by the number of inches to a mile in the scale.

To sum up the analysis of this list of maps :—

* Switzerland (an exceptional case, on account of its mountainous character) has the following intervals : for the $\frac{1}{44,800}$ map, 10 metres ; for the $\frac{1}{50,000}$ map, 30 metres.

The modern topographical map will be on a scale of about $\frac{1}{100,000}$ or from ½ inch to 1 inch to 1 mile.

The hills will be represented by a combination of contours and shade.

The intervals of the contours will be given by the rule*

$$\frac{50}{\text{No. of inches to the mile.}}$$

The map will be printed in colours.

**The Layer System.**—Another means of showing the relief or hill features of a tract of country is by "layers" or "tints." In this system the height above sea-level, between any two selected contours, is indicated by a special tint or colour. Thus, a light tint, say of yellow, will represent all the ground between sea-level and the 200-ft. contour; a somewhat darker tint will show the ground between the 200 and 400-ft. contours, and so on. Mr. Bartholomew's excellent tinted half-inch maps of the United Kingdom (based on the Ordnance Survey) are well known.

A few official maps on this system have been published. Amongst them may be mentioned:—

Austria, $\frac{1}{750,000}$, 11 tints, 9 of brown, 2 of green.

Sweden, $\frac{1}{500,000}$, 13 tints, light yellow to dark brown.

No system can rival this as a means of showing broadly on small scales the absolute and relative heights of the main features. Its disadvantages are, that it gives a stepped appearance to the map, that the darker shades obscure the detail, and that it gives very little idea of the steepness of the hill slopes.

Medium country on a scale of 2 miles to 1 inch may, perhaps, be considered to be equally well represented by contours and shade or by layers, and the effectiveness of the latter method increases as the scale decreases.

**"Metric" Scales.**—In the British Empire the question sometimes arises as to whether the scale shall be in inches to the mile, or shall be a "metric" scale in which the denominator of the fraction is some definite number of thousands. For instance, shall we choose

½ inch to 1 mile ($\frac{1}{126,720}$), or $\frac{1}{125,000}$ ?

The answer may, as a rule, be given in favour of ½ inch to the mile. This statement at once brings up a definite mental picture of the relationship between the length on the map and that on the ground in terms of two most familiar units. The other involves a considerable effort to realize and is indeed only justifiable where the metric system is in use. When we adopt a metric system and have got into the habit of thinking in centimetres and kilometres, then, and not till then, should our maps be drawn on a metric scale.

Plates XVIII. to XXV. are examples of some of the more important small scale maps of the United Kingdom, France, Germany, the United States, Switzerland and Italy.

---

* Or a 25 metre interval for a scale of $\frac{1}{100,000}$.

a 50 ,, ,, ,, ,, ,, $\frac{1}{200,000}$.

Plate XX.

## BIZERTA
(FRENCH TOPOGRAPHICAL MAP)

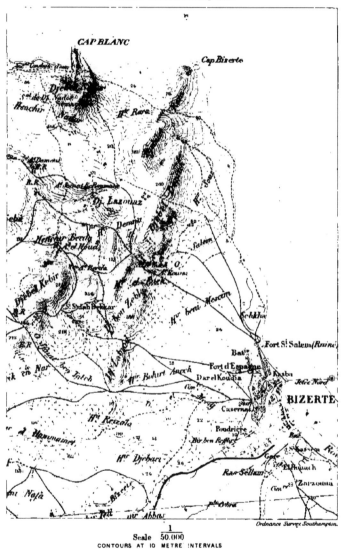

Scale $\frac{1}{50,000}$
CONTOURS AT 10 METRE INTERVALS

Plate XXI.

U. S. A.

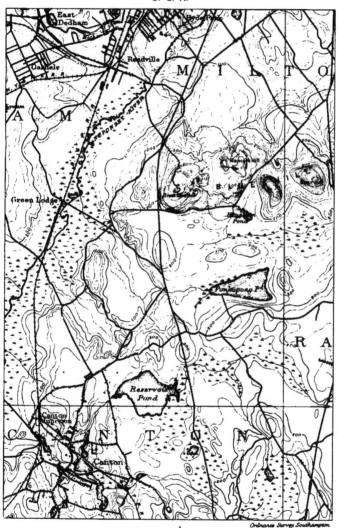

Scale $\frac{1}{62,500}$
CONTOURS AT 20 FEET INTERVALS

Plate XXII.

GERMANY

Scale $\frac{1}{100,000}$

Plate XXIII.

FRANCE

Scale $\frac{1}{80.000}$

Plate XXIV.

SWITZERLAND

Scale $\frac{1}{50,000}$

Plate XXV.

ITALY

Scale 1/100 000

Ordnance Survey Southampton.

CONTOURS AT 50 METRE INTERVAL

Plate XXVI.

## ST. HELENA
Scale - 2½ Inches to 1 Statute Mile.

## CHAPTER XIV.

### FIELD ASTRONOMY.

#### SECTION I.—GENERAL CONSIDERATIONS.

**The Heavens.**—The appearance of the heavens on a clear night is that of a vast concave sphere on which are placed the moon, planets, and stars. This great concave sphere is called in astronomy the *Celestial Sphere*.

The rising and setting of celestial bodies are produced by the revolution of the *earth* on an axis; the North and South Poles of the heavens are on the prolongation of this axis.

Returning to the representation of the heavens as a great concave sphere, imagine, at its centre, the earth, a small sphere of appreciable dimensions; and while the celestial sphere is stationary let the earth revolve on its polar axis. Now, although appearing to be about the same distance from the earth, the distances of the celestial bodies vary enormously. Thus, the mean distance from the earth of the moon is 238,833 miles, of the sun is 92,900,000 miles, and of the nearest fixed star is about 25,000,000,000,000 miles; while the distances to the planets are very variable quantities depending upon our mutual positions in our several orbits.

**Reduction of Observation to a Common Datum-point.**—A second in arc is equivalent to $\frac{1}{206,265}$; therefore, at any point on the moon, the angle subtended by two points a mile apart on the earth's surface may be as much as one second of arc.

Hence the necessity in observations of the sun, moon, and planets—stars are too far off to be affected—of reducing the observations taken to what they would be at the centre of the earth, so that astronomical tables may be of universal use while they are at the same time kept within reasonable bounds in size; and this involves a knowledge of the shape and size of the earth.

Besides the *diurnal motion*, there is the *annual motion* of the earth to be taken into account, and this leads us to the consideration of the Solar System.

#### THE SOLAR SYSTEM.

**The Solar System** is composed of the sun, the planets (of which there are about 600 very small ones and 8 large ones) and their accompanying satellites.

Some of the planets are accompanied by secondary planets or bodies called *satellites* which revolve round their particular planets just as the system of planets revolves round the sun. The planets and their satellites are opaque, non-luminous bodies, and only shine by the light of the sun being reflected from them, and thus they can only be seen when this reflected light can be seen, or when they come between an observer and some luminous body like the sun.

The Solar System may be more particularly described as consisting of:—

1. The Sun the central body.
2. The four inner planets—Mercury, Venus, Earth, and Mars.
3. The small planets or asteroids, of which nearly 600 are known to be revolving between the orbits of Mars and Jupiter.
4. The four outer planets—Jupiter, Saturn, Uranus, and Neptune.
5. The satellites of the planets; the Earth has one called the *Moon*; Mars has two; Jupiter has six; Saturn, nine; Uranus four; and Neptune, one.

**Kepler's Laws.**—The laws of planetary motion are three in number, as follows:—

1. The orbit of each planet is an ellipse, having the sun in one focus.
2. As the planet moves round the sun, its radius vector (*i.e.*, the line joining it to the sun) passes over equal areas in equal times.
3. The square of the time of revolution of each planet is proportional to the cube of its mean distance from the sun.

When a planet is furthest from the sun it is said to be in *aphelion*; when nearest in *perihelion*.

The major axis of a planet's orbit is called the *apse line*, or *line of apsides*, and that of the

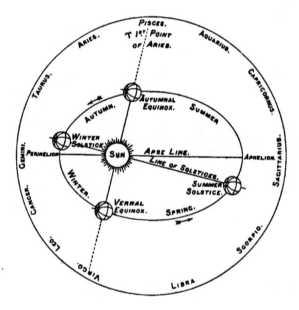

Fig. 46.

earth's orbit is known to be revolving in space 11″·78 annually (as measured at the sun's centre), in the direction of the earth's orbital motion.

**Motion of Planets in the Solar System.**—All the planets revolve round the sun in the same direction, viz., contrary to the motion of the hands of a watch when viewed the north side of the plane of the earth's orbit. Nothing certain is known of the axial rotation of Mercury, Venus, Uranus, Neptune, or the minor planets. The other planets rotate on their axes in the same direction as they revolve round the sun. The satellites move round their respective planets in the same direction, except the satellites of Uranus, the satellite of Neptune and the newly discovered ninth satellite of Saturn; the motions of these are retrograde.

The axis of each planet is inclined to the plane of its orbit at a nearly constant angle, it has a proper secular motion, but practically remains for a considerable interval of time parallel to itself during the motion of the body through space.

The orbits of all the planets are not in the same plane, but those of the principal planets lie in planes within 9° on either side of the plane of the earth's orbit, which plane is called *the ecliptic*.

## THE MOON.

**Motion of the Moon.**—The moon moves round the earth in a path inclined at 5° 8′ 40″ to the ecliptic, and takes 27d. 7h. 40m. to make a complete revolution with reference to a star. The movement of the moon in the heavens is from west to east.

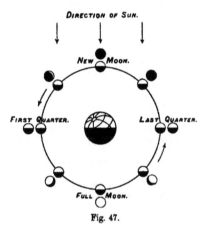

Fig. 47.

**The Phases of the Moon** are shown in fig. 47, where the unshadowed portion of the outer row of moons show its successive appearance as seen from the earth.

## THE STARS.

The great number of stars visible even to the naked eye renders it impossible to distinguish each by a particular name, and thus the stars have been arranged in groups, called *constellations*, for the purpose of more readily distinguishing them, and each group or constellation is given a name.

An idea of **the position of the various constellations** can be best obtained by reference to a celestial globe, map or "star chart."

To distinguish the stars of a constellation from one another, Bayer, in 1603, assigned to each of the stars in every constellation a letter of the Greek alphabet, the brightest being $\alpha$, the next $\beta$, and so on. The name of the constellation was put after each letter. When the Greek letters were exhausted recourse was had to the Roman alphabet, and these two alphabets fully answered the purpose at the time. The order of the letters *was* intended to indicate the relative brilliancy of the stars in the same constellation, without reference to those in other parts of the heavens; thus $\alpha$ *Aquarii* is a star of the same magnitude as $\gamma$ *Virginis*. Since this system was introduced some of the stars have changed in brightness. Sometimes when there are two near stars of nearly equal magnitude, the numerals 1 and 2 are affixed to them, in their order of right ascension, as $x^1$ and $x^2$ *Tauri*.

Flamsteed gave numbers to the stars observed by him in each constellation; the numbers were given according to the order of their right ascensions, and have nothing to do with their brilliancy. This method of naming stars is often used.

**Proper Names of Principal Stars.**—Some of the principal stars of the first two magnitudes have also received particular names, thus:—

| α Canis Minoris | is called | Procyon, |
|---|---|---|
| α Lyræ | „ | Vega, |
| α Boötis | „ | Arcturus, |
| α Canis Majoris | „ | Sirius, |
| α Aurigæ | „ | Capella, |
| α Tauri | „ | Aldebaran, |
| | etc. | |

**The Pole Star.**—The star nearest the north pole of the heavens is known as the north polar star. The north polar star at present is α *Ursæ Minoris*, and is about $1\frac{1}{4}°$ from the pole, which distance will diminish in about 150 years to about $\frac{1}{4}°$, and then it will increase again until the present pole star will be too far from the pole to serve as such. Thus the pole star varies; 4,600 years ago α *Draconis* fulfilled this office and in 12,000 years α *Lyræ* will become the pole star. This change of the pole star is due to *precession*.

**Distance of Stars.**—We can best form an idea of the immense distances of the stars by considering the time which light takes to travel from them to the earth. Thus, the time the *light* of the sun *takes to travel* to the earth is $8\frac{1}{4}$ minutes, whilst Vega is distant 30 light-years and Polaris 50 light-years.

The **annual parallax** of a star is the angle subtended at the star by the mean radius of the earth's orbit.

For the nearest star the annual parallax is less than 1″.

Besides all known displacements in the apparent position of a star due to the earth's motion in its orbit, each star appears to have a small motion of its own which is called **proper motion**.

Stars have been classified into *orders* or *magnitudes* of brightness and these "**magnitudes**" are given in the 'Nautical Almanac,' for all stars there included, under the heads of Mean or Apparent Places of Stars. Those of the sixth magnitude are about the smallest visible to the naked eye. The largest telescopes show stars down to about the 17th magnitude.

NOTATION.

In practical astronomy the following notation is usually adopted:—

| | | | |
|---|---|---|---|
| A. | Azimuth. | L.A.N. | Local apparent noon. |
| E. | Equation of time. | G.A.N. | Greenwich apparent noon. |
| L. | Longitude. | L.M.N. | Local mean noon. |
| R.A. or Æ | Right ascension. | G.M.N. | Greenwich mean noon. |
| R.O. | Referring object. | $h$ | Altitude. |
| T. | Chronometer (or watch) time of observation. | $p$ | Polar distance. |
| | | $t$ | Hour angle. |
| L.S.T. | Local sidereal time. | ϒ | First point of Aries. |
| G.S.T. | Greenwich sidereal time. | $\delta$ | Declination. |
| L.A.T. | Local apparent time. | $\zeta$ | Zenith distance. |
| G.A.T. | Greenwich apparent time. | $\lambda$ | } Latitude. |
| L.M.T. | Local mean time. | $\phi$ | |
| G.M.T. | Greenwich mean time. | | |

## SECTION II.—DEFINITIONS, ETC.

**Celestial Sphere.**—In the same way as circles are made use of as a means of measuring angles, so in spherical trigonometry a sphere is employed. Therefore we may conceive the concave bowl of the heavens as a measuring sphere, with the earth at the centre. This sphere is called the *celestial sphere*, and its diameter may be any size (see HZRN, fig. 48).

**The Axis of the Earth** is that diameter round which it rotates daily. Its extremities are called N. and S. poles. This axis produced cuts the celestial sphere in the N. and S. poles of the heavens (see PV, fig. 49).

**The Terrestrial Equator** is a great circle of the earth whose plane, passing through the centre is perpendicular to the axis. This plane produced cuts the celestial sphere in the *celestial equator*. The terrestrial and celestial equators are in the same plane, referred to generally as the plane of the equator (see EKQ, fig. 49).

**The Zenith** is the point in the celestial sphere vertically above the observer's position (see Z, fig. 48).

**The Horizon** is a plane cutting the celestial sphere at 90° from the zenith (see HBR, fig. 48).

**The Meridian** plane of a place is the plane passing through the poles and that place.

**The Celestial Meridian** is a great circle formed in the celestial sphere by the extension of the meridian plane. It clearly passes through the poles of the heavens and the zenith (see HZPRN, fig. 48); R and H are the N. and S. points of the horizon.

**The Prime Vertical** is the plane through the zenith at right angles to the meridian, and cuts the horizon in E. and W. points.

**The Elevated Pole** is that pole which is above the horizon.

**The Longitude** (L) of a place is the arc of the equator intercepted between the meridian of a place and any standard meridian, such as Greenwich, from which all longitudes are measured E. or W. up to 180° of arc, or 12 hours of time.

**The Latitude** ($\phi$ or $\lambda$) of a place is the inclination of the normal (or plumb-bob) to the plane of the equator, and is N. or S. ( + or - ) according as the place is N. or S. of the equator.

**The Ecliptic** is the great circle in which the celestial sphere is cut by the plane of the earth's orbit. It is therefore the apparent path of the sun among the stars.

**The Obliquity of the Ecliptic** is the angle which the plane of the ecliptic makes with the plane of the equator, and amounts to 23° 27', at present.

The points where the ecliptic and equator intersect are called **the equinoxes**, and the intersection of their planes is called **the line of equinoxes**.

**The Vernal Equinox** is the point at which the sun crosses the celestial equator from south to north (about March 21st).

**The Autumnal Equinox** is the point at which the sun descends—about September 23rd—from the northern to the southern side of the equator.

**The Solstices** are the points of the ecliptic 90° from the equinoxes, and are distinguished as summer and winter (*vide* fig. 46).

## ASTRONOMICAL COORDINATES.

In order to fix the position of any heavenly body at a given instant, it is necessary to employ some system of coordinates. For field astronomy two such systems are employed :—

(1.) $\begin{cases} \text{Altitude.} \\ \text{Azimuth.} \end{cases}$

(2.) $\begin{cases} \text{Declination,} \\ \text{Hour Angle,} \end{cases}$ or $\begin{cases} \text{Declination.} \\ \text{Right Ascension.} \end{cases}$

### (1.) ALTITUDE AND AZIMUTH.

**The Altitude** ($h$) of a heavenly body is the arc of a great circle through the zenith and the heavenly body intercepted between that body and the horizon (see SB in fig. 48).

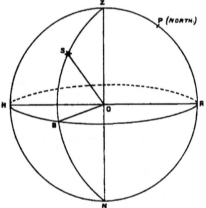

*The plane of the paper is supposed to coincide with the plane of Meridian.*

Fig. 48.

It is clear that the angle SOB would also represent the altitude, and this angle can, whenever the heavenly body is visible, be observed with the theodolite.

Great circles passing through the zenith are called **vertical circles**.

**The Zenith Distance** ($\zeta$) is the complement of the altitude, *i.e.*, $90 - h$ (see ZS in fig. 48).

In working out certain computations which involve *meridional* zenith distance, care must be taken with the algebraical sign of the zenith distance. It is usual to call

Meridional zenith distance ($\zeta$) $\pm$ when the heavenly body is $\frac{S}{N}$ of zenith.

**Culmination.**—When a heavenly body attains its highest altitude it is said to be at its upper culmination. And at its lowest altitude it is at its lower culmination. The latter, in many cases, would not be visible, as it would occur below the horizon.

A star is said to be **at elongation** when it is in such a position that the angle at S in the astronomical triangle PSZ becomes a right angle, *i.e.*, when the plane of the vertical circle ZS becomes tangential to the diurnal path or circle of the star.

**The Azimuth** (A) of a heavenly body is the angle between the meridian plane of the observer, and the vertical plane passing through the body. For astronomical purposes azimuth is measured the nearest way from the meridian plane. Thus in fig. 48 the azimuth of S = angle ROB, = angle PZS.

(2.) DECLINATION AND HOUR ANGLE.

**The Declination** (δ) of a heavenly body is the arc of a great circle through the poles and the body intercepted between the plane of the equator and the heavenly body. It is

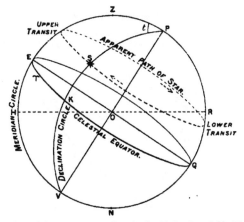

*The plane of the paper is supposed to coincide with the plane of Meridian.*

Fig. 49.

measured from the equator to the poles in degrees 0° to 90° and termed N. or S. ( + or − ) according as the body lies north or south of the equator (see SK, fig. 49).

The declination of the sun is given for every day of the year on pages I and II of each month in the 'Nautical Almanac,' and that of the principal stars (pp. 324 to 452) for every 10th day of the year.

An inspection of these tables will show that the sun's declination varies from approximately 23½° N to 23½° S., but that the stars vary very slightly during the year.

**The Polar Distance** (p) of a heavenly body is the complement of its declination, *i.e.* (90 − δ), it is styled N. or. S. polar distance (N.P.D.) or (S.P.D.) according as it is measured from the N. or S. pole (see P S in fig. 49).

**The Hour Angle** (t) of a heavenly body is the angle at the pole between the *meridian* of a place and the declination circle through the heavenly body (see angle EPS, fig. 49). It is usually measured westward from the meridian from 0° to 360°, or 0h. to 24h.; one hour being equivalent to 15° of arc. From fig. 49 it is clear that this angle is the same as the arc of the equator intercepted between the meridian and declination circle, also measured westward.

In this system, in place of the hour angle the right ascension of the heavenly body may be used.

### RIGHT ASCENSION.

**Right Ascension** (R.A. or Æ) of a heavenly body is the arc of the equator intercepted between the first point of Aries and a declination circle through the body (see ♈ K, fig. 49). It is reckoned in time *eastwards* from 0 to 24 hours.

The R.A. and declination ($\delta$) of all the principal heavenly bodies are tabulated in the 'Nautical Almanac,' which gives their values at particular instants of time for the Greenwich meridian *as if they were seen from the centre of the earth.*

### ASTRONOMICAL TRIANGLE.

A heavenly body at any moment forms a triangle with the zenith and elevated pole (see fig. 50), the sides being arcs of the meridian ZP, the vertical circle ZS, and the declination circle PS.

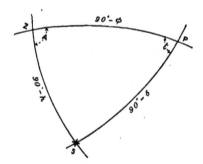

Fig. 50.

Now in fig. 49 the arc EZ is the latitude $\phi$ of the observer, hence PZ is equal to $90 - \phi$, or co-latitude. The latitude is always known approximately, or can be found from one of the methods described later.

The side PS, or polar distance, is equal to $90 -$ declination, and its value can be found from the 'Nautical Almanac.'

The side ZS, or zenith distance, is equal to $90° -$ altitude, and $h$ the altitude can always be observed, whenever the heavenly body is visible, with the theodolite.

The angle at Z is the azimuth.
The angle at P is the hour angle ($t$).*
The angle at S is the parallactic angle.

Given any three of these, the others can be deduced. For formulæ, &c., see Section III.

NOTE.—It will be easily seen from fig. 49 that **the latitude of a place is equal to the declination of the zenith, or the altitude of the pole.**

* More correctly, as drawn, $360° -$ hour angle.

## Hour Angle and Time ($t$).

When any point of the celestial sphere is on the meridian of an observer, that instant is designated the *transit* of that point over the meridian, also the *meridian passage* and *culmination*. In one complete revolution of the sphere about its axis, every celestial body crosses the meridian twice. It is therefore necessary to distinguish between the two transits. The meridian is bisected by the poles of the heaven. The transit over that portion which contains the observer's zenith is the UPPER TRANSIT, and the transit over the other portion is the lower transit. At the upper transit the hour angle (expressed in time) of a heavenly body is 0 hours; at the lower transit its hour angle is 12 hours (see fig. 49). A knowledge of the hour angle of a heavenly body at any moment will, on account of the uniform rate at which it varies, at once give the time that it will take to reach the meridian if it be on the east side of it, or the time elapsed since it crossed the meridian if it be on the west side. In other words, if a clock be so regulated as to go 24 hours between two successive transits of any heavenly body and to read 0h. 0m. 0s. at its upper transit, *the hour angle of the heavenly body* ($t$) *would be the time shown by the clock.*

A **sidereal day** is the interval of time between two successive (upper) transits of the first point of Aries, *i.e.*, the true vernal equinox, over the same meridian. It is divided into 24 hours, each hour into 60 minutes, and each minute into 60 seconds. When $\Upsilon$ is on the meridian the sidereal time is 0h. 0m. 0s. and it is *sidereal noon* at that place, and the sidereal time at any instant is the hour angle of $\Upsilon$ at that instant, reckoned from the meridian westward, 0 to 24 hours.

From the definition of right ascension it is clear that all stars will succeed each other across the meridian in the order of their right ascensions.

An **apparent or solar day** is the interval between two successive transits of the sun's centre across the same meridian, and **apparent time** (AT) at any instant is *the hour angle of the sun* measured westwards throughout the circle expressed in time ($15° = 1$ hour).

Now if the orbit of the earth were a circle with the sun in the plane of the equator and in the centre, this apparent day would give us a constant interval of time; but since the orbit is an ellipse and since it is inclined to the equator, these intervals are continually varying day by day; and to obtain a measure of the day that is uniform and at the same time corresponds fairly well with the recurrence of day and night, the average length of all the solar days of the year is taken and used as the *mean solar day*. This is the day to which ordinary clocks and watches are regulated; and it is adjusted so as to bring 12 noon as near to the time when the sun is on the meridian as possible for every day of the year.

**Mean Time.**—The *mean solar day* is thus regulated by a fictitious or *mean* sun, and is the interval between two successive transits of the *mean sun's centre* across the same meridian; the interval being divided into 24h. 0m. 0s. Mean time (MT) *at any instant is the hour angle of the mean sun's centre at that instant.*

**Astronomical Time.**—Since the ordinary division of the civil day into two periods of 12 hours each is somewhat ambiguous and inconvenient for astronomical purposes, the day (apparent, mean, or sidereal) is conceived by astronomers to commence at noon, and is divided into 24 hours, the hours being subdivided as usual into minutes and seconds.

**Equation of Time.**—As the mean sun is an entirely fictitious assumption, we have no phenomenon by which we can directly observe its progress. We are therefore reduced to observing the motion of the real sun, and by knowing the interval between the real and fictitious sun at any instant, we are enabled to arrive at the mean time at that instant. This interval is called the *equation of time* ($E$); and is defined as the difference between apparent and mean time, or the angle at the pole between the hour circles passing through the centres of the true and mean sun, expressed in time.

**The variation in the equation of time** is due to the two following causes:—

(1) The obliquity of the ecliptic;
(2) The eccentricity of the earth's orbit.

*Firstly.*—Neglecting the effect of the eccentricity of the earth's orbit (*i.e.*, supposing the orbit a circle, and therefore, by Kepler's Laws, that the earth moves uniformly along it), and for convenience imagining the sun to move round the earth with a uniform velocity, then (see fig. 51) let a mean sun, M, be made to move along the equator with the same velocity as the true sun, T, moves in the ecliptic, the two passing through ϒ (the vernal equinox) at the same instant. Now when the true sun is at T the mean sun will be at M, where ϒT = ϒM; and if T¹ be the point where the hour circle through T cuts the plane of the equator, then T¹M will measure the angle between the hour circles of T and M, and therefore the equation of time.

A little consideration will show that T and M will coincide (*i.e.*, will lie in the same hour circle) only at the equinoxes and solstices, and that from equinox to solstice M will be ahead of T, and behind it from solstice to equinox.

The maximum difference between the true and mean suns, on account of the obliquity of

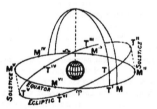

Fig. 51.

the ecliptic, is about 10 minutes, and occurs about the 5th February, 7th May, 8th August, and 8th November.

*Secondly.*—Neglecting the obliquity of the ecliptic, Kepler's second law tells us that the sun would appear to move fastest when the earth is at perihelion (see fig. 46) and slowest when at aphelion. Hence a true sun would be ahead of a mean sun from perihelion to aphelion, and *vice versâ* on the return journey.

Any point of the orbit may be taken as the starting point of the two suns; but one has been chosen which will cause the equation of time at any period of the year to be nearly a minimum. With the point so chosen, the greatest difference due to the elliptical form of the earth's orbit is about 7 minutes.

A combination of the above two causes of variation gives the following data:—The equation of time vanishes (*i.e.*, the hour circles, passing through the mean and true suns, coincide) four times in the year, and one of them having been fixed on the 24th December, the others fall necessarily on the 15th April, 15th June, and 31st August; its maximum positive and negative values as applied to AT, and their respective amounts are as shown below, and occur about the dates given against them.

|  |  | minutes. |
|---|---|---|
| 10th February | ... ... ... ... ... | + 14·26 |
| 14th May | ... ... ... ... ... | − 3·51 |
| 26th July | ... ... ... ... ... | + 6·16 |
| 2nd November | ... ... ... ... ... | − 16·21 |

The value of $E$ is given in the 'Nautical Almanac' for every day of the year, at apparent and mean noon at Greenwich.

We have seen (p. 131) that the sun's hour angle ($t$) at any particular instant and place is the apparent time at that place.

Now the difference of the hour angles of any celestial body at two places at the same instant is equal to the difference of longitude between those places. It follows therefore that the difference of the solar times of two places at one instant is equivalent to their difference of longitude.

Thus if
$T_0$ = Greenwich time,
$T$ = local time,
$L$ = longitude of place,

then $T_0 = T \pm L$ ($+$ or $-$ according as longitude is west or east); this equation being true *whatever time is used* whether solar or sideral, as long as $T_0$ and $T$ are *both of one kind* of time.

### MEAN AND APPARENT TIMES.

In converting apparent into mean time, or *vice versâ*, we use the formula :—

$M.T. = A.T. \pm E$,

where M.T. = the mean time,
A.T. = the apparent time,
$E$ = the equation of time ($+$ or $-$ according as it is additive to, or subtractive from, the apparent time).

*Example* 1.—In longitude 30° 45′ W., on December 7th, civil date, the L.M.T. chronometer marks 3h. 14m. 13·5s. (A.M.), and is fast 25m. 18·7s.; what is the G.M.T. ?

|  | h. | m. | s. |
|---|---|---|---|
| L.M.T. chronometer = | 3 | 14 | 13·5 A.M. |
| Chronometer correction = − |  | 25 | 18·7 |
| Local mean time, Dec. 7th = | 2h. | 48m. | 54·8s. A.M. |

|  | h. | m. | s. |
|---|---|---|---|
| L.M.T., Dec. 6th = | 14 | 48 | 54·8 |
| Longitude $\left(\text{in time} \frac{30° \ 45'}{15}\right) =$ + | 2 | 3 | 0 |
| ∴ the G.M.T. = Dec. 6th | 16h. | 51m. | 54·8s. |

And the civil time is Dec. 7th, 4h. 51m. 54·85s. A.M.*

*Example* 2.—In longitude 60° W., local apparent time May 24th, 1903, 3h. 12m. 10s. what is the local mean time ?

|  | h. | m. | s. |
|---|---|---|---|
| L.A.T. = May 24th | 3 | 12 | 10 |
| Longitude in time = + | 4 | 0 | 0 |
| ∴ G.A.T. = May 24th | 7h. | 12m. | 10s. |

The value of $E$ here required is that of the Greenwich instant corresponding to the local time instant, 3h. 12m. 10s.

* In the further examples, unless A.M. or P.M. is inserted, astronomical time will always be understood.

From the 'Nautical Almanac' of 1903 we get $E$ at G.A.N., May 24th, $= 3$m. $27\cdot64$s., and the hourly variation $= +0\cdot2$ seconds.

Hence at the given G.A.T. instant

$$E = 3\text{m.} \ 27\cdot64\text{s.} + (0\cdot2 \times 7\cdot21) = 3\text{m.} \ 29\cdot08\text{s. and is subtractive.}$$

|  | h. | m. | s. |
|---|---|---|---|
| Thus since L.A.T. = May 24th | 3 | 12 | 10 |
| and $E =$ |  | 3 | 29·08 |
| ∴ L.M.T. = May 24th | 3h. | 8m. | 40·92s. |

## THE YEAR.

The unit of measurement for large intervals of time is the *year*.

The *solar year* (sometimes called Tropical or Equinoctial) is measured between two successive transits of the sun through the mean vernal equinox, that is to say, the interval in time taken by the earth in describing the whole of its orbit around the sun.

There are not, however, an exact number of *mean solar days* in the year, the following being given by Bessel, the accepted authority:—

One solar year = 365·2422 mean solar days.

## MEAN AND SIDEREAL TIMES.

Since in the time the earth has described the whole of its orbit round the sun, the latter has made one transit less over any given meridian than the first point of Aries—

∴ 366·2422 sid. days = 365·2422 mean solar days.
∴ 1 „ = 0·99726957 „
or 24 hours S.T. = 23h. 56m. 4·091s. M.T.
∴ 1 hour S.T. = 1h. M.T. – 9·8296s. M.T. ... (i.)
and 1 hour M.T. = 1h. S.T. + 9·8565s. S.T. ... (ii.)

Equation (i.) shows the *retardation* of M.T. on S.T. and (ii.) the *acceleration* of S.T. on M.T.

Tables to facilitate the conversion of the *intervals* of one species of time into *intervals* of the other are given in the 'Nautical Almanac.'

*Example.*—Convert 18h. 4m. 51s. S.T. units into M.T. units.

|  | h. | m. | s. |
|---|---|---|---|
| By 'Nautical Almanac,' p. 576, 18h. S.T. = | 17 | 57 | 3·07 M.T. |
| 4m. „ = |  | 3 | 59·34 „ |
| 51s. „ = |  |  | 50·86 „ |
| ∴ M.T. equivalent required = | 18h. | 1m. | 53·27s. |

## CONVERSION OF MEAN INTO SIDEREAL TIME AND *Vice Versâ*.

In practical astronomy we are constantly requiring to compute the S.T. corresponding to a given M.T., or *vice versâ*.

In such a conversion, it is necessary to refer to some *epoch* which is given in terms both of mean time and of sidereal time.

Two such epochs are given in the 'Nautical Almanac,' viz.:—The S.T of mean noon (p. II.) and the M.T. of sidereal noon (p. III.), *i.e.*, the M.T. of transit of the first point of Aries both at Greenwich.

The *interval* of the known time, which has elapsed since the epoch, is then to be converted into an *interval* of the other time by means of the tables (see example), and the result added to the time of the epoch (given in 'Nautical Almanac') will complete the computation.

Now, as the epochs in the 'Nautical Almanac' are given for Greenwich time and date, if the instant we wish to convert is given in local time, it is necessary to reduce the given local time to its equivalent Greenwich instant.

This process thus depends on the value of L.S.T. of L.M.N., and this can always be obtained from the G.S.T. of G.M.N. (given on page II., 'Nautical Almanac') by correcting the latter for the divergence in time between the M.T. clock and the S.T. clock, which has occurred in the interval elapsed between the M.N. at Greenwich and the M.N. at the locality.

The difference in *rates* between the two times (M.T. and S.T.) may be taken at 9·86 secs. per hour (see equation (ii.) p. 134), and the elapsed interval in time is the difference in longitude, therefore the above correction = 9·86 secs. × longitude in hours, and is additive in W. longitude and subtractive in E. longitude.

### Example I.—Conversion of Sidereal Time into Mean Time.

On the 1st of May, 1898, in longitude 15° E. the L.S.T. is 20h. 41m. 12·65s. What is the L.M.T.?

|  |  | h. | m. | s. |
|---|---|---|---|---|
| Sidereal time to be converted | = | 20 | 41 | 12·65 |

G.S.T. of G.M.N.         = 2h 37m 44·58s  
Correction for acceleration at 9·86s. per hour of long. = − 9·86  
+ for W. − for E.

L.S.T. of L.M.N. = 2h 37m 34·72s

Sidereal interval from L.M.N. = 18h 3m 37·93s

M.T. equivalents:
- hours = 17 57 3·07
- mins. = 2 59·51
- secs. = 36·90
- ·93 = ·93

M.T. interval from L.M.N., or local M. time = 18h 0m 40·41s

### Example II.—Conversion of Mean Time into Sidereal Time.

On the 1st June, 1898, in longitude 60° W. the L.M.T. is 10h. 15m. 20·73s. What is the L.S.T.?

|  |  | h. | m. | s. |
|---|---|---|---|---|
| Mean time to be converted | = | 10 | 15 | 20·73 |

S.T. equivalents:
- 10 hours = 10 1 38·56
- 15 mins. = 15 2·46
- 20 secs. = 20·05
- ·73 = ·73

Sidereal interval from L.M.N. = 10h 17m 1·80s

G.S.T. of G.M.N., 1st June = 4h 39m 57·81s  
Correction for acceleration at 9·86s. per hour of long. = + 39·44  
+ for W. − for E.

L.S.T. of L.M.N. = 4h 40m 37·25s

Local sidereal time = 14h 57m 39·05s

### SIDEREAL AND APPARENT TIMES.

In computing sidereal time from apparent time, the A.T. may be first converted into M.T. and from this the S.T. computed; or a more direct way may be used by means of the sun's apparent R.A.

Thus
$$\text{for the sun } \quad ST = RA + t = RA + AT,$$

where $t$ is "the hour angle."

## SECTION III.—THE APPLICATION OF SPHERICAL TRIGONOMETRY TO PRACTICAL ASTRONOMY.

It will be seen that at any moment the sun or star in question will form with the zenith and elevated pole a spherical triangle, of which the sides will be portions of great circles of the celestial sphere.

The sides of the triangle have already been shown to have values as follows:—PZ, the

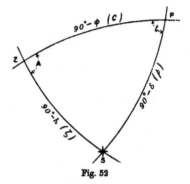

Fig. 52

co-latitude $(c) = 90 - \phi$; ZS, the zenith distance $(\zeta) = 90 - h$; and PS, the polar distance $(p) = 90 - \delta$.

Also the angle at P = hour angle, $(t)$ required for time, and the angle at Z = azimuth.

Knowing any three of the above elements, the remainder can be deduced.

Then by the general formula for a spherical triangle, we get

$$\tan \frac{t}{2} = \sqrt{\frac{\sin(S' - p)\sin(S' - c)}{\sin S' \sin(S' - \zeta)}},$$

$$\tan \frac{A}{2} = \sqrt{\frac{\sin(S' - c)\sin(S' - \zeta)}{\sin S' \sin(S' - p)}},$$

where

$$S' = \frac{p + c + \zeta}{2}.$$

Put for $c$ its value $(90 - \phi)$, and for $\zeta$, $(90 - h)$; then if $S = \dfrac{h + \phi + p}{2}$, we obtain the following :—

(i) For time : $\tan \dfrac{t}{2} = \sqrt{\overline{\sin (S - h) \sec (S - p) \operatorname{cosec} (S - \phi) \cos S}}$,

(ii) For azimuth : $\tan \dfrac{A}{2} = \sqrt{\overline{\sin (S - h) \sin (S - \phi) \sec (S - p) \sec S}}$.

**Solution of Right-angled Spherical Triangles.**—The formulæ for the solution of right-angled spherical triangles are very easily written down by "Napier's Rules of Circular Parts."

In a right-angled spherical triangle there are 5 so-called "circular parts," viz., 2 sides, the complements of the two angles, and the complement of the hypotenuse, the right angle being left out of consideration.

These are written down in the 5 divisions of a circle in the order in which they occur in the triangle; any one may be taken as the middle part, then the two on either side are called the adjacent parts, and the other two the opposite parts; then the rules are—

Sine middle part = product of tangents of the adjacent parts,
 ,,    ,,    ,,    =    ,,    cosines of the opposite parts.

Thus in a triangle ABC where C is the right angle, we have the following figure :—

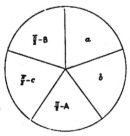

Fig. 53.

**Star at Elongation.**—Another method of obtaining azimuth is to observe a star at "Elongation," *i.e.*, when the angle S of the triangle PZS is a right angle.

In this case we have

$$\cos t = \tan p \cot c,$$

or (i)

$$\cos t = \tan p \tan \phi.$$

Again

$$\sin p = \sin c \sin A,$$
$$= \cos \phi \sin A,$$

whence (ii)

$$\sin A = \sec \phi \sin p.$$

From (i) the time of elongation can be computed and (ii) gives the azimuth of the star.

**Sun or Star on the Prime Vertical.**—It is sometimes desirable to observe a heavenly body when it is on or near the Prime Vertical. In this case the angle at Z will be a right angle.

$$\cos t = \cot p \tan c,$$

or (i)

$$\cos t = \tan \delta \cot \phi,$$

also

$$\cos p = \cos \zeta \cos c,$$

or

$$\sin \delta = \sin h \sin \phi,$$

whence (ii)

$$\sin h = \sin \delta \operatorname{cosec} \phi.$$

Here (i) gives the time of crossing the Prime Vertical and (ii) gives the altitude of the body at that instant. See plates XXXIII, XXXIV.

## SECTION IV.—ELEMENTS FROM THE 'NAUTICAL ALMANAC.'

The 'Nautical Almanac'[*] supplies us with the data necessary for the solution of problems in field Astronomy.

When the time of observation, for which the quantities are required, is exactly one of the instants for which the required quantity is put down in the 'Nautical Almanac,' then we have only to take out this quantity; but as a rule this time falls between two of the instants in the 'Nautical Almanac,' and, as a mere inspection of this book shows that the quantities in it are continually changing, we see that a reduction is necessary when the data are required for any instant of time differing from that for which the quantities are registered. This reduction is done by *interpolation*, and to facilitate this interpolation, the 'Nautical Almanac' contains the rate of change, or alteration of the quantities.

When dealing with with the sun and stars, interpolation by first differences, called *simple interpolation*, is generally sufficient.

*Example.*—Let it be required to find the declination of the sun at 6 o'clock p.m. on a certain day, when at noon $\delta = + 12° 10' 0''$, and is increasing at the rate of 9 seconds an hour.

The result will be $12° 10' 0'' + (9'' \times 6) = 12° 10' 54''$.

We have seen that difference of longitude corresponds with difference of time; therefore as the data given in the 'Nautical Alamanac' are for Greenwich noon they have to be corrected for any other place according to its longitude.

Generally for obtaining a quantity for a certain *local time* we reduce that time to its corresponding *Greenwich* time and take out the quantity accordingly.

The following should be specially noted:—

    For each month page I., 'Nautical Almanac,' gives data for G.A.N.
        „    „    page II.,    „    „    „    „    G.M.N.

The R.A. of a star does not practically alter in 24 hours, therefore the R.A. of a star given in the 'Nautical Almanac' for any date is accepted as its R.A. at any other locality on the same date, or within 24 hours of it.

---

[*] The 'Nautical Almanac' is sometimes called the 'Astronomical Ephemeris' (ἐφημέριος, daily); an ephemeris is properly a table of daily positions of the heavenly bodies.

**The data commonly required in field astronomy** are as follows :—

(1) **Sun's R.A. and $\delta$ at time of observation.** To find either of these assume your longitude as nearly as possible, turn it into time, and add it to your L.M.T. if you are W. of Greenwich, subtract it if E.

This result is the G.M.T. of observation for which R.A. or $\delta$ can be found in the 'Nautical Almanac,' page II, of each month.

*Example.*—In longitude 15° E. required the R.A. of sun at 10 hours on May 1st, 1903.

$$\begin{array}{rl} \text{L.M.T.} = & 10 \text{ hours} \\ \text{Correction for longitude} = & -\ 1\ ,, \\ \hline \text{G.M.T.} = & 9\ ,, \end{array}$$

|  | h. | m. | s. |
|---|---|---|---|
| R.A. of sun at G.M.N., May 1st | = 2 | 30 | 6·23 |
| Correction due to G.M.T., *i.e.*, 9 hours after noon = 9 × 9·52s. = |  | 1 | 25·68 |
| ∴ R.A. at time of observation = | 2 | 31 | 31·91 |

The variations in 1 hour for R.A., $\delta$, and E, are only given on page I 'Nautical Almanac' for apparent time quantities. These figures, however, are equally applicable to the mean time quantities on page II.

(2) **Equation of Time.** Find G.T. of observation as above, and take out E, by simple interpolation, from the column on page I or II of the month in 'Nautical Almanac,' according as it is required for A.T. or M.T. Particular attention must be paid to the heading of the columns of the 'Nautical Almanac,' from which the sign + or - will be obtained.

(3) On page II in the last column is G.S.T. of G.M.N.; its hourly variation is 9·86 seconds (see p. 134).

On page III of the month in 'Nautical Almanac' is a column headed *Transit of the First Point of Aries*. This is identical with G.M.T. of G.S.N.; its hourly variation is 9·83 seconds (see p. 134).

(4) **The R.A. and $\delta$ of the stars.** These are given for all the more important stars, under the head "*Apparent Places of Stars*." They are arranged in the order of their right ascensions for every ten days of the year; and eight stars near to each pole are further detailed for every day of the year. Care must be taken to notice the sign N. or S. against the head of the column for declination. It signifies north or south of the equator.

It sometimes happens that there are two transits of a star in one day, since the sidereal day is shorter than the solar day. When this occurs on any day on which the R.A. and $\delta$ are given in the 'Nautical Almanac' both are given in small figures, and when the double transit occurs *on an intermediate day*, the date alone on which it takes place is given in small figures. In the latter case the procedure for interpolation is fully described in the "*Explanations*" at the end of the 'Nautical Almanac.'

The quantities given for R.A. and $\delta$ in the 'Nautical Almanac,' under the heading "*Mean Places of Stars*," should *not be used in computations*.

**The Tables I., II., and III.** (near the end of 'Nautical Almanac'), for determining the *latitude* by *observation of the Pole Star out of the meridian*, may be used to derive quantities from, by simple interpolation.

The computation is as follows :—

From the observed altitude of *Polaris*, corrected for refraction, &c., is subtracted 0″ 1′ 0″.

To this reduced altitude the further corrections are applied, (1) Table I. with argument LST of observation, (2) Table II. with argument LST and altitude, (3) Table III. LST and date.

The resulting altitude will be the latitude of the place. For example, see computation Form No. 8.

It is to be observed that throughout the 'Nautical Almanac' the astronomical and not the civil day is used. The day of the month is assumed to begin at *mean* noon of the corresponding civil day, and to end at *mean* noon next day. The only exception to this rule is on page I. for each month, where the civil day is used, these tables being for apparent noon.

Great care should be taken that the proper Greenwich *date* is used in taking out data.

CORRECTIONS APPLIED TO OBSERVED ALTITUDES.

Before observed angles of altitude can be used in calculation the following corrections must be applied to them :—

(1) For level error of theodolite (see p. 19).
(2) For refraction.
(3) For semi-diameter (only necessary with single observations).
(4) For parallax.

These corrections should be applied in the order given, and the corrected angle taken in each case as argument for the next correction.

**Refraction.**—Since the rays of light from a heavenly body are bent downwards on passing through the atmosphere, a correction is necessary to obtain the true altitude. This correction depends on the altitude and the state of the atmosphere.

Table XIV. gives the amount of this correction.

**Semi-diameter.**—If the observation taken has been of either limb only of the sun, the angle must be corrected for semi-diameter to reduce it to the centre of the sun.

The *semi-diameter of the sun* is given on page II., 'Nautical Almanac,' but it is preferable to avoid the use of this correction, by taking a series of observations, half the number to the upper limb, and half the number to the lower, the mean of which will give the angle to the centre.

**Parallax.**—Astronomers, in order to work to a common basis, reduce all observations to the same point, viz., the earth's centre. The reduction from the centre to the surface is called diurnal parallax.

Diurnal parallax is the angle subtended at a celestial body by lines radiating to the centre of the earth and to the observer.

Stars have no appreciable diurnal parallax, owing to their enormous distance from the earth.

**Horizontal Parallax.**—When a heavenly body is on the observer's horizon, its diurnal parallax is clearly at its maximum value, and is called *horizontal parallax*. As the altitude increases, parallax diminishes, until it vanishes at the zenith.

The correction for parallax required for application to an observation above the horizon is called *parallax in altitude*. This amount is equal to Horizontal Parallax × cosine altitude.

**Parallax of Sun.**—Table XV. gives values of the sun's *parallax in altitude*.

SECTION V.—INSTRUMENTS, ETC.

For astronomical observations in the field, a theodolite is almost invariably used. A 6-inch micrometer transit theodolite is very suitable.

A theodolite fitted with micrometer microscopes is a great improvement on a vernier instrument. Star sights are useful in enabling the observer to direct the telescope on the star he wishes to observe.

Approximate Details of Certain Theodolites.

| Instrument. | Weight of, with Box. | Weight of Stand. | Price. | Telescope Powers. | Readings of Verniers. | Average Bubble Value. | Details. |
|---|---|---|---|---|---|---|---|
| | lbs. | lbs. | £. s. d. | | | | |
| 3-inch Transit Theodolite (Vernier) | 11 | 5 | 22 10 0 | about 12 | 1' | 1 div. = 1' | |
| 5-inch Transit Theodolite (Vernier) | 26 | 10 | 31 0 0 | 12 to 18 | 30" | 1 div. = 30" | |
| *5-inch Transit Theodolite with Micrometer Microscopes to horizontal and vertical circles | 39½ | 12 | 42 0 0 | 16 to 25 | 10" (estimate to 1") | Vert. bubble 1 div. = 5", plate bubble 1 div. = 30" | With transit-axis level and aluminium star sights. |
| 6-inch Transit Theodolite (Vernier) | 35 | 10 | 36 10 0 | 12 to 18 | 20" | 1 div. = 20" | With transit-axis level. |

A transit theodolite has already been described, and instructions given for its adjustment and setting up (p. 15).

### Notes on the Theodolite when used in Field Astronomy:—

1. All altitudes should be taken on both faces.
2. Note whether the eye-piece used reverses horizontally or vertically, or both.
3. When observing the sun, the dark glass should be used.
4. For observations to stars, it is necessary to light up the field of the telescope. This may be done in two ways: (1) By using the axis lamp, which fits on to a small bracket, and reflecting through the axis of the theodolite, lights up the field. The result generally is far from satisfactory. A more certain method is as follows: (2) A piece of white paper, about 1 inch wide, is fitted to the object end of the telescope by means of an elastic band and bent slightly over the object glass. On this paper is directed the light from any small lamp held at the side of the theodolite. The reflection from the paper then lights up the interior of the telescope. A bicycle lamp answers very well for the purpose.

Care must be taken to throw just enough light on the field to show up the cross-wires without obscuring the star.

The 3-inch transit theodolite, given in the list, is a very good little instrument, but is recommended more for rapid triangulation than for astronomical work. However, for rapid exploration, azimuths and latitudes may, with care, be obtained to within 30".

### TIME-KEEPERS.

The time-keepers used in field astronomy are either **box chronometers** or **chronometer watches**.

A box chronometer is invaluable for stationary work or work on ships, but is not suited for land journeys.

* Where weight is not a consideration of the first importance, a 6-inch Transit Micrometer Theodolite costing about £50, and weighing about 70lbs., is to be preferred, especially for astronomical work.

A keyless half-chronometer watch is recommended for work in the field. These watches are less expensive than pocket chronometers, are less liable to get out of order, and more are easily repaired. They may be carried in the pocket under conditions of rough usage short of actual violence. The watches used by the R.G.S. have backs and faces to screw off, and caps to fit on the winding button. Such a watch costs from £20 to £30.

Both box chronometers and chronometer watches may be regulated to keep either sidereal or mean time.

STAR ATLAS, MAPS AND CHARTS.

A set of star charts is given at the end of this book.

From these, or similar charts, the *relative* positions of the various stars are seen at once, and then their positions in the heavens can be found by referring these positions to imaginary lines joining well-defined stars or groups of stars and following these lines in the heavens.

In using a star chart, the observer may be supposed to be standing (as he is in reality) at the centre of the celestial sphere, in which case the chart approximately represents the stars as seen when looking upwards. When the chart is therefore placed on the table, the stars will appear in their proper relative positions, but the E. and W. points will be reversed with regard to the *true* E. and W.

All observations should be worked out on set forms, a sufficiency of which is an important aid to the topographer. The following books are required for working out observations:—

    An Angle Book (for booking observations in the field).
    The 'Nautical Almanac,' or other Ephemeris, for current and *next* years.
    Book of logarithms.*
    Astronomical Tables (refraction, &c.—see Tables XIV.-XVI.).
    Computation forms (see Forms K-P).
    Star Charts (see Plates XXX.-XXXIII.).


SECTION VI.—OBSERVATIONS.

Observations in field astronomy consist of determinations of the altitude of the sun or star for latitude and time, or for azimuth, the horizontal angle between some "referring object" and the sun or star, in addition, at the same instant.

**A Star** through the theodolite appears as a bright point. It is only necessary therefore to bisect the star with the centre horizontal cross-wire of the theodolite when taking altitudes, and with the vertical wire when taking horizontal angles. Use the same part of the wire on each face.

**The Sun**, however, is a large body, and hence, in order to make all observations possible of comparison, the centre must be observed. As however we cannot tell the precise instant the wire of the theodolite crosses the centre, we are reduced to observing one edge, or "limb" as it is called. For altitudes, the time of contact of the horizontal wire with the upper or lower limb as seen through the theodolite is noted. In the same way, when taking the horizontal angle between some "referring object" and the sun, the right or left limb is observed by making a contact with the vertical wire.

---

\* It saves an enormous amount of time and trouble to use a logarithm book which gives the logarithms of the trigonometrical functions for each second of arc, *e.g.*, Shortrede's log tables.

In order to do away with the necessity for applying a correction for the sun's semi-diameter, an equal number of observations on opposite limbs are made; that is, "upper" and "lower" for altitudes, and "right" and "left" for azimuth angles.

It is advisable to record the limbs in the Angle Book as they *appear* through the telescope, and state which way the eye-piece reverses. This last, however, is not always necessary, if the "pairing" of observations is strictly adhered to.

As already explained, collimation errors are eliminated by taking an equal number of observations on each "face" of the theodolite. No altitude obtained from a theodolite on one "face" only is of any value.

## DETERMINATION OF TIME.

### Time by Single Altitudes.

We have seen that time, either sidereal or solar, is dependent on the hour angle of the 1st point of Aries or of the sun. Since we are able by means of astronomical observations of the sun, to compute its hour angle, the solar time is readily found; and so with a star.

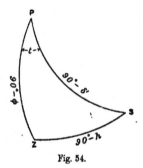

Fig. 54.

Time is known when we have determined the hour angle of the sun or a star. If we adjust a theodolite so that the centre wire describes the plane of the meridian, the determination of the instant of passage of sun or star (with a knowledge of the R.A.) gives the simplest way of computing the time. But this is rarely possible in the field. However, since we are able by an observation of altitude of sun or star, when some distance from the meridian, to deduce its hour angle, we have a ready means of determining the time at the instant of observation.

An examination of the astronomical triangle PSZ (see fig. 54) shows us that the value of the hour angle ZPS depends to a great extent upon the side ZS; for, of the other sides PZ is constant, and PS varies only slightly with the change of declination during the time taken in making an observation. Now ZS is the complement of the altitude, and thus in a position of the heavenly body such as is expressed in the above figure the hour angle changes in accordance with the changes of the altitude.

Obviously, so far as concerns accuracy of observation, the most favourable period to observe altitudes for time will be when they are changing as rapidly as possible, *i.e.*, when for a given change of altitude the corresponding alteration in time is as small as possible; for as in all observations certain imperfections and inaccuracies are to be expected, we must choose a period for our observations when these errors will have the least injurious effect on our results.

**Selection of Stars.**—*In selecting a star for the determination of time by single altitudes choose one on or near the Prime Vertical (*i.e.*, nearly E. or W.). It is easy to show that small errors in the observed altitude, or in the assumed latitude, produce the least effect upon the computed time when the star is on the Prime Vertical. *See Table on p.* 160.

For convenience of observation, and to avoid excessive refraction, the star should have more than 15° and less than 45° altitude.

When a star is on the Prime Vertical, it necessarily follows that its declination must be of the same name (*i.e.*, additive for N. and subtractive for S. lat.), and less than the latitude of the place of observation.

The diagrams on Plates XXXIV. and XXXV. show at a glance the Hour Angle of a heavenly body when on the Prime Vertical, and its Altitude at that instant. The position of a star at any instant can be found from a knowledge of its R.A. (*i.e.*, L.S.T., of transit) and declination. The 'Nautical Almanac' and star charts should therefore always be consulted.

Example:—It is required to take observations at about 8 P.M. in lat. 40° N. Now on looking up the G.S.T. of G.M.N. (p. II., 'N.A.') and correcting it at the rate of 9·86 secs. per hour of longitude (W. additive, E. subtractive), we find the L.S.T. of L.M.N. to be 10h. 25m. 30s., and therefore the sidereal time of 8 P.M. is 18h. 25m. 30s. The star chart shows a large star with declination 25° N. and R.A. 14h. 55m. 30s.

Referring to Plate XXXIV., it will be seen that with this declination and in latitude 40°, a star will cross the P.V. 3h. 45m. before and after meridian transit. The star will therefore be on the P.V. in the west at 18h. 40m. 30s. L.S.T., that is at 8.15 P.M. local watch time.

The time at which the observation should be made is thus easily found, and it is seen from Plate XXXIII. that the star has a suitable altitude at that instant; in this case 41°.

At, or near the equator, the above rules cannot be applied, but any star whose meridional zenith distance ($\pm \phi$) - ($\pm \delta$) does not exceed 15° may be used.

**Pairing Observations.**—When taking observations for time (and also for azimuth), one set should be taken on an east star, and one on a west star. Stars should therefore be selected beforehand to enable the observer to make the observations consecutively.

Time should be found by either of the methods hereafter described, to an accuracy of less than one second of time.

**Chronometer Error.**—The object of the observation is to obtain and record *the error of the chronometer*, which for any instant is the difference between the time it shows and the time it should show, this error being styled *fast* when it is ahead of proper time, and *slow* when it is behind.

The error of the chronometer thus obtained is that *at the mean of the times of observation*. This moment should be noted to the nearest minute.

**Data Required.**—For this purpose, besides our observed altitude, we must know two other elements; on of these must be PZ, the co-latitude; the other PS, the polar distance, which can be derived from the 'Nautical Almanac.'

A small error in the assumed latitude, and even a considerable error in the assumed longitude, would produce no practical error in the result.

---

* The selection of a star near the Prime Vertical is a counsel of perfection. Practically all that is required is to choose a star well to the East or West and between 15° and 45° in altitude.

The following equation is used in the computation for time (see p. 136):—

$$\tan \frac{t}{2} = \sqrt{\cos S \sin (S - h) \operatorname{cosec} (S - \phi) \sec (S - p)},$$

where

$$S = \frac{h + \phi + p}{2}.$$

Lat. $\phi$ always positive and $p$ reckoned from the *elevated* pole.

NOTE.—If a number of observations of the same star at the same place are to be individually computed, it will be most readily done by the fundamental equation

$$\cos t = \frac{\sin h - \sin \phi \sin \delta}{\cos \phi \cos \delta},$$

here $\phi$ and $\delta$ are constant for all the observations, and it only remains to compute $\sin h$.

**Instructions for Observing** (*General*).—For precision, several altitudes will be taken, and the mean of these, with the corresponding mean of the times noted, will be taken for computation. If the time assumed is much out, the computation must be repeated with new functions obtained by means of the corrected time.

The most favourable position for the observer for taking this observation is on the equator, the worst at either of the poles.

When the latitude is undetermined, a small error in its assumed value has no appreciable effect on the results of the observation if the stars taken are near the prime vertical (see p. 160).

To eliminate instrumental errors, errors of refraction and assumed functions, observations must be taken in pairs on either side of the zenith; in the case of the sun, by morning and afternoon observations, in that of the stars, by two stars, one east and one west of the meridian.

*With Theodolite.*—When observing the sun, adjust the instrument, turn the telescope on the sun and make contact with the upper or lower limb on the centre horizontal wire, noting the time of observation and the readings of both ends of the level and of the vertical circle; next change face and make contact with the other limb, noting readings as before.

The procedure for observing a star is the same, except that the observations are taken by allowing the star to make its own passage across the centre horizontal wire near to the vertical wire. Four[*] observations complete a set, half on each face of instrument, thus, F.R., F.L., F.L., F.R.

### Time by Equal Altitudes.

If a star, the declination of which is constant during the observation, be observed at the same altitude, before and after transit over the meridian, it is obvious that the hour angles of the star at both instants of observation must be equal to each other though on opposite sides of the meridian, and that the mean of these instants will give the actual time that the star is on the meridian.

As a star rises in altitude we may record an instant (say $t$), which by our time-keeper corresponds to the time that the star reaches a certain altitude, then, by waiting till the star has crossed the meridian and descends to the same altitude, and again recording the instant (say $t'$), we can find the moment of transit, which should have been $\frac{t + t'}{2}$. A comparison of this time with the R.A. of star would determine for us the error of our time-keeper.

This method of equal altitudes is independent of an accurate knowledge of latitude or declination, or of errors in the graduation of the instrument. It is (with a star) very simple

---

[*] Clearly, all that is absolutely necessary is F.R., F.L.

in its procedure, but requires a clear sky for a considerable interval of time, and for altitudes less than about 45° it is liable to variations on account of changes in temperature, and, therefore, of refraction.

When the sun is used with this method, a correction has to be made for the change of declination that occurs between the times of observation before and after the passage over the meridian.

Having obtained the mean of the two times of observation, it may be corrected to the time of apparent noon, *i.e.*, time of transit, by the following formula:—

$$\frac{\pm t \times d\delta}{15} (\tan \phi \operatorname{cosec} t - \tan \delta \cot t),$$

where

$t$ = half the interval between the two observations,
$d\delta$ = the change of declination in one hour,
$\delta$ = the declination of the sun when on the meridian.

This correction is positive when the sun is leaving the elevated pole, negative when it is approaching it.

In observation by equal altitudes of both stars and sun, instead of one observation only being made each side of the meridian, it is necessary, for precision, to obtain a set as in observing for absolute altitudes. The instrument must be set forward and readings taken at intervals of 10, 20, or 30 minutes of arc, on the upward journey of the heavenly body; and to these readings the vernier can be easily adjusted on the downward journey of the sun or star. The mean time of each set of observations will then be used as if they were the instants corresponding to one pair of altitudes only.

## COMPARISON OF WATCHES.

Suppose it is intended to compare a chronometer watch ($a$) and a chronometer watch ($b$). The observer A watches the seconds hand of ($a$), while B watches ($b$). When the seconds hand of ($a$) appears to cross a definite division, A makes a rap on the table, B notes the time of this by ($b$) to the nearest $\frac{1}{10}$th second.

B then signals in a similar manner to A, who notes the time of the signal.

They then change watches and go through the same procedure.

The mean of these four determinations will give a very fair comparison.

## COMPARISON OF BOX CHRONOMETERS.

In order to get a good comparison, it is desirable that at least one of the chronometers should be sidereal, if the others are mean, and *vice versâ*.

Box chronometers always tick half seconds.

As the difference of the two times, sidereal and mean, is approximately 10 seconds an hour, and the clocks tick half seconds, a coincidence will occur every 3 minutes.

Place the two chronometers on the same table, and take up the time on one, counting: one, half, two, half, three, &c. When the coincidence appears exact, while still counting, note the time shown by the other chronometer.

## COMPARISON OF WATCH AND CHRONOMETER.

Owing to the fact that a watch beats five times in 2 seconds, it is impossible to get a satisfactory coincidence. It is therefore best to compare them in the same way as two watches.

## DETERMINATION OF AZIMUTH.

To determine the direction of the meridian line or the azimuths of any number of points from a station, it is usual to select some well defined terrestrial object suitably situated to

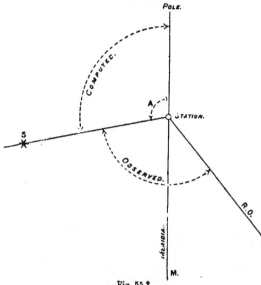

Fig. 55.*

which all azimuth angles can be referred. Such an object is called "**The Referring Object**" (R.O.).

In all **astronomical observations for azimuth** the procedure adopted is, by means of an observed altitude, to *compute* the azimuth (A) of the sun or star (S) (see fig. 55), and to *observe*, at the same time, the azimuthal angle between the R.O. and the sun or star. From these two angles (one *computed* and the other *observed*) it is simple to deduce the azimuth of the R.O.

The computed azimuth is always measured from the elevated pole, the shortest way.

If we can determine the precise instant a heavenly body is on the meridian, the determination of azimuth is simple, and can be done by measuring the horizontal angle at that instant between the heavenly body and the R.O.

But, when on the meridian, a heavenly body is moving at the greatest possible rate in azimuth, so this method is impossible, and it is necessary to select a time for observing when a small error in altitude or time will have the least possible effect on the result.

Stars should be observed for azimuth at or near elongation, and sun or stars on or near the prime vertical. As regards the latter, practically all that is required is to choose a star well to the east or west and between 15° and 45° in altitude.

* Azimuths, used for the purposes of triangulation, as described in Chapter III., are measured from the South by the West. Thus, if the deduced azimuth of the R.O., in fig. 55, were 150°, the azimuth used for triangulation would be 330° (the North being the elevated pole in this case).

## AZIMUTH BY OBSERVATION OF SUN OR STAR EAST OR WEST.

As regards the **selection of stars** for observation by this method, the condition that they are to be well to the E. or W. necessitates their selection being made in the same way as those required for the determination of time (*q.v.*).

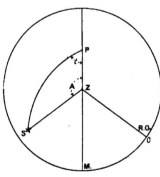

Fig. 56.

For the purposes of the determination of azimuth the theodolite, having two circles, a horizontal (or azimuthal) and a vertical (or altitude) circle, has particular advantages.

In fig. 56, S represents the sun or star, O the referring object, and P the pole.

We can *observe* the angle SZO with the theodolite, and there only remains for us to *compute* the angle SZP, *i.e.*, A or the azimuth of the sun or star.

In the astronomical triangle we have the following quantities:—

$$PS = 90 - \delta,$$
$$PZ = 90 - \phi,$$
$$ZS = 90 - h,$$
$$\text{Angle SPZ} = t.$$

It is required to compute the value of the angle SZP; this can be done with either of the two following sets of data:—

(*a*.) The three sides as above.
(*b*.) The two sides $(90 - \delta)$, $(90 - \phi)$ and included angle "*t*."

ASTRONOMICAL FORM 3 utilises the data (*a*) in the following equation, and can be employed with either sun or star observations:—

$$\tan \frac{A}{2} = \sqrt{\sec S \sec (S - p) \sin (S - \phi) \sin (S - h)},$$

where

$$S = \frac{h + \phi + p}{2}, \text{ and A is the angle SZP.}$$

**Instructions for Observing for Azimuth.**—The R.O. selected should be suitable for intersection with the cross-wires, and as far off as possible, so that solar focus may be used At night the R.O. will be a light not less than 400 yards off.

In all cases care must be taken that the lower plate remains clamped throughout the set of observations.

It is not advisable or necessary that the R.O. should read exactly zero.

In all azimuth observations an endeavour should be made to take as many observations on one "swing" as on the other "swing."

The observed altitudes and azimuths are simultaneously recorded.

The times are not absolutely necessary except for sun observations (to correct $\delta$) but are usually recorded, as then a fair value for time (which is always useful) can be computed on Astronomical Forms 1 or 2 from the observed altitudes.

The procedure in observing **a star** for azimuth is as follows :—

Note the reading of the R.O. on the horizontal arc and turn the telescope on the star; take an observation with the intersection of the centre cross-wires and note the time, the readings of both ends of the bubble, and of the vertical and horizontal arcs. Now reverse the telescope, and take another observation of the same star, noting time, bubble readings and vertical and horizontal arcs, as before; then read the R.O. This completes one set of observations on a single star.

The procedure for obtaining *simultaneous* azimuth and altitude readings **with the sun** is as follows :—

Place the *horizontal* wire in *advance* of the limb to be observed and allow it to make its own contact; just before the moment of contact keep the *side* limb to be observed in contact with the *vertical* wire by means of the slow motion screw; just as the contact with the horizontal wire is completed remove the hand from the vertical wire slow motion screw, note the time and then read level error, vertical and horizontal angles.

The following order of observations should be followed :—

> Read R.O.
> Observe (as above) the right upper limb.
> Reverse the telescope.
> Observe (as above) the left lower limb.
> Read R.O.

This completes one set of observations.

A rough diagram should always be drawn in the angle book, showing the relative positions of the observer, sun or star, and referring object, and also the meridian line.

In rapid triangulation it is frequently of the greatest importance to determine the azimuth of some side of the triangulation. In such cases, it is often convenient to determine this by observations of the sun in the afternoon. Three sets should be taken, and each set worked out independently, so as to obtain a check.

### Azimuth by Circum-Polar Star at Elongation.

**A Circum-polar Star** useful for this observation is one with a polar distance of less than 10°. As already explained, a star is at elongation when the angle at S in the triangle PZS is a right angle.

The time of elongation is computed from the formula $\cos t = \tan \phi \tan p$. Then the angle between the R.O. and the star is observed at this time.

The azimuth of the star can be computed for the same moment from the formula $\sin A = \sec \phi \sin p$.

It is sufficient to take an observation on each face within a few minutes of the computed time of elongation. Only the times and horizontal angles need be recorded. Thus the observations would be :—Face right; R.O., star. Face left; star, R.O. With a quick observer these might be repeated.

The means of the angles between the R.O. and the star can then be taken, provided that in ordinary latitudes no time differs from that of elongation by 5 minutes either way. This gives a range of 10 minutes for observing.

The angle A is then added to or subtracted from this observed angle to obtain the azimuth of the R.O. A rough diagram should be drawn.

The altitude of the star at elongation is given by, $\sin h = \sin \phi \sec p$.

The method is an easy one, both as regards observing and computing, and gives good results, provided the time limits are not exceeded.

The computation is so simple that no example is thought necessary.

## ROUGH METHODS.

The foregoing observations occupy considerable time in both observing and computing, and a shorter and simpler method of obtaining azimuth is sometimes necessary.

There are two such methods of obtaining an approximate azimuth by circum-polar stars, at culmination, and at elongation; these are as follows:—

*Circum-polar Star at Culmination (rough).*

Compute the chronometer time of upper or lower transit of Polaris, or some other known circum-polar star. As this time approaches, bisect, with the central wire of the telescope, the R.O. and read the horizontal arc. Then at the computed instant of the star's culmination, bisect the star with the central wire and read the arc. Observe the R.O. again. The difference between the reading of the star and the mean of readings of R.O. will give the azimuth of R.O., E. or W.

Repeat the process with other circum-polar stars if required.

*Circum-polar Star at Elongation (rough).*—Assume the elongation to be *about* 6 hours from the time of transit, easily found from the R.A. given in the 'Nautical Almanac.'

As nearly as possible *at this time* take a set of observations to obtain the horizontal angle between the R.O. and the star.

Now the azimuth of a star at elongation is obtained from the relation $\sin A = \sin p \sec \phi$.

A fair value for the azimuth of the R.O. is obtained, therefore, by adding or subtracting the angle A from the *greatest* of the observed angles, according as the elongation is E. or W.

In the northern hemisphere there are two stars, $\alpha$ and $\delta$, *Ursæ Minoris*, which come to elongation at intervals of approximately 6 hours from each other. One of these can therefore always be found at elongation in the first half of the night. In the southern hemisphere there are two suitable stars, $\beta$ *Hydri* and $\beta$ *Chamæleontis*, each about 12° from the pole, and of the third and fourth magnitude respectively.

A plan of the constellation *Ursa Minor* is given in fig. 57, which may assist in the identification of the several stars.

Fig. 57.

URSA MINOR.

The elements of all these seven stars, except $\eta$ which is a double star, are given in the 'Nautical Almanac.'

# DETERMINATION OF LATITUDE.

## GENERAL REMARKS.

In fig. 58, in which the plane of the paper is supposed to coincide with the plane of the meridian:—

It is evident that the *latitude of a place* is equal to the *declination of the zenith* EOZ.

It is also the *altitude of the pole* NOP, for in fig. 58, ZE = latitude of the place = $\phi$, and

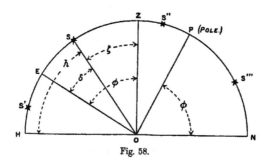

Fig. 58.

therefore PZ = 90 - $\phi$, but PN = 90 - PZ. Hence PN or PON, which is clearly the altitude of the pole, is equal to $\phi$.

Let $h$ = altitude of a heavenly body.

$\zeta$ = meridian zenith distance. This is $\pm$ when the heavenly body is $\frac{S}{N}$ of zenith.

$\delta$ = declination
$\phi$ = latitude of place } These are $\pm$ when $\frac{N}{S}$ of equator.

Then it is clear from the figure that $\phi = \zeta + \delta$.

If the heavenly body be—

    at S    $\zeta$ is +ve, $\delta$ is +ve.
    If at S'    $\zeta$ is +ve, $\delta$ is -ve.
    ,, S''    $\zeta$ is -ve, $\delta$ is +ve.
    ,, S'''    $\zeta$ is -ve,* and $\delta$ is 180—true $\delta$.

In the first two cases the altitude ($h$) is observed while facing south.

    ,, last    ,,    ,,    ,,    ,,    ,, north.

Now suppose the observations are made in *south* latitude (see fig. 59). Then if the heavenly body be at S'''' we have $\phi = \zeta + \delta$, whence giving $\zeta$ and $\delta$ their proper signs, $\phi$ if south must come out negative.

From the above results it will be seen we get the following general formula giving $\phi$, $\zeta$, and $\delta$, their proper signs:—

$$\phi = \zeta + \delta \quad \ldots \quad \text{(i.).}$$

\* $\delta$ is here measured from E through the zenith and elevated pole, and is equivalent to the angle S''' OE, that is to say when a star is taken at its *lower* transit the $\delta$ is obtained by taking the supplement of the value given in the 'Nautical Almanac.'

We have thus an easy method of determining the latitude, viz. :—by measuring the altitude of any celestial body at the exact moment when it is on the meridian, δ being obtainable from the 'Nautical Almanac,' if we know the approximate Greenwich time of observation.

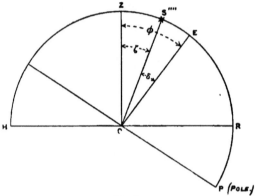

Fig. 59.

### Latitude by Meridian Altitudes.

This is the simplest way of determining latitude, though not the most accurate.

First, calculate the *time of transit* of the sun or star (see p. 153).

Then, having adjusted the theodolite, commence observing a minute or so before the time of transit, and take a series of altitudes, F.R., F.L., F.L., F.R., F.R., F.L., &c., noting the times. The mean of the pair of observations (F.R. and F.L.) nearest to the time of transit is to be taken as the observed meridian altitude. Or, if the error of the watch is not accurately known, the pair which gives the *greatest* altitude must be taken. (In the case of the sun, the correction for semi-diameter must first be applied, in order to ascertain this.)

Correct the meridian observed altitude, so obtained, for level error, refraction (and parallax in case of the sun), and from this obtain the true meridian zenith distance ($\zeta$). Then, as before :—

$$\phi = \zeta + \delta \quad \text{(taking care of the signs).}$$

To get good results from meridian altitudes, the following rules should be observed :—

(1) No altitude should be less than 40°.

(2) Equal numbers of North stars and South stars should be observed.

(3) The North and South stars should be balanced; *e.g.*, if a North star of altitude 40° is observed, try and find a South star of the same altitude.

It is clear that a sun altitude will be unbalanced, and that a latitude by meridian altitude of the sun can only be an approximation.

### Latitude by Circum-Meridian Altitudes.

When a number of altitudes are observed very near the meridian, they are called *circum-meridian* altitudes.

This method is an accurate and satisfactory one.

Each altitude reduced to the meridian gives nearly as accurate a result as if the observation were taken on the meridian.

It can be shown by investigating the formula that near the meridian, the correction varies as the *square* of the hour angle, and not simply in proportion to the time. Hence it is that near the meridian we cannot reduce a number of altitudes by taking their mean to correspond to the mean of the times, as is done (in most cases without sensible error) when the observations are remote from the meridian. We are obliged, therefore, to reduce *each* altitude separately and take the mean of all the results.

(*a*) The zenith distance should not be greater than 50° and under no circumstances less than 10°.

(*b*) The hour angles should not, as a rule, exceed 10 minutes.

(*c*) The closer the meridian (and therefore the more rapidly the observations are taken) the better will be the results.

(*d*) Equal numbers of North stars and of South stars should be observed.

(*e*) The North and South stars should be balanced.

**Selection of Stars.**—When selecting stars, care must be taken to choose those which transit at a convenient time. The stars chosen should have zenith distances less than 50°, and greater than 10°. This can be found from the equation

$$\zeta = \phi - \delta$$

by giving $\phi$ an approximate value and carefully looking to the signs of $\zeta$, $\delta$, and $\phi$. The altitude ($h$) of the star at transit will be $90 - \zeta$.

**Time of Transit.**—The time of meridian transit is obtained, in the case of a star, from the R.A. given in the 'Nautical Almanac,' which is the same thing as the L.S.T. of transit. This last can be reduced to L.M.T. from a knowledge of the G.S.T. of G.M.N., p. II., 'Nautical Almanac' (see Astronomical Form 4).

With the sun, the L.M.T. of transit is obtained from the *equation of time* (p. I., 'Nautical Almanac') corrected for longitude. If a sidereal clock is used, the L.S.T. of transit will be the R.A. of the sun, as given in the 'Nautical Almanac,' corrected for longitude (see Astronomical Form 5).

PRACTICAL HINTS ON OBSERVING FOR LATITUDE.

**Programme of Work.**—Before commencing observations for latitude, an observer should make out a programme of the observations he intends to take, determining the approximate time of each by his watch, and selecting such stars as will enable him to carry out his programme at convenient hours.

As the travelling rate of the chronometer can rarely be depended on, observations for time should always form part of the programme; these can be taken at any convenient time while waiting for the culmination of latitude stars.

For example, in latitude 40° N. on the 1st January, observations are required for latitude. The sidereal time at mean noon on that day is 18 h. 50 m., and supposing we wish to begin at dark and complete the observations before 10.30 P.M., we must choose stars for latitude which culminate between 0 h. and 5 h. 30 m. sidereal time.

The zenith distance of a star at culmination is found from the equation $\zeta = \phi - \delta$, where the signs of the several quantities must be carefully attended to.

Then knowing that the latitude is approximately 40° 0', and looking out the declination of α Cassiopeia from the 'Nautical Almanac,' we obtain the approximate zenith distance of this star thus :—

$$\phi = + 40° \quad 0'$$
$$\delta = + 55 \quad 58$$
$$\phi - \delta = \zeta = - 15° \ 58' \text{ north of the zenith.}$$

Similarly we shall find for—

　　　　　　α Arietis　　ζ = + 17° 3' south,
　　　　　　ε Cassiopeiæ　ζ = − 23　9 north,
　　　　　　α Tauri　　　ζ = + 23 42 south,
　　　　　　Polaris　　　ζ = − 48 45 north,
　　　　　　β Orionis　　ζ = + 48 19 south,

forming three pairs of approximately equal north and south zenith distances.

As regards the times of transit:—If the watch is M.T., look out in the 'Nautical Almanac' the G.S.T. of G.M.N., and correct this for longitude at the rate of 9·86 seconds per hour (W. +, E. −). This will give the L.S.T. of L.M.N. Now the R.A. of the star is its L.S.T. of transit, hence the difference of the R.A. and the L.S.T. of L.M.N. gives the sidereal interval from L.M.N., which can then be reduced to a M.T. interval and used with a M.T. watch (corrected for any known error). If the watch shows S.T., no computation is necessary, except to apply any known error.

The programme can then be drawn up as below, entering altitudes instead of zenith distances:—

| Time. | Star. | Mag. | Aspect. | Altitude. | Remarks. |
|---|---|---|---|---|---|
| h.　m. |  |  |  | ° ′ |  |
| 5　44 | α Cassiopeiæ | Var. | N. | 74　2 | For latitude circum- |
| 6　30 | Polaris | 2·2 | N. | 41　15 | meridians. |
| 6　56 | ε Cassiopeiæ | 3·6 | N. | 66　51 |  |
| 7　11 | α Arietis | 2·0 | N. | 72　58 |  |
| 9　39 | α Tauri | 1·0 | N. | 66　18 |  |
| 10　19 | β Orionis | 0·3 | N. | 41　41 |  |

E. and W. stars suitably placed for time observations can be observed between the observations to α Arietis and α Tauri.

**Instructions for Observing Latitude.**—The observer should commence observing altitudes about 15 minutes before the transit, reversing the telescope after the *first* observation and then after *each pair* of observations.

In the case of the sun, altitudes are taken to the upper and lower limbs alternately.

The observations should be continued till it is quite evident that the sun or star is falling, and that some 10 minutes has elapsed since transit.

As the altitudes are changing very slowly contacts have to be *made* by the slow motion screw.

At the time of observing, then, set the vertical circle to the proper altitude (*h*) and turn the telescope a little to the east of the meridian; the star should then be in the field of the telescope. The altitudes of the star on both faces are recorded in the *Angle Book*, and the star is easily found by setting the vertical circle to these altitudes approximately.

The most favourable series of observations can afterwards be selected to compute from, as no greater accuracy can practically be obtained by computing from a large number of observations in preference to six or eight well situated, *i.e.*, near the meridian.

It is not necessary to have an *equal* number of observations on either side of the meridian, but the better they are *balanced* the better will be the results; for example, three observations on one side of the transit and five on the other may be said to be well balanced.

**The Chronometer Error** or correct local time at the mean of the observed times must be known, and is usually obtained by an independent set of observations (see p. 143).

The chronometer error has to be applied to the *true time* of transit to obtain the *chronometer time of transit*.

The *Rate* of chronometer is practically only required when the chronometer or watch used is unreliable, or when the *interval* between the time of transit and the time at which the chronometer error was determined exceeds one or two hours.

*Example.*—Error of watch determined by set of observations for time at 20h. 5m. 0s. was found to be 2m. 20s. *slow.*

*Rate* of watch ascertained to average about 1·3s. per hour, *gaining.*

*Mean* of times of transit 24h. 10m. 0s. on the same day.

We have therefore:—

Watch gained in elapsed interval 4·1h. × 1·3s. = 5·3s.

∴ Watch *error at time of transit* = 2m. 20s. less 5·3s.

= 2m. 14·7s. *slow.*

In observation with a star the chronometer error is generally obtained by an independent set of observations, within one hour or less of the circum-meridian altitude observations, and any good ordinary watch may be relied on to keep sufficiently accurate time for such a small interval without any correction for *rate*, but it is advisable to ascertain the rate in case it should be unusually high. No good ordinary watch should gain or lose more than 12 seconds per diem.

**The notation and formulæ** employed in the computations for latitude by circum-meridian altitudes are as follows:—

$\phi$ = latitude N. or S.   $\zeta_0$ = mean zen. dist.   $\delta$ = dec. at mean of times of obsn.

$\phi_1$ = approx. lat.   $\zeta_1$ = approx. mer. zen. dist.

$m = \dfrac{*2 \sin^2 \frac{t}{2}}{\sin 1''}$.   $Z_0$ = correct. do. do. do.

$m_0$ = mean of the values of $m$ (computed separately for each hour angle).

$$\phi_1 = \delta + \zeta_1 \text{ and } \phi = \delta + Z_0.$$

$$Z_0 = \zeta_0 - Am_0 \text{ and } A = \dfrac{\cos \phi_1 \cos \delta}{\sin \zeta_1}.$$

As regards signs, treat A and $m_0$ as +ve. $Z_0$ must be less than $\zeta_0$.

In this method of reducing circum-meridian altitudes the computation of the correction $Am_0$ can be repeated a second time with the computed values for the latitude and zenith distance, instead of the *approximate* ones $\zeta_1$ and $\phi_1$, and a closer result obtained.

Astronomical Forms 4 and 5 provide for the computation of latitude from circum-meridian altitudes of the star and sun respectively.

### Latitude by Pole Star.

When the correct local time and the approximate longitude are known the latitude may be found from observations of the altitude of the *Pole Star* at any time.

To obtain the correct local time it is usual to take an independent set of observations for *time*, either immediately before or after the observations for latitude.

The following formula is employed:—

$$\phi = h - p \cos t + \tfrac{1}{2} p^2 \sin 1'' \sin^2 t \tan h.$$

The hour angle $t$ is to be deduced from the sidereal time, and the star's right ascension by the formula—

$$t = \text{S.T.} - \text{R.A.}$$

Calculation Form 8 provides for the reduction of observations by these formulæ, and also by Tables I., II., and III. at the end of the 'Nautical Almanac.' The latter tables are useful when great accuracy is not required; but when precision is necessary it is best to work out the

* See Table XVI.

formula given above, as interpolation in the tables can only be accurately done by means of second differences—a somewhat lengthy calculation.

### Latitude by Rough Methods.

A useful method of finding latitude which does not entail a knowledge of the declination of any celestial body is the following:—

The latitude of a place is equal to the altitude of the pole at that place. Hence if we take a circum-polar star and observe its altitude at both upper and lower transit, the mean of these altitudes (corrected for refraction, &c.) will give at once a value for the latitude of the place. Since, however, two such observations at upper and lower transit must necessarily be separated by a period of 12 hours, it is not always practicable to make them.

A single observation of the altitude of a star at time of its transit gives the latitude readily by the formula—

$$\phi = \zeta + \delta,$$

but it must be recollected that when the star is observed at its lower transit, $\delta$ then is taken as the supplement of the declination given in the 'Nautical Almanac.'

Another rough method has already been described on p. 152.

## SECTION VIII.—GENERAL PRACTICAL HINTS IN FIELD ASTRONOMY.

(*a*.) Before taking observations for *time* or *azimuth*, suitable stars, one E., and one W., may be selected from the star chart, or else stars may be picked up east and west of the observer in the sky. It is necessary of course to know the names of the stars used. In every case where observations for *time* are taken, comparison of watches should be made, either before or after observation.

(*b*.) For *latitudes*, the altitude and chron. time of transit of the star must be worked out previous to observation (see p. 153 and Ex. Form 6).

(*c*.) Before commencing work, an observer should make out a programme of the observations he intends to take, determining the approximate time of each by his watch, and selecting such stars as will enable him to carry out his programme at convenient hours.

(*d*.) Observations should always be made in pairs, with face right and left alternately, so as to eliminate collimation and level errors.

The same principle should be adopted in selecting stars for observation; thus for *latitude*, pairs of stars of opposite and approximate equal zenith distance should be observed, and for *time* and *azimuth* pairs of stars east and west.

(*e*.) A booker should always be employed if possible.

(*f*.) Observations should be made in the following sequence:—
  1st. The observer carefully sets up and adjusts the theodolite.
  2nd. The observer informs the booker what the observation is for, and the name of the star used. The booker enters these in the angle book and also the following data: the approximate watch correction, the thermometer and barometer readings, the value of one division of the vertical circle level, the place, date and referring object when required.
  3rd. The observer gets the star into the field of the telescope, telling the booker to "stand by." On making a contact the observer says "up."[*]
  4th. The booker takes the time and records it.
  5th. The observer reads out the level readings, always giving that at the eye end of the theodolite first. Thus "E, 7·4 ; O, 5·2."
  6th. The observer reads out the vertical angles, C and D verniers.
  7th. The observer reads out the horizontal angles, A and B verniers.

[*] Some observers consider it better to use a word beginning with a dental consonant, such as "top."

(*g.*) The booker must always repeat the readings after he has written them down, to show that he has got them correctly.

The booker should remind the observer each time he is about to observe, what "face" should be used, and for sun observations what limb (upper, lower, right or left) he is to take. To simplify this, the limbs and faces to be used should be written down in the angle book in their proper order for the particular observation, before observing.

(*h.*) The only occasion where both vertical and ho izontal angles are simultaneously recorded is for azimuth observations, Computation Form 4. The method of making the contacts has already been explained, p. 149.

(*i.*) After observing, everything required by the angle book form should be inked in and the book signed. This is to ensure that no necessary data are omitted.

## IDENTIFICATION OF STARS.

When making observations of the stars for time, latitude, or azimuth, it is very helpful to be able, by a mere inspection of the heavens, to pick out the star previously determined on for observation, or at any rate the constellation in which that star exists.

This involves a knowledge of the relative positions of the larger stars in the different constellations.

Star charts are, as a rule, on stereographic or globular projections, and do not represent satisfactorily the relative positions of stars.

For this reason, a short description of the positions of some of the principal stars is given.

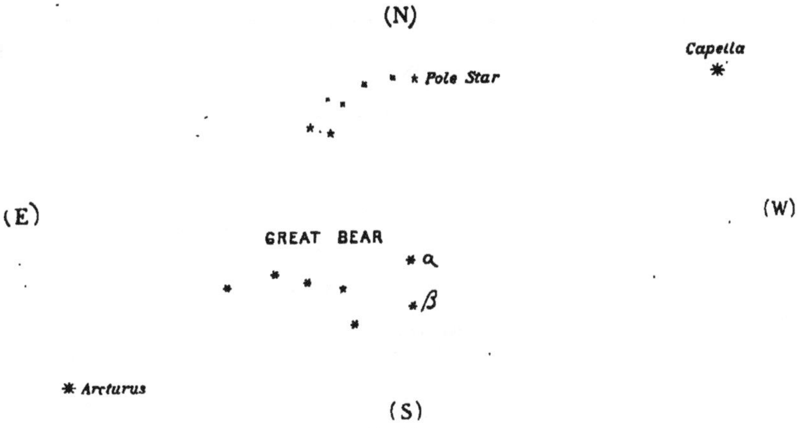

Fig. 60.

It must be understood that, when speaking of a straight line between two stars, what is meant is the portion of a great circle of the celestial sphere intercepted between, and passing through, those stars. A little consideration will show that, owing to the position of the observer being at the centre of the sphere, a diagram of what is seen will have the E. and W. points interchanged.

The angular distance between two stars can be very fairly accurately estimated, if one compares it with half the distance from the point in the heavens above the head to the horizon, which of course represents 45°.

(1.) Referring first to the constellation "Ursa Major" or "Great Bear," the two stars $\alpha$ and $\beta$ point nearly to *Polaris* or the POLE STAR. The Pole Star itself forms the end of the tail of the "Little Bear" or "Ursa Minor."

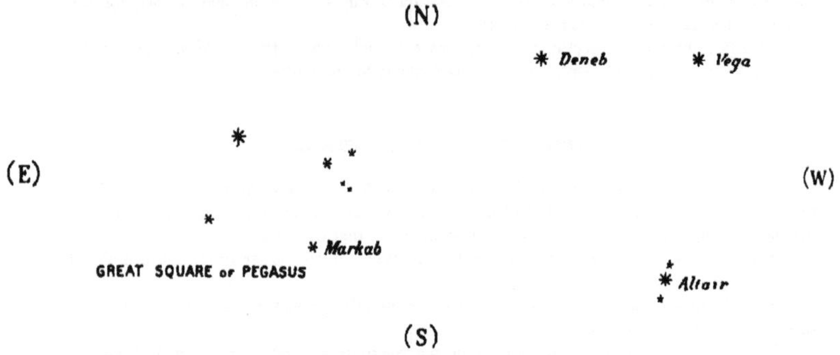

Fig. 61.

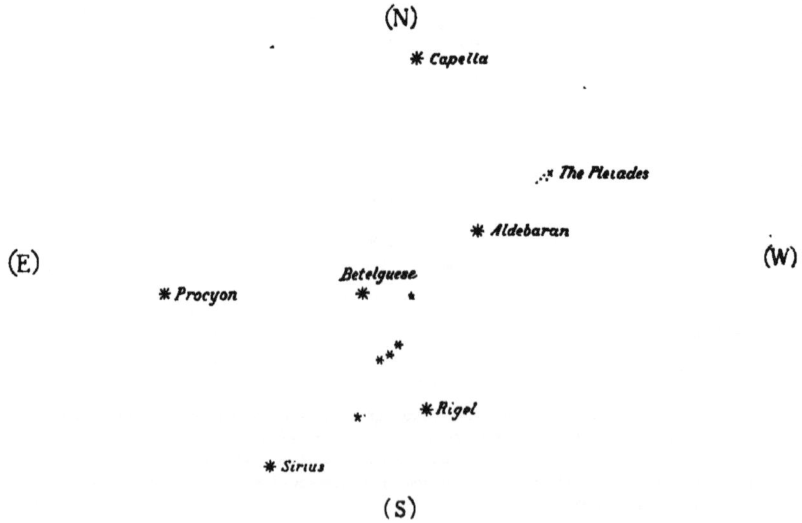

Fig. 62.

The stars α and β of the " Great Bear " are generally known as the pointers.

(2.) If the curve formed by the tail of the "Great Bear" be continued away from the pointers, the bright star ARCTURUS or α *Bootis* will be seen at about 30° from the end of the tail.

(3.) A line drawn from the Pole Star perpendicular to the line from the Pointers passes through CAPELLA or α *Aurigæ*, which is distant about 50° from the Pole Star.

(4.) A line from the Pole Star through the last star but one of the tail of the Great Bear will, if produced, pass through SPICA or α *Virginis*. This star will be found to be about 30° beyond Arcturus.

(5.) In a line with the pointers, on the opposite side of the Pole Star, and distant about 60° from it, will be seen the "Great Square of Pegasus" formed by three stars of that constellation, and one, nearest the Pole, belonging to "Andromeda." The star at the opposite corner is called MARKAB.

(6.) Midway between the "square" and the Pole Star will be noticed the constellation "Cassiopeia," formed by five stars in the shape of a W.

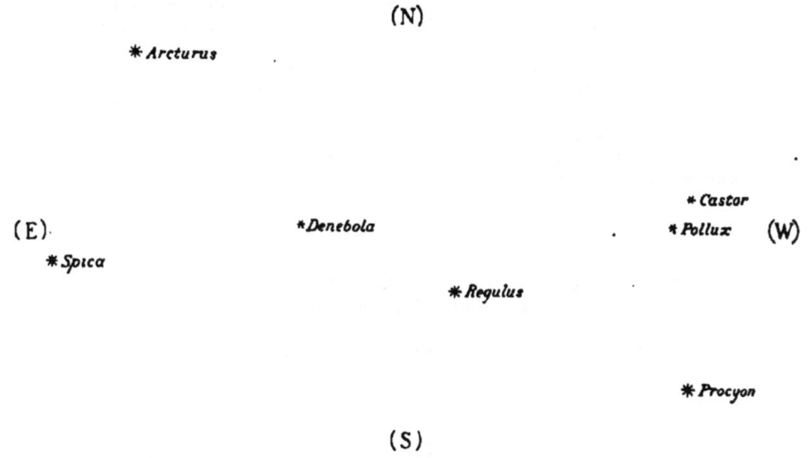

(N)

✷ *Arcturus*

✷ *Castor*
(E) ✷ *Denebola*　　✷ *Pollux*　(W)
✷ *Spica*

✷ *Regulus*

✷ *Procyon*

(S)

Fig. 63.

(7.) A diagonal drawn from the S.E. to the N.W. corner of the square, if produced, will pass through DENEB or α *Cygni* at about 40° distance from the square.

(8.) If that line be continued another 25° still further, and at 10° to the right of it, VEGA or α *Lyræ*, a large white star in the Lyre, will be seen.

(9.) At about 35° to the south of Deneb and Vega will be seen ALTAIR or α *Aquilæ*. This star is situated between two smaller stars, and forms, with Deneb and Vega, what is very nearly an isosceles triangle, with Altair at the apex.

(10.) A line from the Pole Star through Capella will touch RIGEL, a star in "Orion," distant about 65° from Capella. This line also passes between two other bright stars, ALDEBARAN or α *Tauri*, 10° to the west, at about 30° from Capella; and BETELGUESE in Orion, 10° to the east, at about 40° from Capella

(11.) A line from Aldebaran through "Orion's Belt" passes close to SIRIUS or α *Canis Majoris*, the brightest star in the heavens, and at about an equal distance on the other side of the belt. This star, with Rigel, Betelguese, and Aldebaran, forms a trapezium, having "Orion's Belt" in the centre of it.

(12.) At about 50° to the east of Betelguese will be seen PROCYON or α *Canis Minoris*. This star, with Betelguese and Sirius, forms a very nearly equilateral triangle.

(13.) About 30° due north of Procyon will be seen CASTOR, which, together with POLLUX, 5° to the S.E. of Castor, are the two well-known second-magnitude stars in "Gemini."

(14.) At a distance of 45° eastwards is seen REGULUS or α *Leonis*. This star forms the apex of an isosceles triangle with Castor and Procyon.

(15.) Continuing the line from Procyon to Regulus, at about 30° beyond the latter will be seen DENEBOLA, a second-magnitude star also in Leo.

(16.) It will be noticed that Denebola forms an equilateral triangle with Arcturus and Spica. The positions of these last two have already been referred to.

(17.) 30° from the Southern celestial pole is the well-known constellation the Southern Cross, the cross being formed of 4 stars (2 first mag., 1 second, 1 third). The star nearest to the pole is α, that furthest away is γ. When the line α, γ is in a vertical plane these stars are approximately on the meridian.

(18.) 40° south of Sirius is α Argûs (CANOPUS), a first-magnitude star.

(19.) Canopus, the Southern Cross, and ACHERNAR (α Eridani), form a right-angled triangle with the right angle at Canopus. Achernar is about 40° distant from, and transits about 5 hours before, Canopus.

(20) CANOPUS, ACHERNAR, and FOMALHAUT (α Piscis Australis) are in a line and nearly equi-distant, being about 40° apart.

### Effect on determination of Time (by single altitudes) of error of 10″ in observed Altitude. See p. 144.

| Azimuth from N. | 90°. | 75° or 105°. | 60° or 120°. |
| --- | --- | --- | --- |
|  | seconds. | seconds. | seconds. |
| Latitude 60° | 1·33 | 1·38 | 1·54 |
| 40° | 0·87 | 0·90 | 1·00 |
| 20° | 0·71 | 0·73 | 0·82 |
| 0° | 0·67 | 0·69 | 0·77 |

### Effect on determination of Time (by single altitudes) of error of 10″ in assumed Latitude.

| Azimuth from N. | 90°. | 75° or 105°. | 60° or 120°. |
| --- | --- | --- | --- |
|  | seconds. | seconds. | seconds. |
| Latitude 60° | 0·00 | 0·36 | 0·77 |
| 40° | 0·00 | 0·24 | 0·50 |
| 20° | 0·00 | 0·19 | 0·41 |
| 0° | 0·00 | 0·18 | 0·38 |

# EXAMPLES

OF

# ASTRONOMICAL OBSERVATIONS.

## Specimen Page of Angle Book.

Page 1

Instrument ..................
Observations for.... Time
To Sun or ~~Star~~ ..................
Place ...R.E. Institute..................
Date ...24.01 (Civil)..................

Bar. ...29".98..................
Ther. ...50..................
Value of 1 Div. of Level ...20'..................

~~Star~~ for M T Chron. No. ...21..................  Rate. ..........h. m. s.
Approx. Chron. Error on ~~Star~~ for M T = 0. 8. 15 slow

Approx. Long. E. ~~or W.~~ ......h. m. s. / 0 . 2 . 0
Referring Object (R.O.) ..................
Magnetic Bearing ..................

| Apparent Limb or Object. | Face. | VERTICAL ANGLES. | | | | | | Level. | | Chron. Times, T. | | | HORIZONTAL ANGLES (for Azimuth only). | | | | | | | |
|---|---|---|---|---|---|---|---|---|---|---|---|---|---|---|---|---|---|---|---|---|
| | | C Vernier. | | | D. | | Means of C and D. | | | | | | A Vernier. | | | B. | | Means of A and B. | Angle Diff. of R.O. and Star. | Means of Pairs (with R.O.). |
| | | ° | ' | " | ° | ' | " | E. | O. | h. | m. | s. | ° | ' | " | ° | ' | " | | |
| O | L. | 23 | 26 | 0 | 26 | 30 | 26 | 15 | 5 | 7 | 20 | 13 | 45 | | | | | | | |
| O | R. | 65 | 31 | 0 | 42 | 0 | 24 | 27 | 30 | 0 | 10 | 20 | 15 | 20 | | | | | | | |
| O | R. | 62 | 33 | 0 | 31 | 0 | 27 | 8 | 30 | 4 | 7 | 20 | 17 | 15 | | | | | | | |
| O | L. | 25 | 0 | 0 | 0 | 0 | 26 | 0 | 0 | 0 | 12 | 20 | 18 | 24 | | | | | | | |

"WITH Sxs."
Eye-piece reverses :—
Horizontally...No.................
Vertically...Y...................

NOTE.—The red figures are those which would be entered at the time of observing. The blue figures as soon after observing as possible.

Observer ......A. B. C..................
Recorder ......X. Y. Z..................

(5455)

## Specimen Page of Angle Book.

Page 2.

Instrument ..................
Observations for........ Time
To ...... Star, β Ophiuchi
Place R E Institute
Date 22.4.01 (Civil)

Bar. 29".80
Ther. 46°
Value of 1 Div. of Level 5"

5-7-(or M T) Chron. No. 10 .......... Rate.
Approx. Chron. Error on ...... (or M T) = 0 h. m. s. 5.40 slow

Approx. Long. E. ...... h. m. s. 0 . .
Referring Object (R.O.) ..................
Magnetic Bearing ..................

### VERTICAL ANGLES.

| Apparent Limb or Object. | Face. | C Vernier. | | | D. | | | Means of C and D. | | | Level. | | Chron. Times, T. | | |
|---|---|---|---|---|---|---|---|---|---|---|---|---|---|---|---|
| | | ° | ' | " | ° | ' | " | ° | ' | " | E. | O. | h. | m. | s. |
| * | L | 214 | 49 | 0 | 33 | 48 | 0 | 13 | 14 | 20 | 10 | 1h | 46 | 12 |
| | R | 15 | 40 | 40 | 0 | 41 | 0 | 15 | 40 | 30 | 15 | 11 | 10 | 44 | 32 |
| | R | 15 | 53 | 21 | 24 | 54 | 0 | 15 | 53 | 55 | 19 | 12 | 12 | 37 | 2 |
| | L | 313 | 58 | 47 | 40 | 53 | 0 | 16 | 16 | 17 | 10 | 20 | 16 | 51 | 43 |

### HORIZONTAL ANGLES (for Azimuth only).

| A Vernier. | | | B. | | | Means of A and B. | | | Angle Diff. of R.O. and Star. | | | Means of Pairs (with R.O.) | | |
|---|---|---|---|---|---|---|---|---|---|---|---|---|---|---|
| ° | ' | " | ° | ' | " | ° | ' | " | ° | ' | " | ° | ' | " |
| | | | | | | | | | | | | | | |
| | | | | | | | | | | | | | | |
| | | | | | | | | | | | | | | |
| | | | | | | | | | | | | | | |

NOTE.—The red figures are those which would be entered at the time of observing. The blue figures as soon after observing as possible.

"WYTH Box."
Eye-piece reverses:—
Horizontally ..................
Vertically ..................

Observer ...... A. B. C.
Recorder ...... X. Y. Z.

## Specimen Page of Angle Book

Page 3

Instrument ..................
Observations for... Azimuth
To Sun or Star ..................
Place ...... R.E.I. Paper ..................
Date ..... Sun, Oct. (Civil) ..................

Bar. 29°.55
Ther. 70°.2
Value of 1 Div. of Level ... 10″

S-Tier M T) Chron. No. ..... 15
Approx. Chron. Error on S-T-for M T) =

Approx. Long. E-or-W. 0 h. 2 m. 9 s.
Referring Object (R.O.) Gillingham
Church Trig. Station
Magnetic Bearing ..................

h. m. s.
0 2 9 low

| Apparent Limb or Object. | Face. | VERTICAL ANGLES. | | | | | | | | Level. | | Chron. Times, T. | | | HORIZONTAL ANGLES (for Azimuth only). | | | | | | | | | |
|---|---|---|---|---|---|---|---|---|---|---|---|---|---|---|---|---|---|---|---|---|---|---|---|---|
| | | C Vernier. | | | D. | | | Means of C and D. | | E. | O. | h. | m. | s. | A Vernier. | | | B. | | | Means of A and B. | | Angle Diff. of R.O. and Star. | Means of Pairs (with R.O.). |
| | | ° | ′ | ″ | ° | ′ | ″ | ° | ′ | | | | | | ° | ′ | ″ | ° | ′ | ″ | ° | ′ | ″ | ° | ′ | ″ | ° | ′ | ″ |
| R.O. | L | 211 | 41 | 11 | | 41 | 23 | | 41 | 17 | 7½ | 7½ | | | | | | | | | | | | | | |
| ⊙ | L | | 18 | 6 | | 18 | | | | 14 | 2½ | 4 | 19 | 51 | 21 | 25 | 25 | 12 | 35 | 35 | 23½ | 22½ | 25 | 23·5 | | 50 | | |
| ⊙ | R | 227 | 18 | | | 46 | | | 41 | | 7½ | | 19 | 51 | 9½ | 27 | 25 | 54 | 26 | 26 | 25½ | 27½ | 26 | 8·5 | 48 | 48 | | 4 | |
| R.O. | R | | | | | | | | | | | | | | | 47 | 21 | 22 | 24 | 24 | 24 | 47 | 24 | 17 | | | 40 | | 41 |
| | | | | | | | | | | | | | | | 45 | 25 | 42 | 35 | 35 | 32 | 45 | 35 | 37 | | | | | |

"WITH SUN."
Eye-piece reverses :—
Horizontally... Yes
Vertically... No

NOTE.—The red figures are those which would be entered at the time of observing. The blue figures as soon after observing as possible.

Observer .... A. B. C.
Recorder .... N. Y. Z.

# Specimen Page of Angle Book.

Page

## Specimen Page of Angle Book.

Page ;

Instrument .................................
Observations for, Latitude .................
To Sun or Star ............................
Place M.E. Institute .......................
Date  M.I.C., Civil ........................

Bar. 30 ·00 ..............
Ther. 72° ...............
Value of 1 Div. of Level 15′ ......

S T (or M T) Chron. No. ............... Rate ..........
Approx. Chron. Error on S T (or M T) = 0 0 30 fast

Approx. Long. E. or W.  h. m. s.
Referring Object (R.O.) ...........
Magnetic Bearing ..................

| Apparent Limb or Object. | Face. | VERTICAL ANGLES. | | | | | | Level. | | Chron. Times, T. | | | HORIZONTAL ANGLES (for Azimuth only). | | | | | | |
|---|---|---|---|---|---|---|---|---|---|---|---|---|---|---|---|---|---|---|
| | | C Vernier. | | | D. | | | Means of C and D. | | | E. | O. | h. | m. | s. | A Vernier. | | | B. | | | Means of A and B. | | | Angle Diff. of R.O. and Star. | | | Means of Pairs (with R.O.). | | |
| | | ° | ′ | ″ | ° | ′ | ″ | ° | ′ | ″ | | | | | | ° | ′ | ″ | ° | ′ | ″ | ° | ′ | ″ | ° | ′ | ″ | ° | ′ | ″ |
| O | L. | 239 | 16 | 8 | 13 | 16 | 13 | 59 | 48 | 10·5 | 7 | 2 | 12 | 59 | 14·5 | | | | | | | | | | | | | | | |
| O | R. | 289 | 20 | 5 | 30 | 30 | 16 | 60 | 9 | 45·5 | 4 | 5 | | 62 | 35 | | | | | | | | | | | | | | | |
| O | R. | 299 | 30 | 40 | 8 | 28 | 30 | 60 | 0 | 30 | 8 | 6 | | 65 | 11 | | | | | | | | | | | | | | | |
| O | L. | 230 | 20 | 0 | 28 | 20 | 19 | 30 | 20 | 19 | 7 | 3 | | 67 | 5 | | | | | | | | | | | | | | | |
| O | L. | 229 | 18 | 40 | 10 | 48 | 40 | 30 | 48 | 40 | 6 | 3 | | 68 | 48 | | | | | | | | | | | | | | | |
| O | R. | 299 | 40 | 10 | 10 | 40 | 19 | 60 | 19 | 50 | 5 | 5 | | 71 | 2 | | | | | | | | | | | | | | | |
| O | R. | 299 | 12 | 30 | 42 | 11 | 47 | 59 | 47 | 54 | 5 | 2 | | 72 | 14 | | | | | | | | | | | | | | | |
| O | L. | 240 | 17 | 30 | 10 | 13 | 7 | 60 | 17 | 30 | 7 | 2 | | 74 | 23 | | | | | | | | | | | | | | | |

" WTTH SUN."
Eye-place traverse -
Horizontally ....... Yes.
Vertically ........ No.

NOTE.—The red figures are those which would be entered at the time of observing. The blue figures as soon after observing as possible.

Observer ...A.B.C. ..................
Recorder ...X.Y.Z. ..................

## Specimen Page of Angle Book

Page 6

Instrument: ..........................................
Observations for ... Latitude
To S̶i̶g̶m̶a̶ Star ... Polaris
Place ... R.E. Staff Quarters
Date ... 1.10.97

Bar ... 20'·27
Ther. ... 42°
Value of 1 Div. of Level ... 13″·4

S̶ T̶r̶u̶e̶ (M T) Chron. No. ... 12 ............... Rate ..........
Approx. Chron. Error on S̶ T̶ (or M T) = 0 h. 3 m. 24 s. fast

Approx. Long. E. or W. ... 0 h. 2 m. 9 s.
Referring Object (R.O.) ..........................
Magnetic Bearing ..........................

### VERTICAL ANGLES.

| Apparent Limb or Object. | Face | C Vernier. | | | D. | | | Means of C and D. | | | Level. | | Chron. Time, T. | | |
|---|---|---|---|---|---|---|---|---|---|---|---|---|---|---|---|
| | | ° | ′ | ″ | ° | ′ | ″ | ° | ′ | ″ | E. | O. | h. | m. | s. |
| * | L. | 231 | 37 | 15 | 27 | 10 | | 51 | 37 | 12·5 | 7 | 5 | 7 | 3 | 54 |
| | R. | 308 | 20 | 29 | 20 | 12 | | 51 | 39 | 49·5 | 4 | 8 | 8 | 8 | 12 |
| | R. | 308 | 19 | 38 | 19 | 23 | | 51 | 40 | 29·5 | 4 | 8 | | 11 | 0 |
| | L. | 231 | 41 | 0 | 40 | 51 | | 51 | 40 | 33·5 | 7 | 5 | | 16 | 33 |

### HORIZONTAL ANGLES (for Azimuth only).

| A Vernier. | | | B. | | | Means of A and B. | | | Angle Diff. of R.O. and Star. | | | Means of Pairs (with R.O.) | | |
|---|---|---|---|---|---|---|---|---|---|---|---|---|---|---|
| ° | ′ | ″ | ° | ′ | ″ | ° | ′ | ″ | ° | ′ | ″ | ° | ′ | ″ |
| | | | | | | | | | | | | | | |

"Wyrm Scr."
Eye-piece reverses:—
  Horizontally..........................
  Vertically..........................

NOTE.—The red figures are those which would be entered at the time of observing. The blue figures as soon after observing as possible.

Observer ... A. B. C.
Recorder ... X. Y. Z.

# EXAMPLES

OF

## ASTRONOMICAL COMPUTATIONS.

TIME by Altitudes of Sun.

| From Page 1, Angle Book. | | | | | | | | |
|---|---|---|---|---|---|---|---|---|
| | | ° | ′ | ″ | | h. | m. | s. |
| Vert. Angles | | 23 | 26 | 15 | Times | 20 | 13 | 45 |
| | | 24 | 27 | 30 | | 20 | 15 | 20 |
| | | 27 | 5 | 30 | | 20 | 17 | 15 |
| | | 25 | 0 | 0 | | 20 | 18 | 34 |
| Totals | | 99 | 59 | 15 | | 81 | 4 | 54 |
| Means | | 24 | 59 | 49 | | 20 | 16 | 13½ |

Level Correction.*

| | E. | O. | | Correction. |
|---|---|---|---|---|
| | 5 | 7 | | |
| | 0 | 10 | | $\dfrac{(36-9)}{8} \times 20''$ |
| | 4 | 7 | | $= + \dfrac{27 \times 20''}{8}$ |
| | 0 | 12 | | $= + 1'\ 7''\cdot 5$ |
| Totals | 9 | 36 | | |

$$\begin{array}{rrrr} & ° & ′ & ″ \\ \therefore & 24 & 59 & 49 \\ + & 0 & 1 & 8 \\ \hline & 25 & 0 & 57 \end{array}$$ Correct mean observed altitude.

* With micrometer theodolites the level should be read to one place of decimals.

## Form K.—(S.M.E. Astronomical Form 1.)
### Time by Altitudes of the Sun.

*Those parts in Italics are only necessary when the Chronometer keeps Sidereal Time.*

Place: R. E. Institute　　Lat. N. or S. 51 23 30　　L M T

Date: 2.4.01 (Civil)　　　　　　　　　　　Chron. No. 21　　Bar. 29·9

Page in Angle Book: 1　　Approx. Long. E. or W. 0 2 9 h.m.s.　　Approx. Chron. Corr. on L S T or L M T: 0 8 15 slow　　Ther. 50

### ELEMENTS.

| | h. m. s. |
|---|---|
| *G S T of G M N (Page II., N.A.)* = | |
| *Corr. for Long. at 9·86s. per hour (− E. + W)* = | |
| *L S T of L M N* = | |

FROM ANGLE BOOK:
| | h. m. s. |
|---|---|
| † Mean of Obsd. Times | 20 16 13·5 |
| Approx. Chron. Corr. ± (Slow + Fast −) | + 0 8 15 |
| *L S T of Obs.* (Approx.) | |
| *L S T of L M N* | |
| *Sid. Int. from L M N* | |
| M T Equivalents { Hrs. ...... Mins. ...... Secs. ...... } | |
| L M T of Observation (Approx.) | 20 24 28·5 |
| Long. W.(+) or E.(−) ± | − 0 2 9 |
| G M T of Mean of Observed Times | 20 22 19·5 |
| ‡ Int. in hours from G M N | 3·63 |

### POLAR DISTANCE.

This is reckoned from the *Elevated Pole* $p = 90 \pm \delta$.

For Dec. (δ) at time of observation (see below).

Dec. (δ). N. or S. = 4 42 35·64
　　　　　　　　　　90 0 0
Polar Dist. (p) = 85 17 24·6

If the Lat. (φ) and Dec. (δ) are of the same name, δ is *subtracted* from, but if of different names, *added* to 90°.

### DECLINATION.
Dec. (δ).

At Greenwich Noon　N. or S. 4 46 7·4 (Page II., N.A.)

Hourly Variation (Page I., N.A.) × ‡ Interval in hours from G M N = ± − 0 3 31·7

At mean of Obsd. Times　N. or S. 4 42 35·7

### EQUATION OF TIME.

(Page I., N.A.)　m. s.　± + 3 48·53

± + 0 2·74

± + 3 51·3

### REFRACTION.

| | ′ ″ |
|---|---|
| Refn. due to Alt. | 2 4·1 |
| Corr. for Bar. | ± 0 0 |
| Corr. for Ther. | ± 0 0 |
| Refraction | 2 4·1 |

### PARALLAX.

Hor. Par. (Page I., N.A.) × cos h

Par. in Alt. = 0 8·1

### FORMULA AND COMPUTATION.

$$\tan \frac{t}{2} = \sqrt{\cos s \sin (s - h) \operatorname{cosec}(s - \phi) \sec(s - p)}$$

where $s = \dfrac{h + \phi + p}{2}$

The Lat. (φ) is *always* positive and p is reckoned from the *Elevated Pole*.

FROM ANGLE BOOK.

| | ° ′ ″ |
|---|---|
| Obsd. Mean Altitude = | 25 0 57 |
| Refraction = | − 0 2 4·1 |
| Semi-diam. = | ± |
| Parallax = | + 0 0 8·1 |
| True Alt. (h) = | 24 59 1 |
| Lat. (φ) = | 51 23 30 |
| Polar Dist. (p) = | 85 17 24·6 |
| | 2 ) 161 39 56 |
| s = | 80 49 58　Log cos　9·2022606 |
| s − p = | 4 27 27　Log sec　10·0013156 |
| s − φ = | 29 26 28　Log cosec 10·3081511 |
| s − h = | 55 50 57　Log sin　9·9178009 |
| | 2 ) 19·4295282 |

Log tan $\dfrac{t}{2}$ = 9·7149141

| | ° ′ ″ |
|---|---|
| $\dfrac{t}{2}$ = | 27 24 56·4 |
| t in arc = | 54 49 52·8 |
| | h. m. s. |
| t in time = | 3 39 19·5 |
| If before noon subtract from | 24 0 0 |
| L A T of Observation = | 20 20 40·5 |
| Equation of Time ± = | + 0 3 51·3 |
| L M T of Observation = | 20 24 31·8 |

S. T. Equivalents { Hrs. ...... Mins. ...... Secs. ...... } *Not required with M.T. Chron.*

*Sid. Int. from L M N* =
*But L S T of L M N* =
*L S T of Obs.* =

† Chron. T. of Obs. = 20 16 13·5
　　　　　　　　　　h. m. s.
Chron. Slow or Fast ± on L S T or L M T = 8 18·3

(5455)　　　　Z

Time by Altitudes of an E. Star.

From Page 2, Angle Book.

|  |  | ° | ′ | ″ |  | h. | m. | s. |
|---|---|---|---|---|---|---|---|---|
| Vert. Angles |  | 15 | 11 | 14 | Times | 10 | 46 | 12 |
|  |  | 15 | 40 | 50 |  | 10 | 48 | 32 |
|  |  | 15 | 53 | 55 |  | 10 | 50 | 2 |
|  |  | 16 | 1 | 17 |  | 10 | 51 | 43 |
| Totals |  | 62 | 47 | 16 |  | 43 | 16 | 29 |
| Means |  | 15 | 41 | 49 |  | 10 | 49 | 7·25 |

Level Correction.

| | E. | O. |
|---|---|---|
| | 20 | 10 |
| | 15 | 15 |
| | 19 | 12 |
| | 10 | 20 |
| Totals | 64 | 57 |

Correction.

$$\frac{(57 - 64) \times 5''}{8}$$

$$= -\frac{7 \times 5''}{8}$$

$$= -4''$$

$$
\begin{array}{rrr}
° & ′ & ″ \\
15 & 41 & 49 \\
- \quad 0 & 0 & 4 \\
\hline
15 & 41 & 45
\end{array}
$$ Correct mean observed altitude.

FORM L.—(S.M.E. Astronomical Form 2.)

## Time by East and West Stars.

```
STAR......β. Ophiuchi............E. or W.        Lat. (φ) N. or S.  51  21  30    Bar. ...29..8..
Place ..R.E. Institute.............              Approx. Long. E. or W.  2  9    Ther. ...42....
Date...27.4.21...............                                                    m.  s.
Page in Angle Book...2......    –F– or M T Chron. No....19.....  Rate........   Approx. Corr. 5..40. slow.
```

$$\tan \frac{t}{2} = \sqrt{\cos s \sin (s-h) \operatorname{cosec}(s-\phi) \sec(s-p)}$$

where $s = \frac{h+\phi+p}{2}$ with $\phi$ always taken positive and $p$ reckoned from the *Elevated Pole*.

$h$ = Altitude.    $p$ = Polar Distance.

### COMPUTATION.

|  |  | ° | ′ | ″ |
|---|---|---|---|---|
| FROM ANGLE BOOK. |  |  |  |  |
| Mean Obsd. Alt. | = | 15 | 41 | 45 |
| Corr. for Refn. | = − | 0 | 3 | 24·2 |
| True Alt. (h) | = | 15 | 38 | 20·8 |
| Lat. (φ) | = | 51 | 23 | 30 |
| Polar Dist. (p) | = | 85 | 23 | 33·3 |
|  | 2 | 152 | 25 | 24·1 |

### REFRACTION.

|  |  | ′ | ″ |
|---|---|---|---|
| Refn. due to Alt. | = | 3 | 21·8 |
| Corr. for Bar. | ± − | 0 | 1·4 |
| Corr. for Ther. | ± + | 0 | 0·4 |
| Refraction | = | 3 | 24·2 |

### POLAR DISTANCE.

|  |  | ° | ′ | ″ |
|---|---|---|---|---|
| Decl. (δ) N. or S. | ± = | 4  | 36 | 26·7 |
|  |  | 90 | 0 | 0 |
| Polar dist. (p) or 90 ± δ | = | 85 | 23 | 33·3 |

NOTE.—"p" is reckoned from the *Elevated Pole*. ∴ if the Lat. and Dec. are of the *same* names, *subtract* from, and if of *different* names, *add* to 90°.

|  |  | ° | ′ | ″ |
|---|---|---|---|---|
| s | = | 76 | 12 | 42 |
| s − p | = | 9 | 10 | 51 |
| s − φ | = | 24 | 49 | 12 |
| s − h | = | 60 | 34 | 22 |

| Log cos | = | 9·3771801 |
| Log sec | = | 10·0052901 |
| Log cosec | = | 10·3772007 |
| Log sin | = | 9·9390054 |
|  | 2 | 19·6997865 |

| Log tan $\frac{t}{2}$ | = | 9·8498932 |

|  |  | ° | ′ | ″ |
|---|---|---|---|---|
| $\frac{t}{2}$ | = | 35 | 17 | 23·2 |
| $t$ in arc 5 | 70 | 34 | 46·4 |
| 3 | 14 | 6 | 57·3 |

```
                                                 h.  m.  s.
GST of GMN         =    1  49  23·55      t in time ±     4  42  19·1
 (Page II., N.A.)                         (− E. + W.)
Corr. for Long. at 9·86} ± − 0 0 0·35     R A of Star    17  54  37·19
 per hour (− E. + W.) }                   L S T of Obsn.  22  36  56·3
                                         *L S T of L M N   1  49  23·20
*LST of LMT        =    1  49  23·2       Sid. Int. from L M N  20  56  33·1
```
```
                                                            h.  m.  s.
                                          (Hrs.              9  56  21·7611
                        M T Equivalents   (Mins.             0  55  50·8257
                                          (Secs.             0   0  54·8490
                                                             0   0   0·0897
                        L M T of Obsn.                      10  53  7·48
FROM ANGLE BOOK: Chron. Time of Obsn.                       10  49  7·25
                                                              m.  s.
                        Chron. Fast or Slow on } ± +          6   0·2
                        L M T or L S T        }
```

## Azimuth by the Sun.

From Page 3, Angle Book.

|  | ° ′ ″ |  | h. m. s. |
|---|---|---|---|
| Vert. Angles | 31 41 17 | Times | 19 51 21 |
|  | 32 41 14 |  | 19 54 9 |
| Means | 32 11 15·5 |  | 19 52 45 |

                                                        h. m. s.

Chron. watch is approx.    0  3  29 slow

L M T of observation is    19  56  14

Level Correction.

| | E. | O. | | Correction. |
|---|---|---|---|---|
| | 3·5 | 7·5 | | $\dfrac{(11\cdot5 - 11) \times 10''}{4}$ |
| | 7·5 | 4·0 | |  |
| Totals | 11 | 11·5 | | $= +\,1''\cdot25$ |

                                              ° ′ ″

                                        32  11  15·5

                                +   0   0   1·3

                                        32  11  16·8   Correct mean observed altitude.

175

FORM M.—(S.M.E. Astronomical Form 3.)

## Azimuth by Sun or Star with the Theodolite.

| | | | |
|---|---|---|---|
| SUN or STAR .......... | Lat. ($\phi$) N or S. 51 23 30 | R O Gillingham Church |
| Page in Angle Book 3 | Apprx. Long. E or W. h. m. s. 0 2 9 | (Trig. Point) |
| Place R.E. Institute (Parapet) | Bar. 29"·55 | |
| Date 21.7.01 (A.M.) | Ther. 76°·2 | Mag. Bearing .......... |

$\operatorname{Tan} \frac{A}{2} = \sqrt{\operatorname{Sec.} s \operatorname{Sec.} (s - p) \operatorname{Sin} (s - \phi) \operatorname{Sin} (s - h)}$ and $s = \frac{h + \phi + p}{2}$.

A = Horizontal Angle between the *Elevated Pole* and the Sun or Star.

In this formula the latitude ($\phi$) should be taken with the *positive* sign whether N. or S. and Polar Distance ($p$) is then to be reckoned from the *Elevated Pole*.

### ELEMENTS.

|   |   |   ′   ″ |   |   |   h. m. s. |
|---|---|---|---|---|---|
| Refn. due to Alt. | = | 1 32·38 | G S T of G M N (Page II., N.A.) | = | |
| Corr. for Bar. ± | = − | 0 1·4 | Corr. for Long. at 9·86 sec. per hour (+ W. − E.) | | |
| Corr. for Ther. ± | = + | 0 4·8 | L S T of L M N | = | |
| Refraction | = | 1 26·1 | L S T of Obsn. | = | |
| | | | Sid. Int. from L M N | | |
| Hor. Par. (Page I., N.A.) × Cos. Alt | | | M T Equivalents | | Hrs.......... Mins.......... Secs.......... |
| Parallax in Alt. | = | 0 7"·4 | L M T of Obsn. (If M T Chron. used this is taken direct from Angle Book). | = | 19 56 14 |
| | | | Long. E. (−) or W. (+) | ± = − | 0 2 9 |
| **COMPUTATION.** | | | G M T of Obsn. Int. from G M N in Hrs. | = | 19 54 5 4·10 |

|   |   |   °   ′   ″ |   |   |   °   ′   ″ |
|---|---|---|---|---|---|
| FROM ANGLE BOOK. | | | | | |
| Mean Obsd. Alt. | = | 32 11 16·8 | Decl. ($\delta$) at G M N (N. or S.) | ± = + | 19 45 51 |
| Corr. for Refraction | = − | 0 1 26·1 | Hour. Var. (Page I., N.A.) × Int. from G M N | ± = + | 0 2 20 |
| Corr. for Parallax | = + | 0 0 7·4 | Dec. ($\delta$) of Sun or ☆ at time of Obsn. | ± = + | 19 48 11 |
| True Alt. ($h$) | = | 32 9 58 | | | 90 0 0 |
| Latitude ($\phi$) | = | 51 23 30 | Polar Dist. ($p$) or 90 − (± $\delta$) | = | 70 11 49 |
| Polar dist. ($p$) | = | 70 11 49 | | | |
| | 2 | 153 45 17 | NOTE.—"$p$" is reckoned from the elevated pole, ∴ if $\phi$ and $\delta$ are of *different* names "$\delta$" will be *negative* and $p = 90 + \delta$. | | |
| $s$ | = | 76 52 38 | Log sec | = | 10·6439005 |
| $s − p$ | = | 6 40 49 | Log sec | = | 10·0029586 |
| $s − \phi$ | = | 25 29 8 | Log sin | = | 9·6337547 |
| $s − h$ | = | 44 42 40 | Log sin | = | 9·8472542 |
| | | | | 2 | 20·1278680 |
| | | | Log tan $\frac{A}{2}$ | = | 10·0639190 |
| | | | $\frac{A}{2}$ | = | ° ′ ″ <br> 49 12 11·5 |
| | | | A | = | 98 24 23 |
| Angle between R O and ☆ (or Sun) (from Angle Book) | | | | = | 1 49 41 |
| † AZIMUTH of R O from *Elevated Pole* | | | | = | 96 34 42 |

† NOTE.—Azimuth for trigonometrical purposes should be measured from the South (by West). It is best ascertained from the above data by consideration of a figure showing the relative positions of Star, Pole and R O.

LATITUDE by Circum-meridian Altitude of Star.

|  |  |
|---|---|
| October 7th, 1897, "ε Pegasi," R A = | 21h 39m |
| G S T of G M N = | 13  6 |
|  | 8 33 |
| Chron. Correction | + 0  4 |
| Approx. Chron. Time of Transit | 8 37 |

NOTE.—This computation is made before observation and guarded in the angle book.

From Form 4. Chron. Time of Transit ($T_0$) = 8h 36m 37s·5

| From Angle Book, Page 4, Chron. Times (T). | Hour Angles ($T - T_0$). | Values of m (Vide Table XVI.). | Face. | From Angle Book, Page 4, Vert. Angles. |
|---|---|---|---|---|
| h. m. s. | m. s. | ″ |  | ° ′ ″ |
| 8 29 42 | − 6 55·5 | 94·1 | R. | 48 1 33 |
| 8 31 58 | 4 39·5 | 42·6 | L. | 48 0 31·5 |
| 33 39 | 2 58·5 | 17·4 | L. | 48 1 14·5 |
| 36 26 | 0 11·5 | 0·1 | R. | 48 2 51 |
| 38 5 | + 1 27·5 | 4·1 | R. | 48 2 51·5 |
| 40 49 | 4 11·5 | 34·5 | L. | 48 0 46 |
| 42 39 | 6 1·5 | 71·3 | L. | 48 0 6 |
| 45 10 | 9 32·5 | 178·7 | R. | 48 0 30·5 |
| Totals |  | 442·8 |  | 10 24 |
| Means |  | 55·3 |  | 48 1 18 |

Level Correction.

| E. | O. |
|---|---|
| 6 | 6 |
| 8 | 4 |
| 8 | 4 |
| 5 | 7 |
| 5 | 7 |
| 8 | 4 |
| 8 | 4 |
| 6 | 6 |
| Totals 54 | 42 |

Correction.

$$= \frac{(42 - 54) \times 15''}{16} = -11''\cdot 2$$

$$\begin{array}{r} ° \phantom{0}′ \phantom{00}″ \\ 48 \phantom{0}1 \phantom{0}18 \\ - \phantom{0}0 \phantom{0}0 \phantom{0}11\cdot 2 \\ \hline 48 \phantom{0}1 \phantom{0}6\cdot 8 \end{array}$$ Correct mean observed altitude.

## Form N.—(S.M.E. Astronomical Form 4.)

### Latitude by Circum-Meridian Altitudes of Star.

STAR .... ε Pegasi  
Place .... R.E. Staff Quarters.  
Date .... 7.10.1897  

Approx. Long. E. or W. 30°7′27″  
Bar. ........  
Ther. 42°  

h. m. s. 0 2 9  

M T or S T  
Approx. Chron.  

Page in Angle Book ............  
No. ........ 4  
Error ± Fast 1m. 21s  
Rate ± ........

---

### COMPUTED CHRON. TIME OF TRANSIT ($T_0$).
(See Note).

|  |  | h. m. s. |
|---|---|---|
| R A of Star, i.e., Sid. Time of Transit | = | 21 39 11·37 |
| G S T of G M N (Page II., N.A.) | = | 13 5 34·11 |
| Corr. at 9·86s. per hr. (+ W. − E. long.) ± | = | − 0 0 0·5 |
| Sid. Int. of Transit from L M N | = | 8 33 37·6 |
| L S T of L M N | = | 13 5 33·76 |
| M T Equivalents | Hrs. | 7 56 41·36 |
|  | Mins. | 0 32 54·59 |
|  | Secs. | 0 0 37·60 |
| L M T of Transit | = | 8 32 13·15 |
| Chron. Error on S T or M T (Fast + Slow −) | = | + 0 4 24 |
| Chron. Time of Transit ($T_0$) | = | 8 36 47·5 |

(Omitted with S T Chron.)

---

NOTE.—Chron. T of Transit and mean value of "$m$" are preferably computed on a separate sheet of paper. If a MT Chron. is used and some of the Hour Angle intervals are great (say over 10 minutes), all of them should first be converted into Sid. Time intervals before taking out the values of "$m$" from the tables.

### NOTATION AND FORMULÆ.

$\phi$ = Lat. N. or S.  $\zeta_0$ = Mean Zen. Dist.  
$\phi_1$ = Approx. Lat.  $\zeta_1$ = Approx. Mer. Zen. Dist.  
$\delta$ = Dec. at Mean of Times of Obsn.  $Z_0$ = Correct Mer. Zen. Dist.

$m = \dfrac{2 \sin^2 \frac{t}{2}}{\sin 1''}$ (Values found in Table XVI. See above note.)

$m_0$ = Mean of Values for $m$.

$\phi_1 = \delta + \zeta_1$  and  $\phi = \delta + Z_0$.

These Eqns. hold good in all cases if the signs of the functions $\phi$, $\delta$ and $\zeta$ are carefully attended to as follows:—

Lat. and Dec. are $\pm$ when $\frac{N}{S}$ of Equator.

Zenith. Dist. is $\pm$ when the Star is $\frac{S}{N}$ of Zenith.

$Z_0 = \zeta_0 - A m_0$  and  $A = \dfrac{\cos\phi_1 \cos\delta}{\sin\zeta_1}$.

---

### MEAN ZEN. DIST. ($\zeta_0$).

|  |  | ′ ″ |
|---|---|---|
| Refraction due to Obsd. Mean Alt. | = | 0 52·3 |
| Corr. for Bar. | = | + 0 0·8 |
| Corr. for Ther | = | + 0 0·8 |
| Refraction Corrected | = | 0 54·6 |

|  |  | ° ′ ″ |
|---|---|---|
| Obsd. Mean Alt. (From Angle Book) | = | 48 1 6·6 |
| Refraction (see above) | = | − 0 0 53·9 |
| True Mean Alt. (A) | = | 48 0 13 |
|  |  | 90 0 0 |
| $\zeta_0$ Mean Zen. Dist. = (90 − A) | = | + 41 59 47 |

### APPROX. LATITUDE ($\phi_1$).

|  |  | ° ′ ″ |
|---|---|---|
| From Angle Book, Greatest Obsd. Alt. | = | 48 2 3 |
| Refraction (use that for $\zeta_0$) | = | − 0 0 1 |
| Approx. Mer. Alt. | = | 48 2 2 |
|  |  | 90 0 0 |
| $\zeta_1$ (Approx. Mer. Zen. Dist.) | = | + 41 58 ... |
| Dec. ($\delta$) | = | + 9 24 24·5 |
| Appx. Lat. $\phi_1 = \delta + \zeta_1$ | = | + 51 23 ·5 |

---

### COMPUTATION FOR TRUE LATITUDE ($\phi$).

|  |  |  |
|---|---|---|
| Log cos $\phi_1$ | = | 9·7955·2 |
| Log cos $\delta$ | = | 9·9941·2 |
| Log cosec $\zeta_1$ | = | 10·17477 |
| Log A | = | 9·96421 |
| Log $m_0$ ($m_0 = 55\frac{2}{3}$) | = | 1·74275 |
| Log A $m_0$ | = | 1·7069m |
| A $m_0$ | = | 50′·93 |

|  |  | ° ′ ″ |
|---|---|---|
| A $m_0$ | = | 0 0 50·93 |
| $\zeta_0$ or Mean Zen. Dist. | = | 41 59 47·0 |
| $Z_0$ or ($\zeta_0$ − A $m_0$) | = ± | 41 58 56·07 |
| Dec. ($\delta$) N. or S. | = + | 9 24 33·6 |
| Lat. $\phi = \delta + Z_0$ | = | 51 23 29·7 N. or S. |

## LATITUDE by Circum-meridian Altitudes of Sun.

|  |  | h. | m. |  |
|---|---|---|---|---|
| Equ. of Time, 15th July, 1897 | = + | 0 | 6 | (Page I., N.A.) |
| Chron. Correction | = + | 0 | 1 |  |
| Approx. Chron. Time of Transit | = | 0 | 7 |  |

NOTE.—This computation is made before observation and guarded in the angle book.

h. m. s.
From Form 5.  Chron. Time of Transit $(T_0)$ = 0  6  40·7.

| From Angle Book, Page 5, Chron. Times (T). | Hour Angles $(T - T_0)$. | Values of $m$ (Vide Table XVI). | From Angle Book, Page 5, Vert. Angles. |
|---|---|---|---|
| h.   m.     s. | m.     s. | " | °    '    " |
| 23   59   14·5 | −   7   26·2 | 108·6 | 59   46   10·5 |
| 23   62   35   | 4   5·7 | 33·0 | 60   9   47·5 |
|      65   11   | 1   29·7 | 4·4 | 60   0   36 |
|      67   5    | + 0   24·3 | 0·3 | 60   20   19 |
|      68   48   | 2   7·3 | 8·8 | 59   48   40 |
|      71   2    | 4   21·3 | 37·2 | 60   19   50 |
|      72   14   | 5   33·3 | 60·6 | 59   47   54 |
|      74   23   | 7   42·3 | 116·5 | 60   17   50 |
| Totals 193   0   32·5 |  | 369·4 | 480   31   7 |
| Means   0   7   34 |  | 46·2 | 60   3   53·4 |

Level Correction.

| | E. | O. |
|---|---|---|
| | 7 | 2 |
| | 4 | 5 |
| | 3 | 6 |
| | 7 | 2 |
| | 6 | 3 |
| | 5 | 5 |
| | 5 | 5 |
| | 7 | 2 |
| Totals | 44 | 30 |

Correction.

$$= \frac{(30 - 44) \times 15''}{16} = -13''\cdot1$$

°     '     "
60    3    53·4
−     0    0    13·1
────────────
60    3    40·3   Correct mean observed altitude.

FORM O.—(S.M.E. Astronomical Form 5.)
## Latitude by Circum-Meridian Altitudes of Sun.

| | | |
|---|---|---|
| Place M.E. Institute | Appx. Long. E. or W. 30°    h. m. s.   0 2 9 | Page in Angle Book 5 |
| Date 15.7.97 (Civil) | Bar. ....   Ther. 72° | MT or ST Chron. { No. 1   Error ± 0 0 56·8 Fast.   Rate ± |

### COMPUTED CHRON. TIME OF TRANSIT ($T_0$).

| With Sid. Time Chron. | | | With Mean Time Chron. | | | † NOTE.—Mean Value of "$m$" is preferably computed on a separate sheet of paper. |
|---|---|---|---|---|---|---|
| | h. m. s. | | | h. m. s. | | |
| Sun's R A (Page I., N.A.)   = | | | L A T of L A N   = | 0 0 0 | | Values for "$m$" are found in Table XVI |
| Corr. for Long. (*i.e.*, Long. in Hrs. × Hrly. Var. of R A) (If Long. E. − W. +)   ± = | | | Eqn. of Time (Page I., N.A.)   ± = | + 0 5 43·91 | | If a ST chron. is used and some of the Hour Angle Ints. are great (say over 10 mins.) all of them should first be converted into Mean Time Ints. before taking out value of "$m$" from the tables. |
| Chron. Error on S T (Fast + Slow −)   ± = | | | *Corr. of Do.   ± = | − 0 0 0·009 | | |
| | | | L M T of L A N (True MT of Transit)   = | 0 5 43·9 | | |
| | | | MT Chron. Error (Fast + Slow −)   ± = | + 0 0 56·8 | | |
| Chron. Time of Transit ($T_0$)   = | | | Chron. Time of Transit ($T_0$)   = | 0 6 40·7 | | |

### MEAN OF HOUR ANGLES ($T-T_0$).

* This Corr. = Long. in Hours × Hourly Var. of Eqn. of Time (Page I., N.A.).
In E. Longs. if Eqn. of T is *increasing* numerically, this Corr. should be given a contrary sign to that of the Eqn. of T, and *vice versâ*.
In W. Longs. if Eqn. of T is *increasing* numerically, the Corr. is given the same sign as the Eqn. of T, and *vice versâ*.

| FROM ANGLE BOOK. | | | |
|---|---|---|---|
| | | h. m. s. | |
| (a) Mean of Obsd. Times (T)   = | | 0 7 31 | |
| (b) Chron. T of Transit ($T_0$)   = | | 0 6 40·7 | |
| (c) Mean of Hour Angles ($T-T_0$)   ± = | | + 0 0 53·3 | |

NOTE.—(c) is ± if (a) is later/earlier than (b).

### NOTATION AND FORMULÆ.

$\phi$ = Latitude N. or S.    $\zeta_0$ = Mean Zen. Dist.
$\phi_1$ = Approx. Lat.    $\zeta_1$ = Appx. Mer. Zen. Dist.

$$m = \frac{2\sin^2 \frac{t}{2}}{\sin 1''} \quad (\text{See Note †}) \quad Z_0 = \text{Correct Mer. Zen. Dist.}$$

$\delta$ = Dec. at Mean of Times of Obsn.
$m_0$ = Mean of Values of $m$ (for each hour angle).
$\phi_1 = \delta + \zeta_1$ and $\phi = \delta + Z_0$.

These Eqns. hold good in all cases if the signs of the functions $\phi$, $\delta$ and $\zeta$ are carefully attended to as follows:—

Lat. and Dec. are $\pm$ when $\frac{N}{S}$ of Equator.

"Zen. Dist." is $\pm$ when the Sun is $\frac{S}{N}$ of the Zenith.

$$Z_0 = \zeta_0 - A m_0 \quad \text{and} \quad A = \frac{\cos\phi_1 \cos\delta}{\sin\zeta_1}$$

### TIME INT. OF OBSN. FROM G M N.

| | | h. m. s. |
|---|---|---|
| G M T of G A N (Eqn. of T, p. I., N.A.) = | | 0 5 43·91 |
| Long. in Hrs. × Hourly Var. of Eqn. of T (see Note *)   ± = | | − 0 0 0·009 |
| L M T of L A N (or Corr. Eqn. of T)   = | | 0 5 43·9 |
| Long. (Add if W., subtract if E.)   = | | − 0 2 9 |
| G M T of L A N   = | | 0 3 34·9 |
| Mean of Hour Angles (see above)   ± = | | + 0 0 53·3 |
| Int. of Mean of Obsn. Times from G M N   = | | 0 4 28·2 |
| Do. do. in hours   = | | ·074 |

### COMP. FOR MEAN ZEN. DIST. ($\zeta_0$).

| FROM ANGLE BOOK. | | | | | |
|---|---|---|---|---|---|
| | ° ′ ″ | | ″ | | |
| Obsd. Mean Altitude   = | 60 3 40·3 | | 0 32·0 | × | 4·35″ |
| Refraction   = | − 0 0 32·0 | | 0 1·7 | | |
| Parallax   = | + 0 0 4·35 | | +·4 | | |
| | | | | Refrn. due to Alt. Corr. for Bar. Corr. for Ther. | Refraction Hor. Par. × cos $\phi$ ∴ Parallax in Alt. |
| True Mean Altitude   = | 60 3 12·7   90 0 0 | | | | |
| $\zeta_0$ Mean Zen. Dist.   ± = | + 29 56 47·7 | | | | |

### COMP. FOR TRUE LATITUDE. ($\phi$).

| | | | |
|---|---|---|---|
| | | Log cos $\phi_1$ = | 9·75526 |
| | | Log cos $\delta$ = | 9·96883 |
| | | Log cosec $\zeta_1$ = | 10·30193 |

### COMP. FOR APPX. LATITUDE ($\phi_1$).

| FROM ANGLE BOOK. | ° ′ ″ |
|---|---|
| Greatest Obsd. Alt.   = | 60 20 |
| Refraction (use Refrn. for $\zeta_0$)   ± = | − 0 0·5 |
| Semi-Diam.   ± = | |
| Approx. Mer. Altitude   = | 60 4   90 0 0 |
| $\zeta_1$ Approx. Mer. Zen. Dist.   = | + 29 56 |
| Dec. ($\delta$)   = | + 21 27 35·1 |
| Appx. Lat. $\phi_1 = \delta + \zeta_1$   = | 51 23 |

| | | |
|---|---|---|
| Log A   = | | 10·06599 |
| Log $m_0$ ($m_0$ = 46·2″)   = | | 1·66464 |
| Log A $m_0$   = | | 1·73063 |
| A $m_0$   = | | 53·78 Secs. |
| | | ° ′ ″ |
| A $m_0$   = | | 0 0 53·78 |
| $\zeta_0$   = | | + 29 56 47·7 |
| $Z_0$ or ($\zeta_0 - A m_0$)   = | | + 29 55 53·9 |
| Dec. $\delta$   = | | + 21 27 35·1 |
| Lat. $\phi = \delta + Z_0$   = | | 51 23 29·0 N. or S. |

### DECLINATION.

| | | ° ′ ″ |
|---|---|---|
| Dec. at G M N (Page II., N.A.) N. or S.   = | | 21 27 36·9 |
| Hourly Var. (Page I., N.A.) 24·09 × ·07   } ± = | | − 0 0 1·8 |
| Int. in Hrs. from G M N | | |
| Dec. ($\delta$) at Mean of Obsn. Times N. or S.   = | | 21 27 35·1 |

(5455)      2 A

LATITUDE by Altitudes of Polaris.

From Page 6, Angle Book.

|  | ° | ′ | ″ |  | h. | m. | s. |
|---|---|---|---|---|---|---|---|
| Vert. Angles | 51 | 37 | 12·5 | Times | 7 | 3 | 54 |
|  | 51 | 39 | 39·5 |  | 7 | 8 | 12 |
|  | 51 | 40 | 29·5 |  | 7 | 11 | 0 |
|  | 51 | 40 | 55·5 |  | 7 | 16 | 33 |
| Totals | 206 | 38 | 17 |  | 28 | 39 | 39 |
| Means | 51 | 39 | 34·2 |  | 7 | 9 | 54·8 |

Level Correction.

| | E. | O. |
|---|---|---|
| | 7 | 5 |
| | 4 | 8 |
| | 4 | 8 |
| | 7 | 5 |
| Totals | 22 | 26 |

Correction.

$$+ \frac{(26 - 22) \times 15''}{8}$$

$$= + \frac{4 \times 15''}{8}$$

$$= + 7''\cdot 5$$

|  | ° | ′ | ″ |
|---|---|---|---|
|   | 51 | 39 | 34·2 |
| + | 0 | 0 | 7·5 |
|   | 51 | 39 | 41·7 | Correct mean observed altitude.

FORM P.—(S.M.E. Astronomical Form 6.)

## Latitude by Altitudes of "Polaris."

| | | | | | |
|---|---|---|---|---|---|
| Place... R.E. Staff Quarters. | Approx. Long. E. or W. | | h. m. s. 0 2 9 | Bar....30".27. | |
| Date ....7.10.97... | | No. ....12........ | | Ther.....52°. | |
| Page in Angle Book....6... | L S T or L M T Chronometer | Error ...3m. 24s. fast... | | | |
| | | Rate...... | | | |

### ELEMENTS.

| | h. m. s. |
|---|---|
| G S T of G M N (page II., N.A.) = | 13 5 34·1 |
| Corr. for Long. at 9·86 s. per hour (− E. + W.) ± = | − 0 0 0·3 |
| L S T of L M N = | 13 5 33·8 |

*Omitted if S T Chron. is used.*

### FROM ANGLE BOOK.

| | h. m. s. |
|---|---|
| Mean of Obsd. Times = | 7 9 34·8 |
| Chron. Corr. on L M T or L S T (Slow + Fast −) ± = | − 0 3 24 |
| L M T of Obsn. = | 7 6 30·8 |
| S T Equivalents { Hrs. 7 Mins. 6 Secs. 30·8 } = | 7 1 9<br>0 0 1<br>0 0 30·8 |
| Sid. Int. from L M N = | 7 7 40·8 |
| L S T of L M N (see above) = | 13 5 33·8 |
| * L S T of Obsn. = | 20 13 14·6 |

*This is omitted if S T Chron. is used.*

### REFRACTION.

| | ° ′ ″ |
|---|---|
| Refn. due to Obsd. Mean Alt. (Angle Book) = | 0 46 |
| Corr. for Bar. ± = + | 0 0·12 |
| Corr. for Ther. ± = + | 0 0·73 |
| Refraction Corrected = | 0 47·1 |

### TRUE ALTITUDE (h).

| | ° ′ ″ |
|---|---|
| Obsd. Mean Alt. (Angle Book) = | 51 39 41·7 |
| Corr. Refraction = | − 0 0 47·1 |
| True Altitude (h) = | 51 38 54·6 |

### COMPUTATION BY TABLES IN NAUTICAL ALMANAC.

| | ° ′ ″ |
|---|---|
| True Alt. (h) (see above) = | 51 38 55 |
| Subtract | 0 1 0 |
| Reduced Alt. = | 51 37 55 |
| With argt. L S T° 20h 13m.15s. 1st Corr. ± = − | 0 16 42 |
| Approx. Latitude = | 51 21 13 |
| With argt. L S T° and True Alt. 2nd Corr. = + | 0 0 ·7 |
| With argt. L S T° and date 3rd Corr. = + | 0 1 21 |
| Latitude N. or S. = | 51 23 34 |

### COMPUTATION BY FORMULA.

$$\phi = h - p\cos t + \frac{p^2}{2}\sin^2 t \sin 1'' \tan h.$$

This computation gives more accurate results than those obtained by use of N.A. Tables.

### POLAR DISTANCE (p).

| | ° ′ ″ |
|---|---|
| Dec (δ) = | 88 15 50·8<br>90 0 0 |
| p (or 90° − δ) = | 1 11 9·2 |
| p (in seconds) = | 4149 |

### HOUR ANGLE (t).

| | h. m. s. |
|---|---|
| * L S T of Obsn. = | 20 13 14·6 |
| R A of Star = | 1 22 26 |
| t in time (S T − R A) ± = + | 18 50 48·6 |
| If t is negative subtract from 24 hrs. = + | |
| | ° ′ ″ |
| t in arc = | 282 42 8·55 |

| | |
|---|---|
| Log p (p in seconds) = | 3·6482624 |
| Log cos t = | 9·3421987 |
| Log 1st Corr. = | 2·9901611 |
| 1st Corr. (in seconds) = | 978·3 |
| Log p = | 3·6482624 |
| Log sin t = | 9·9892386 |
| Log p sin t = | 3·6375010 |
| Log p² sin² t = 2 × (Log p sin t) = | 7·2750020 |
| Log tan h = | 10·1017066 |
| Log sin 1″ † = | 4·6855749 |
| Ar. comp. Log 2 † = | 1·6989700 |
| Log 2nd Corr. = | 1·7612535 |
| 2nd Corr. (in seconds) = + | 57″·7 |

| | ° ′ ″ |
|---|---|
| True altitude = + | 51 38 54·6 |
| 1st Corr. (opposite sign to cos t) ± = − (if t is in 2nd or 3rd Quadrant cos t is −) | 0 16 18·3 |
| | 51 22 36·3 |
| 2nd Corr. (always added) = + | 0 0 57·7 |
| Latitude N. or S. = | 51 23 34 |

† log sin 1″ = 6̄·6855749

Ar. comp. Log. = 1̄·6989700.

(5455)

## CHAPTER XV.

### LATITUDES AND AZIMUTHS.

ON expeditions and boundary commissions it is frequently impossible to keep up a regular triangulation. Recourse may then be had to the device known as "latitudes and azimuths."

Suppose A and B to be the two mutually visible points. If a latitude is observed at A and the azimuth AB is determined, and then subsequently the latitude of B is determined, it is clear that provided that there is a considerable amount of northing and southing in the line AB, values can be obtained for the length AB, and the difference of longitude between A and B, either graphically by drawing on a plane-table, or better, by means of a calculation.

**Degree of Accuracy.**—Obviously no very great degree of refinement can be obtained by this method. We will analyse the error which will result in a fairly favourable case. Let the azimuth of AB be about 45°, let latitudes be observed at A and B with a probable error of 3″. Then the probable error in the range of latitude is $3\sqrt{2}$, and the probable error in the length of AB = $3\sqrt{2} \times \sqrt{2}$ = 6″ or 600 feet. Now suppose AB to be about 40 miles, the

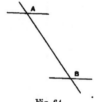

Fig. 64.

ratio of error in the length of AB = $\dfrac{600}{40 \times 5,280}$, or $\dfrac{1}{352}$. Now, although this is a large error as compared with trigonometrical errors, it is smaller than would usually result from any rapid system of traversing.

**The most favourable case** is that of a journey nearly north and south with long rays. Latitudes and azimuths should rarely be attempted with rays of a greater azimuth than 45° from N. or S.

Supposing the explorer on a rapid march is fixing his framework by this means, when he observes a latitude and azimuth he should take a round of angles to all likely hills anywhere near to his probable line of march; very often he cannot predict the hill which he is likely to visit next. Even with this precaution he may find either that he cannot visit or cannot identify with sufficient accuracy any hill included in his round of angles from his first station. He must then give up the idea of identifying the hills previously observed to and must observe a latitude and take an azimuth from any convenient hill *back* to his first station, on which he should, if possible, have erected some cairn or pole or mark.

Latitudes and azimuths were freely used on the march of the Afghan Boundary Commission, as is shown on Plate VIII.

They were also much employed in the exploration of the southern frontier of Abyssinia in 1902–03.

**Computation.**—Let A and B be two stations at which latitudes have been observed and at A the azimuth of AB has been determined.

Let the difference of the two latitudes in seconds $= \Delta\lambda$.
Let the azimuth of AB at A $= $ A.
It is required to find—

(i.) The length of the line AB $= c$;
(ii.) The difference of longitude of A and B in seconds of arc $= \Delta L$.

Let the azimuth at the middle point of AB $= A + \dfrac{\Delta A}{2} = a$.

Let the mean latitude of A and B $= \lambda_m$.

Let the difference of longitude determined approximately (from a plane-table, or rough method) $= \Delta'L$.

Then

$$\frac{\Delta A}{2} = \frac{\Delta'L}{2} \sin \lambda_m \quad \ldots \ldots \ldots \ldots (1).$$

Now apply this value of $\dfrac{\Delta A}{2}$ to A (remembering in which way the meridians converge) and thus obtain $a$.

Then for the required difference of longitude

$$\Delta L = \frac{\rho}{\nu} \tan a \sec \lambda_m . \Delta\lambda \quad \ldots \ldots \ldots \ldots (2),$$

where $\log \dfrac{\rho}{\nu}$ is the value of Q in Table I. (for the *mean* latitude).

Also, for the length AB,

$$c = \Delta\lambda . \sec a . \rho \sin 1'' \quad \ldots \ldots \ldots \ldots (3),$$

where $\log \rho \sin 1''$ is the arithmetical complement of P in Table I. (for the *mean* latitude).

## CHAPTER XVI.

### LONGITUDE.

**The Longitude** of a place is the angle made by its meridian plane with some one fixed meridian plane arbitrarily chosen, and is measured by the arc of the equator intercepted between these two meridians.

Longitudes are reckoned east or west of the fixed meridian up to 180°.

The fixed meridian in use in all British territories is that of Greenwich.

Since the earth makes a complete revolution once in 24 hours, it is clear that each meridian is moving round at the rate of 15° an hour; we may therefore describe the relative positions of two meridians in arc or in time. We may say, for instance, that the longitude of a place is 30° E., or 2 hours E.

Now, as has been explained in Chapter XIV., the local time is measured by the angle at the celestial pole between the local meridian and the mean sun; and the local sidereal time is the angle between the local meridian and the First Point of Aries.

It follows that the difference in longitude of two places is the difference, at any instant, of their local times, mean or sidereal.

The longitude of a place may be determined by any one of the following methods:—

(1) Triangulation
(2) Exchange of telegraphic signals
(3) Latitudes and azimuths
(4) Transport of chronometers
} Relative methods.

(5) Occultations of stars by the moon
(6) Moon photographs
(7) ,, culminations
(8) ,, altitudes
(9) Lunar distances
(10) Eclipses of Jupiter's satellites
} Absolute methods.

### TRIANGULATION.

(1) **Triangulation** is by far the best of all these methods, not only on account of its accuracy, but also because the process is itself necessary for the detail survey. Even in wooded countries, such as parts of Burma, it is still usually possible to triangulate; and the accumulated error, even when it is impossible to erect signals, will only amount perhaps to some 5 seconds of arc in 100 miles. The method has been fully explained in Chapter III. If, however, the longitude of the starting-point is not known, it must be determined by telegraph, chronometers, or by some absolute method.

A good example of the combined use of several methods is afforded by the Nyasa-Tanganyika Boundary Survey of 1898. The head of the transcontinental telegraph line had, at that time, reached a point (Nkata) near the shore of Lake Nyasa, about 140 miles south of the starting point of the boundary (which was on the lake) and some 1900 miles from Cape Town.

The difference of longitude between Cape Town observatory and Nkata was determined by exchange of telegraphic signals on three nights, and a good value obtained.

Fourteen chronometers were now transported backwards and forwards between Nkata and

Kambwé near the starting point of the boundary. From this point a continuous triangulation was carried out along the whole length of the boundary zone (over 200 miles) to Lake Tanganyika.

## EXCHANGE OF TELEGRAPHIC SIGNALS.

(2) **Exchange of Telegraphic Signals.**—For a description of the refined determination of differences of longitude by telegraph for geodetic purposes by means of chronographs, the reader is referred to any standard work on geodetic methods; we shall here describe the simple means used to obtain telegraphic differences of longitude with sufficient accuracy for map making.

Let A and B be two stations which are connected by telegraph, and of which it is required to determine the difference of longitude.

Arrangements must first be made to insure direct communication between A and B; this may involve the use of relays. The line must be free of all traffic during certain hours of the night, and there must be a telegraph operator at each end.

At each end there must be a good box chronometer; one of these should be a mean time chronometer and the other a sidereal time chronometer. The reason for this is, that if both were sidereal or both mean time chronometers, the intervals between the beats of the chronometer would be constant, and there would be no coincidences.

The chronometer should be on the same table (in tent, hut, or house) as the telegraph key, and close to it, so that the person sending signals can easily read the chronometer.

Before commencing the exchange of signals, a programme should have been arranged between A and B.

A convenient programme is as follows:—

> Signals to be exchanged on three nights.
> Each night's work to consist of at least one set of signals sent in each direction.
> Unsatisfactory sets to be repeated.
> Each set of signals should be preceded by a rattle on the key about 30 seconds before the set begins. Each set should end with a rattle. A message should be sent at the end of a set to ask if it was received all right.
> Each set of signals to last 5 minutes.
> Each set to begin at 0 seconds. Succeeding signals to be sent at 10 seconds, 20 seconds, 30 seconds, 40 seconds, 0 seconds, 10 seconds, &c., omitting the 50th second.
> When the key is depressed to send a signal it should be kept down for 2 seconds.

**Receiving.**—The chronometers should beat half seconds, and in receiving signals it is necessary to count mentally the beats of the chronometer, thus: one, half, two, half, three, half, &c. If now the signal from the other end should arrive, say, between half and two, the person receiving must judge whether it was 1·6, 1·7, 1·8, or 1·9; if he is doubtful he should say 1·8. It requires a little practice to judge the intervals.

If the chronometers beat half seconds, a coincidence will occur every three minutes; at a coincidence the chronometer beat and the signal will sound together, the receiver will then say "up"; the beat and signal will sensibly coincide two or three times running.

Before and after exchanging signals, observations for time should be taken. These observations should be made either with a portable transit instrument or with a theodolite; if with the former, two N. stars and two S. stars before and the same number after; if with a theodolite, two E. and two W. stars before and the same number after.

The box chronometer should not be moved from the table on which it rests during the operations; any movement will tend to give it an irregular rate.

Its error should be determined by transit instrument or theodolite as laid down, each moment of transit or intersection being referred to the chronometer in one of the following ways :—

By having a recorder at the chronometer, and,
- (i.) Fitting up a temporary telegraph line a few yards long from instrument to hut, and sending signals by key;
- (ii.) By shouting "up" to recorder, or
- (iii.) The time may be transferred by stop watch; this is only suitable for use with a theodolite, which must be near the hut.

The next morning, after an interchange of signals, each station should send a message to the other station :—

"My local mean (or sidereal) time of first signal sent was        " (to tenths of seconds).
"My local mean (or sidereal) time of first signal received was         ."

The list of received signals should now be inspected, there should be a slow progressive increase or decrease in the decimals, any obvious mistake should be corrected.

The difference between the means of the corrected local times of receiving and sending is the difference of longitude in time. This is burdened with two errors in addition to instrumental errors: (1) Rate of transmission of signal along the wire; this is eliminated by sending the same number of times in each direction, and is very small; (2) Personal equation. Every observer records a transit, intersection or signal a little too early or too late. This error is best eliminated by exchange of observers from one end of the line to the other when half the work has been completed. This is not usually practicable in topographical work.

In view of the existence of personal equation, it must be expected that a telegraphic difference of longitude determined as above described will be in error ·3 to ·5 of a second of time, or in equatorial regions 150 to 250 yards.

(3) **Latitudes and Azimuths.**—See Chapter XV.

## TRANSPORT OF CHRONOMETERS.

(4) **Transport of Chronometers.**—Suppose that it is required to ascertain the difference of longitude between two places A and B. Determine the error of a chronometer at A on local time, and also determine the "rate" of the chronometer, $i.e.$, the amount which it gains or loses in a day. Now transport the chronometer to B, and determine its error on local time at B. Since we know the error at A, and the rate, we can also at any instant find the local time at A; compare this with the local time at B at the same instant, the difference of these two local times is the difference of longitude between A and B.*

As an example :—

At A, at 9 P.M., November 3, 1903, the chronometer was 2 minutes 5·6 seconds fast, gaining 1·4 seconds a day;

At B, at 10·45 P.M., November 4, 1903, the same chronometer was 4·2 seconds slow; what is the difference of longitude?

The interval in time between the two determinations is 1 day 1·75 hours = 1·07 day;
The rate = 1·4 seconds, the chronometer has therefore gained 1·4 seconds × 1·07 seconds
= 1·5 seconds.

The difference of longitude is therefore :—

|   | m. | s. |
|---|----|----|
|   | 2  | 5·6 |
| +0 |   | 4·2 |
| +0 |   | 1·5 |
|   | 2  | 11·3 |

and B is east of A 0° 32′ 49·5″.

* It is clear that the difference of the local times is the same as the difference at the same instant of the errors of the same chronometers on local times.

In practice it is found that for land journeys box chronometers give unsatisfactory results, for such work good keyless half-chronometer watches, costing from £24 to £30, are the best.

Experience shows that the rate of a watch at rest is not the same as its rate when being transported; it is therefore desirable, when possible, to ascertain the "travelling rate" of the watch. This may be done by sending the watch out for a march every day, the march to last as long as it is expected marches will be on the actual journey. After 3 or 4 days a fair rate will be obtained. The "travelling rate" in this sense includes the rate when at rest at night.

When every precaution has been taken, it will be found that a single watch is very variable in its rate. It is thus necessary to carry several; five is a good number. Each watch should be known by some distinctive letter or number. One watch should be used for the time observations, and comparisons with all the others should be made before and after the observations.

**Watch Comparisons.**—Watches usually beat 5 times in 2 seconds, and the comparison of two watches by ear is difficult; the following is a useful method:—Let A and B be two observers, X and Y two watches. Let A take watch X, and B watch Y. A should make a sharp tap on the table when the seconds hand of X is exactly, by eye, on some definite second. B should record the position of Y to the nearest tenth. B now signals to A in the same way, and A records. A and B now change watches and repeat the operations. There are now four values for the difference between the watches and the personal equations of the observers are, if constant, eliminated.

**Return Journeys.**—If a watch be transported from a place A to a place B, and then immediately carried back to A, and if its errors on local time, before starting at A, on arrival at B and on return to A be determined, a very good value of the difference of longitude A to B can be found.

For if no halt is made at B, beyond the ordinary night's rest, the travelling rate can be found by dividing the difference between the errors at A (starting and returning) by the number of days which elapsed.

If a halt is made at B, the rate at rest should be determined and applied for the period of the halt.

**Chronometer Journeys by Sea.**—Sometimes it is necessary to determine the difference of longitude of two places on the sea coast, or connected by a navigable river, or on one of the great African lakes. In such a case, as many box chronometers as can be obtained should be carried, and the method of return journeys, as above described, should be adopted.

**Variations in Rate.**—Watch rates are variable, and the only way to eliminate the effects of such variability is to carry a large number of watches. The attached table (p. 188) will serve to show the variations which may be expected, as judged by comparisons of four watches on a journey in E. Africa.

## LONGITUDE BY OCCULTATION OF A STAR.

(5) **Longitude by the Occultation of a Star.**—We now come to the determination of longitude by *absolute* methods. The rapid movement of the moon amongst the fixed stars gives the means of determining, by their relative positions, the Greenwich time at the moment of observation; and the difference, at the same instant, between the Greenwich time and the local time is the longitude, in time, E. or W. of Greenwich.

These *absolute* determinations of longitude depend only upon observations made on the spot, and the best of all absolute methods is undoubtedly the occultation of a star

The *apparent* movement of the moon in the heavens from east to west does not take place at the same rate as that of the stars, but is slower; in other words, the *relative*

TABLE of Daily Comparisons of four watches carried from River Sabi to River Limpopo and back, between September 8th and September 28th, 1892.

### Differences between Watches.

| 1 and 2. | Δ. | 1 and 3. | Δ. | 1 and 4. | Δ. | Dates. |
|---|---|---|---|---|---|---|
| m.  s. |  | m.  s. |  | m.  s. |  |  |
| + 0  29 |  | + 0  13 |  | + 0  34·5 |  | 8th Sept. |
|  | 12 |  | 4 |  | 11·5 |  |
| 17 |  | 9 |  | 23 |  | 9th  ,, |
|  | 6 |  | 8·5 |  | 13·5 |  |
| 9 |  | 0·5 |  | 9·5 |  | 10th  ,, |
|  | 8 |  | 14 |  | 7·5 |  |
| 0 |  | − 0  14·5 |  | − 0  17 |  | 11th  ,, |
|  | 12 |  | 10·5 |  | 16 |  |
| − 0  12 |  | 25 |  | 33 |  | 12th  ,, |
|  | 11·5 |  | 11 |  | 19·5 |  |
| 23·5 |  | 36 |  | 52·5 |  | 13th  ,, |
|  | 12 |  | 11·5 |  | 19·5 |  |
| 35·5 |  | 47·5 |  | 1  12 |  | 14th  ,, |
|  | 13·5 |  | 9·5 |  | 15 |  |
| 49 |  | 57 |  | 27 |  | 15th  ,, |
|  | 10·5 |  | 10·5 |  | 18 |  |
| 59·5 |  | 1  7·5 |  | 45 |  | 16th  ,, |
|  | 10 |  | 9 |  | 17 |  |
| 1  9·5 |  | 18·5 |  | 2  2 |  | 17th  ,, |
|  | 10·5 |  | 8 |  | 16 |  |
| 20 |  | 26·5 |  | 18 |  | 18th  ,, |
|  | 11 |  | 8 |  | 12·5 |  |
| 31 |  | 34·5 |  | 30·5 |  | 19th  ,, |
|  | 15 |  | 3·5 |  | 10·5 |  |
| 46 |  | 38 |  | 41 |  | 20th  ,, |
|  | 7 |  | 16·5 |  | 30 |  |
| 53 |  | 54·5 |  | 3  11 |  | 21st  ,, |
|  | 9 |  | 6·5 |  | 13 |  |
| 2  2 |  | 2  1 |  | 24 |  | 22nd  ,, |
|  | 6 |  | 10·5 |  | 30 |  |
| 8 |  | 11·5 |  | 54 |  | 23rd  ,, |
|  | 1·5 |  | 3·5 |  | 30 |  |
| 9·5 |  | 25 |  | 4  24 |  | 24th  ,, |
|  | 11·5 |  | 4 |  | 16·5 |  |
| 21 |  | 29 |  | 40·5 |  | 25th  ,, |
|  | 10·5 |  | 7·5 |  | 14·5 |  |
| 31·5 |  | 36·5 |  | 35 |  | 26th  ,, |
|  | 3 |  | 16·5 |  | 28·5 |  |
| 34·5 |  | 53 |  | 5  23·5 |  | 27th.  ,, |
|  | 10 |  | 8 |  | 15 |  |
| 44·5 |  | 3  2 |  | 38·5 |  | 28th  ,, |

Comparisons on this journey were only made to the nearest half second.

movement* of the moon among the stars is from west to east at a rate of about one diameter per hour.

It follows, therefore, that in the course of its monthly revolutions round the earth, the moon shuts out for a time, from the view of an observer on the earth's surface, certain of the fixed stars. The disappearance of a star behind the moon's disc is called its immersion, and the re-appearance its emersion; and it follows from what has been said that the immersion will always take place on the eastern limb of the moon, and the emersion on the western.

Data for these phenomena are given in the 'Nautical Almanac,' under the heading "Elements of Occultations," together with the G.M.T. of their occurrence as they would be seen from the earth's centre. The G.M.T. of the phenomena, as observed from any station on the earth's surface, can be computed, having given the latitude and approximate longitude of the station; and thus, by observing the L.M.T. at which the phenomena actually occur, the difference of time, and therefore of longitude, between the station and Greenwich is determined.

To observe an occultation it is desirable to know beforehand the approximate L.M.T. at which it will take place at the observer's station.

This can be done roughly by applying the approximate longitude of the station (in time) to the G.M.T. of conjunction in R.A. of the moon and the star (which is given in the 'Nautical Almanac'), *adding* the longitude if it is east, and *subtracting* if west.

Two hours before the time so found, the telescope should be directed at the eastern limb of the moon, and search should be made for any star (the magnitude being known is a great assistance) approaching it from a distance equal to one or two diameters of the moon.

There can be little difficulty in identifying the star, owing to the rarity with which these phenomena occur, and a star chart is unnecessary for the purpose.

Parallax in R.A. will have a considerable effect on the time at which the occultations will take place, and with a little practice the observer will be able to save himself a good deal of waiting by studying its probable effects.

Thus he will be able to say whether the hour found from the 'Nautical Almanac' will be accelerated or retarded, and roughly how much, simply by ascertaining whether the moon will be to the east or west of the meridian respectively, and how far.

It is evident, however, that this system of waiting considerable periods of time for an occultation (the actual occurrence of which will in many cases be doubtful up to the last moment) will not commend itself to the skilled observer.

A preferable method is to predict beforehand the circumstances of an occultation. This can be done sufficiently correctly, the accuracy of the prediction depending on the exactness of the position of the observer's station assumed in the calculation.

Thus it can be ascertained—1st, whether the occultation will take place at all; 2nd, the L.M.T. of immersion and emersion; 3rd, the exact position of the moon at these instants; 4th, the points on the moon's disc (measured in degrees of arc from the moon's vertex, or from its north point) at which the star will disappear and re-appear; and 5th, whether the immersion and emersion will occur in the moon's bright or dark limb—a most important consideration in actual practice.

A method for predicting the circumstances of an occultation is given in the "explanation" of the 'Nautical Almanac,' and there is also a graphical method devised by Colonel S. C. N. Grant, C.M.G., R.E., which is explained further on.

In 1896 several quantities ($q_0, p^1, q^1$), were added to the "Elements of Occultations" in the 'Nautical Almanac,' and Mr. A. C. D. Crommelin uses these quantities graphically in an

* Of course the moon has a second relative movement among the stars, from N. to S. and *vice versâ*, corresponding to its changes of declination; the latter cause is also responsible for the circumstance that the range of fixed stars which are at one time or another occulted, is such a wide one.

adaptation of Grant's method (published in pamphlet form by the Royal Geographical Society), and thereby considerably reduces the labour required in applying the latter.

One can now, with a little practice, predict the whole of the circumstances of an occultation in a quarter of an hour, and the times of immersion and emersion will be found to be correct within 2 or 3 minutes. An example of such a prediction is appended.

Longitude by the observation of a single occultation cannot ordinarily be depended upon to within less than about 1 second of time (15 seconds of arc, or a quarter of a mile on the equator), and therefore, whenever possible, more than one phenomenon should be observed at the same place. To get full benefit from the accuracy of the method, it is necessary to obtain from some fixed observatory the *observed* declination and right ascension of the moon during the night in question, so as to correct the co-ordinates given (by prediction) in the 'Nautical Almanac'; differences even in the second place of decimals of seconds in these quantities appreciably affect the result of the calculation.

Another disadvantage possessed by this method of ascertaining longitudes consists in the scarcity of suitable occultations that are visible at any one place during the year, and a large percentage of those that take place, in every climate, cannot be observed owing to the state of the weather.

Thus in England, during the winter months, the heavens are generally obscured by clouds, and in the tropics, climatic causes, such as clouds, the dust haze accompanying the Harmattan wind (which makes small stars difficult to see when in close proximity to the full moon), bring about many disappointments.

In the 'Nautical Almanac' about 150 stars are given for each month, the occultation of which is visible at some point or other of the earth's surface.

In considering any one station, however, about half of these must be struck out, owing to the station being outside the limits* of latitude (last column) within which, only, the occultation will be visible.

Even now it by no means follows that because the station is within the stated limits of latitude, the occultation will be visible there, for in many cases among the stars that still remain, the occultations will take place either in daylight or when the moon is below the local horizon, to say nothing of the instances in which the occultations will be "spoiled" by the effects of parallax, both in declination and in right ascension, more particularly the latter. Generally, about five satisfactory occultations only will be found to be visible at any one station each month, and in actual practice, with field instruments, dependence can only be placed on an average of two of these being observed.

In Northern Nigeria, in 1903, the Anglo-French Boundary Commissioners (having two sets of instruments at their disposal) waited two months at one place before a set of three occultations was obtained.

The longitudes when worked out had an outside variation from one another of only ·9 seconds of time. At another place, on the other hand, with the second telescope in use during the same period of time, three occultations were observed in a fortnight, the resulting longitudes having a maximum variation of three seconds.

It should be stated, however, that in all six cases the calculations were worked out without using the *observed* co-ordinates of the moon, those predicted in the 'Nautical Almanac' only being available at the time.

**Occultation of a Planet.**—This observation can be made with field instruments, but it cannot be expected that the result will be as accurate as in the case of a fixed star, owing to the appreciable time taken by the disc of the planet to disappear into or emerge from the moon, and the consequent difficulty of judging the moment at which the *centre* of the planet comes in contact with the moon's limb.

In making this observation accurately the times should be taken of the *immersion* of the

* Owing to the possible accumulation of neglected quantities these limits should be extended by 1°.

first limb of the planet into, and the *emersion* of the second limb from the moon's disc, and a correction can then be applied for the apparent semi-diameter of the planet.

In the calculation also the variations of right ascension and declination of the planet, as well as its parallax, have to be considered, and further complications arise (especially when considering occultations of Jupiter, Saturn and Saturn's ring) in determining the extent of illumination of the planet's elliptical outline.

*Practical Hints.*—For field work a telescope with a 3-in. or 3½-in object-glass will be found suitable, but care should be taken that the stand on which it is fixed is firm, and if possible, as an extra precaution against vibration, the object end of the telescope should be secured to the stand by an arm working in conjunction with that attached to the eye end of the instrument for the purposes of elevation.

During observation the telescope should of course be protected as much as possible from the wind by placing it inside a roofless hut, or else within a building having a door or window facing the moon. It will be found, too, that an eye-piece of comparatively low power is preferable for use :—

(a.) For better definition ;
(b.) To minimise the effects of vibration on the instrument.

A diagonal eye-piece is more convenient than a direct one, as it enables the observer by adjustment of the legs of the telescope to remain seated while waiting ; but it should be remembered that the image thus seen, although upright, is inverted in azimuth, and allowance should be made accordingly, when expecting an emersion.

Stars of which the occultations are predicted are nearly all invisible to the naked eye, being of the 5th, 6th, or 7th magnitude, for the most part.

These are so small that even when seen through a powerful telescope it may be taken as a general rule that an immersion into, or an emersion from the *bright* limb of the moon, even when observed, cannot be depended upon for great accuracy of result. Nor, of course, can observations into or from the dark limb be made until some time after sunset, or before sunrise, especially if the moon is in the west or east respectively—though as a matter of interest it may be stated that occultations of large stars are often observed in observatories in broad sunlight. Generally speaking, also, if the moon is only 15° or less above the horizon, the observation will be impeded by haze.

So much stress has been already laid on the scarcity of suitable occultations at any one place that it need hardly be said, when an opportunity occurs of observing one, every precaution should be taken to ensure success ; for example, observations for time should be made as shortly as possible before *and after* each phenomenon, so that the error of the chronometer on L.M.T. may be known as nearly as possible.

Also, every timekeeper, however experienced, is always liable to make mistakes in booking times ; when possible, therefore, there should be two timekeepers, with separate chronometers, which latter should be compared immediately after the observation.

If this cannot be done the observer should use a stop-watch as a check on the timekeeper, and the procedure would be as follows :—

At the moment that the signal is given by the observer, he should start the stop-watch ; then after the time has been booked by the timekeeper, the observer should go to the chronometer and stop his watch when the seconds hand of the chronometer has exactly reached some minute or half-minute ; the time indicated by the chronometer should then be written down, and the number of seconds shown by the stop-watch subtracted from it.

If the time thus arrived at is the same, within one or two tenths of a second, as that booked by the timekeeper, the latter may be accepted as correct.

It should be borne in mind that an ordinary stop-watch cannot be relied upon after it has been going several minutes.

If an accurate stop-watch is available, it would perhaps be simpler to start it from the chronometer at some exact minute shortly before the phenomenon is due to take place, and to then stop it at the moment the signal is given.

The hours and minutes can then be booked from the chronometer, and the seconds and fractions of a second from the stop-watch.

As regards the calculation of longitudes from occultations, there are several ways of doing this, all depending on the same principle.

The method described in the 'Connaissance des Temps,' for example, is different in form from that given in the 'Hints to Travellers,' which is taken from Raper, and an example of which is appended; the latter will be found to give good results—although formidable in appearance it is not so in reality, and depends almost entirely on accurate arithmetic.

In using it, it should be noted that it is preferable to use five places of decimals in the logs instead of four only, ternary proportional logs to five places being given in Chambers' tables.

In conclusion, the attention of the beginner might with advantage be drawn to two points in this calculation :—

(i.) Care should be taken to *reduce to date* the assumed places of stars given in the "Mean Places of Occultation Stars" in the 'Nautical Almanac,' by adding (arithmetically) the quantities $\Delta\alpha$, $\Delta\delta$ (columns 4 and 5 in "Elements of Occultations," 'Nautical Almanac') to the R.A. and declination respectively given under "Mean Places of Occultation Stars," 'Nautical Almanac.'

It will be noticed that $\Delta\alpha$ is given in seconds of time and $\Delta\delta$ in seconds of arc.

(ii.) Table XVIII. for reducing the observed latitude to geocentric latitude is based on Clarke's first figure (1858), the value for compression being $\frac{1}{294.26}$.

A compression of $\frac{1}{300}$ has been used in the following example ; the value adopted makes an appreciable difference in the resulting longitude.

In the calculation of longitude by occultations, the value for the moon's semi-diameter given on p. III, 'Nautical Almanac,' should be diminished in the proportion of 15' 34"·09 to 15' 32"·65 or 93409 : 93265. The former value is the mean semi-diameter adopted for transits, and the latter that adopted for occultations."

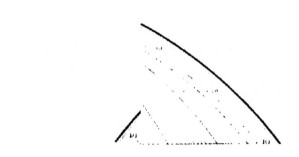

Plate XXVIII.

# COMPUTATION

OF THE

# LONGITUDE OF BUSA

FROM THE

IMMERSION OF 76 LEONIS, ON MAY 6TH, 1903.

## Computation of the Longitude of Buss from the Immersion of 76 Leonis, on May 6th, 1903.

|  |  | h. m. s. | *Error of Watch.* | | m. s. | ) 's Semi-diameter noon 6th | = 15 54·16 (p. III, 'N.A.') |
|---|---|---|---|---|---|---|---|
| Chron. time of Immersion | = | 10 4 18·8 | | | | Do. midnight | = 15 50·54 |
| Watch slow | = + | 0 10 8·0 | By α Leonis (W.) | = | 10 8·04 slow | 12-hourly diff. | 0 3·61 |
| M.T. at Place | = | 10 14 21·8 | By α Coronæ (K.) | = | 10 7·90 slow | | 0 2·96 |
| Approx. Longitude E. | = − | 0 24 7·5 | Mean | = | 10 7·97 slow | 3"·61 × 9·837 + 12 | → 0 2·96 |
| Corrected Green. Date | = | 9 50 14·3 | | | | ) 's Semi-diameter for Green. Date = 15 51·19 |

|  |  | ′ ″ |  |  | ° ′ ″ |  |
|---|---|---|---|---|---|---|
| ) 's Hor. Parall., noon, 6th May | = | 58 16·77 (p. III, 'N.A.') | Latitude | = | 14 25 24 | |
| Do. midnight, do. | = | 58 2·53 | Reduction | = | 0 5 30·9 (Table XVIII) | |
| 12-hourly difference | = | 0 13·24 | Geoc. Latitude | = | 14 19 53·1 | |
| Now 9h. 50m. 14·3s. = 9·837h. | | | | | | |
| 9·837 × 13″·24 + 12 | = | 0 10·85 | ) 's Declination at 9h. | = | 2 25 35·0 (p. VI, 'N.A.') | |
| ∴ ) 's Hor. Parall. for Green. Date | = | 58 4·92 | Correction | **= | 0 8 58·37 | **Var. in 1′ = 10″·716 |
| Correction for Latitude | = − | 0 0 0·68 (Table XVII) | ) 's Reduced Declination | = | 2 16 36·63 N. | 50′ 24 × 10·716 = 8′ 58″·37 |
| ) 's Reduced Hor. Parall. | = | 58 4·24 | | | | |

|  |  | h. m. s. |  |  | h. m. s. |  |
|---|---|---|---|---|---|---|
| | | | Star's R.A. | = | 11 13 56·16 (a) | (a) (a) " Mean Places of Occultation |
| Sid. Time of G.M.N. | = | 2 52 40·13 (p. II, 'N.A.') | Correction | = + | 0 0 1·90 (b) | Stars," 'N.A.' |
| Acceleration (Table XXXI, {9h. | = | 0 1 28·71 | | = | 11 13 58·06 | (b) (b) "Δα, Δδ," in "Elements of |
| 'Hints to Travellers') {50m. | = | 0 0 8·21 | | | | Occultation," 'N.A.' |
| {14s. | = | 0 0 0·04 | Star's Declination | = + | 2 10 56·0 (N.) (a) | |
| Reduced Sid. Time | = | 2 54 17·09 | Correction | = | 0 0 14·9 (b) | |
| M.T. at Place | = | 10 14 21·80 | | = | 2 10 41·1 | |
| R.A. of Meridian | = | 13 8 38·89 | | | | |
| Do. Star | = − | 11 13 58·06 | Star's Declination | = | 2 10 41·1 | (+, ∴ Latitude and Declination are |
| | | | Arcs (A−C) | = − | 0 14 21·6 | same name) |
| Star's Hour Angle | = | 1 54 40·83 W. | | = | 2 25 2·7 | |
| | = | 28° 40′ 12″·45 (in arc) | Arc B | = − | 0 1 52·57 | (−, ∴ Hour Angle is less than 6 hrs.) |
| | | | Prepared Declination | = | 2 23 10·13 | |

| | | | | | |
|---|---|---|---|---|---|
| ☾'s Reduced Hor. Parall. *Prolog. | = | ·49132 | | | ·49132 |
| Cosec. Geoc. Latitude | = | 10·00637 | Sec. | = | 10·01373 |
| Sec. Star's Declination | = | 10·00081 | Cosec. | = | 11·43016 |
| | | | | | 10·05680 |
| | | 1·08800 | | | 1·98201 |
| = *Prolog. of 14′ 21″·84 = Arc A | | | = *Prolog. of 1′ 52″·57 = Arc B | | |

| | | | | | |
|---|---|---|---|---|---|
| | | | | | ·49132 |
| | | | | | 10·01373 |
| | | | | Cosec. | 10·31897 |
| | | | | | ·82402 |
| | | | | | 2 |
| | | | | | 1·64804 |
| | | | Cot Star's Declination | = | 1·58800 (constant) |
| | | | | | 11·41984 |
| | | | | | 4·64988 |
| | | | = *Prolog. of 0′ 0″·24 = Arc C | | |

| | ° ′ ″ | | | | |
|---|---|---|---|---|---|
| Prepared Declination | 2 23 10·13 | Cos. | = | 9·99962 | |
| ☾'s Declination | 2 16 36·63 | Constant | = | 1·17610 | |
| Difference | 0 6 33·5 | | | ·82402 (sum of 3 logs. used to find Arc C) | |
| ☾'s Semi-diameter | 0 15 51·19 | | | 1·99974 | |
| Difference | 0 9 17·69 | ½ *Prolog. | = | ·64352 | = *Prolog. of 1′ 48″·055 Part II. |
| Sum | 0 22 24·69 | ½ *Prolog. | = | ·45240 | |
| | | | | 2·27164 | = *Prolog. of 0′ 57″·783 Part I. |

| | h. m. s. | | |
|---|---|---|---|
| Star's R.A. | = - 0 0 57 78·06 | (Part I., -, ∴ *Immersion*) | |
| | 11 13 0·277 | (Part II., +, ∴ ☾ is W. of Merid.) | |
| | + 0 1 48·055 | | |
| ☾'s R.A. | (i.) 11 14 48·342 | 9 hrs. (p. VI., 'N.A.') | |
| Do. preceding hour | (ii.) 11 12 59·82 | 10 hrs. | |
| Do. following hour | (iii.) 11 15 9·38 | | |

| | | | |
|---|---|---|---|
| Difference between (i.) and (ii.) | = 0 1 48·522 | *Prolog. | = 1·987912 |
| Do. (ii.) and (iii.) | = 0 2 9·56 | Arithmetical complement of Prolog. | = 8·079046 |
| | | Constant | ·477100 |
| | | | ·554058 |
| | | = *Prolog. of 60′ 15″·65 = z | |

| | | h. m. s. |
|---|---|---|
| Hour of (ii.) | = | 9 0 9 |
| z | = | 0 50 15·55 |
| G.M.T. of Immersion | = | 9 50 15·55 |
| L.M.T. of Immersion | = | 10 14 21·80 |
| Difference of Longitude | = | 0 24 6·25 E. |
| | = | 6° 1′ 33″·75 E. (in Arc) |

* To find the Prolog. (Proportional Logarithm) of TIME less than 3 hours or of ARC less than 3 ; proceed thus: From 4·08342 subtract the logarithm of the TIME or ARC reduced to seconds. The result will be the Prolog. required.

NOTE.—For greater accuracy the *observed* places of the moon should be obtained from Greenwich and substituted in the above; and the resulting longitude should be used at the beginning in doing the calculation a second time.

(5455)   2 c

*EXAMPLE OF PREDICTION OF OCCULTATION BY GRAPHIC METHOD.

Let us examine the circumstances of the occultation of 76 Leonis on May 6th, 1903, at Busa (a French military post in the Western Soudan).

Latitude = 14° 25′ 24″ N. (by observation).
Approx. Longitude = 24m. 7·5s. E. of Greenwich (in time).

Data:—

76 Leonis, Magnitude 6·5
$T_0$, G.M.T. of conjunction in R.A. = 9h. 26m. 58s.
Limits of latitude, 44 N. to 25 S.
$q_0 = + 0·1736$
$p' = ·5572$
$q' = - ·1845$
} 'Nautical Almanac,' "Elements of Occultations."

|  | h. | m. | s. |
|---|---|---|---|
| Assumed R.A. 1903 = | 11 | 13 | 56·16 |
| Assumed Declination, 1903 = + | 2 | 10 | 56·0 |
|  | (i.e., N.) | | |

} 'Nautical Almanac,' "Mean Places of Occultation Stars."

Reduction to Date:—

$\Delta a = + 1s. ·90$
$\Delta \delta = - 14″ ·9$
} 'Nautical Almanac,' "Elements of Occultations."

There are two diagrams to be used. Plate XXVII. is †Grant's, and from it we obtain the parallaxes in declination and R.A. of the moon, which are transferred to Plate XXVIII. (Crommelin's), this second diagram being drawn on such a scale that the parallaxes found on Plate XXVII. can be entered on it unaltered. (It is in this that the advantage of Crommelin's adaptation of Grant's method is to be found.) On the second diagram the remainder of the work is done.

Plate XXVII. represents an orthographic projection of earth (OO, being the equator). The horizontal lines are parallels of latitude and the up and down lines hour circles.

In Plate XXVIII. the circle in the centre represents the moon, the radius of which is taken as being constant and 0·2725 of the earth's radius.

**Latitude.**—It will be seen that Busa is situated well within the limits of latitude given in the data.

**Moon.**—From p. IV. of the 'Nautical Almanac' for May, 1903, it will be seen that the meridian passage of the moon is at 8h. 17m. 24s. G.M.T.

Now G.M.T. of occultation = 9h. 26m. 58s., from which it will be seen that the moon will be well up above the local horizon—in fact near, and to the W. of, the meridian.

The effect of the moon being to the W. of the meridian will be to *delay* the occultation, which may be expected to occur about 10 P.M.

**Sunlight.**—Evidently at 10 P.M. the sun will not interfere with the observation.

**Parallax.**—The effects of parallax in declination and in R.A. have now to be considered, and we come to the use of the diagrams.

On Plate XXVII. draw a horizontal line DD′, to represent the latitude of Busa, 14° 25′ 24″ N. (If the place had a south latitude, DD′ would be drawn below OO.)

Draw a straight line through C, the centre of the diagram, to F, the point on the circumference of the circle representing the moon's declination at 9 or 10 P.M. on May 6th. (F would be below OO, if the moon had a S. declination.)

**Parallax in Declination.**—Now from D′, inwards towards D, set off the moon's hour angles at $T_0$, $T_1$, $T_2$, i.e., at the G.M.T. of conjunction in R.A., one hour later and two hours

* See a pamphlet by Mr. A. C. D. Crommelin, published in 1902 by the Royal Geographical Society, entitled: "An Adaptation of Major Grant's Graphical Method of Predicting Occultations to the Elements now given in the 'Nautical Almanac.'"
† Taken from diagram by Colonel S. C. N. Grant, C.M.G., R.E.

later. (Note.—If the moon had been near the western horizon, the effects of parallax in R.A. would have been much greater, and we would have selected the hour angles at $T_1$, $T_2$, $T_3$, instead of the above. Similarly if the moon had been slightly to the *east* of the meridian, we would have substituted the hour angles at $T_{-1}$, $T_0$ and $T_1$. An indication of which hours to select is soon obtained with a little experience in predictions.)

From these points on the line DD' draw perpendiculars to CF, and call their lengths $a$, $b$, $c$.

Thus $a$, $b$, $c$ are the values of the parallax in declination at $T_0$, $T_1$ and $T_2$, and will be transferred direct, a little later, to Plate XXVIII.

**Parallax in R.A.**—For the values of parallax in R.A. set off the same hour angles along the latitude line, but this time from D towards D'; then D$d$, D$e$ and D$f$ will be the values of parallax in R.A. to be transferred to Plate XXVIII.

We have now done with Plate XXVII. It might be noted here, that if the hour angle exceeds 6 hours, in plotting the parallax in R.A. on the latitude line we first move outwards to the circumference of the circle and then count inwards again from that point, for the excess on 6 hours. Also the feet of the perpendiculars $a$, $b$, $c$, will be to the *right* of C if the moon is above the horizon, and to the *left* if it is below.

**Hour Angles.**—To find the hour angles at $T_0$, $T_1$ and $T_2$ we proceed as follows:—

|  | h. | m. | s. |
|---|---|---|---|
| 9h. 26m. 58s. ($T_0$) | 9 | 26 | 58 |
| accelerated for 9h. = | 1 | 28 | ·71 |
| accelerated for 26m. = |  | 4 | ·27 |
| accelerated for 58s. = |  |  | ·16 |
|  | 9 | 28 | 31·14 |
| Sidereal time at G.M.N. (May 6th) = | 2 | 52 | 40·13 |
| Sidereal time at Greenwich = | 12 | 21 | 11·27 |
| East longitude in time = + |  | 24 | 7·5 |
| Sidereal time at Busa = | 12 | 45 | 18·77 |
| = | 12 | 45 | 0 (nearly) |

as seconds cannot be plotted on Plate XXVII.

| Moon's R.A. at $T_0$ = | 11 | 14 | 0 (nearly) |
|---|---|---|---|
| (p. VI., 'Nautical Almanac') |  |  |  |
| ∴ ☽'s hour angle at $T_0$ = | 1 | 31 | 0 W. of the meridian. |

Now the variation in R.A. of the moon in 10 minutes at $T_0$ = 21·59s. (about).
∴ Variation in 1 hour = 129·54s. = 2m. 9·5s.
Whence ☽'s hour angle at $T_1$ = (1h. 31m.) + (1h.)* − (2m. 9·5s.) = 2h. 29m. (nearly).
Similarly ☽'s hour angle at $T_2$ = 3h. 27m. (nearly).
Next, in Plate XXVIII.:—
Make $OG_0 = q_0$†, on the vertical line through O.
Draw $G_0M$ horizontally *to the left* of $G_0$ and equal to $p'$.
Then plot $MG_1 = q'$‡

Join $G_0G_1$, and produce it both ways, marking off equal intervals $G_2G_1$, $G_1G_0$, and $G_0G_{-1}$. Then $G_0$, $G_1$, $G_2$, &c., are the geocentric places of the moon at the intervals $T_0$, $T_1$, $T_2$, &c., respectively.

* − because ☽ is W. of meridian, + if ☽ is E. of meridian.
† When $q_0$ is +, $G_0$ is above O; when $q_0$ is −, $G_0$ is below O.
‡ „ $q'$ „ +, $G_1$ „ M; „ $q'$ „ −, $G_1$ „ M.

Next, to get the *apparent* position of the moon's centre at these times as seen from the station under consideration, we plot at $G_0$, $G_1$, $G_2$, the parallaxes in declination and R.A. obtained in Plate XXVII.

When the points marked off on the latitude line in Plate XXVII. are *above* the "moon line," the parallaxes in declination are plotted *downwards* from $G_0$, $G_1$, $G_2$; when *below*, the parallaxes are plotted upwards.

Similarly, if the hour angles at $T_0$, $T_1$, $T_2$, are *east*, the parallaxes in R.A. are plotted to the left of $G_0$, $G_1$, $G_2$; if *west*, to the right.

Then $L_0$, $L_1$, $L_2$, is the *apparent* path of the moon's centre, as seen at the station.

If this line cuts the circle in the centre of Plate XXVIII., the occultation will take place, and the points I, E, correspond to the immersion and emersion of the star respectively.

To find the times at these instants, select that part of the scale (fig. 2, Plate XXVIII.), on which a distance equal to $L_0$, $L_1$ corresponds to 1 hour.

It will be seen that the distance $L_0I$ corresponds to about 22m. 20s. Whence

|   |   | h. | m. | s. |
|---|---|---|---|---|
| G, time of immersion = |   | 9 | 26 | 58 |
|   | + |   | 22 | 20 |
|   |   | 9 | 49 | 18 |
|   | + |   | 24 | 7·5 |
| and L.M.T. of immersion = |   | 10 | 13 | 25·5 (Long. E.). |

(Note the actual L.M.T. of immersion was observed to be 10h. 14m. 21·8s.)

Treating $L_1$, $L_2$, and $L_1$, E. similarly, we get the L.M.T. of emersion = 11h. 2m. 20s.

**Angles from Moon's Vertex.**—The angles on the graduated arc in Plate XXVIII., at I and E, are the angles from the moon's north point (always reckoned to the *left*) of immersion and emersion, (in the example 68° and 349°) respectively.

It will be easier in the observation to judge these angles by eye, however, if they are reckoned from the moon's vertex.

Now, at the time $T_0$, the line $L_0G_0$ passes through the zenith of the observer, and so by drawing a line through O, parallel to $L_0G_0$, we find that the vertex of the moon at $T_0$ is 60° to the *left* of the north point.

For exactness the line $G_0G_1$ should be divided proportionally (at $g$) as $L_0L_1$ is by the point I, and a line should be then drawn through O parallel to I$g$. We thus get the moon's vertex *at the instant of immersion*, which in the example is 64° to the *left* of the moon's north point.

Whence the angle of immersion of the star is 68° − 64° = 4° to the left of the moon's vertex.

Similarly, the angle of emersion is 349° − 70° = 279° to the left of the moon's vertex.

Thus the occultation will occur as shown in fig. 65. It will of course appear, if seen through the diagonal eye-piece, as shown in fig. 66.

**Moon's Limb—Illumination.**—Further, we can foretell whether the immersion and emersion will take place in or from the bright or dark limb of the moon.

On p. XII., 'Nautical Almanac' for May, 1903, under "Phases of the Moon," we find that the moon is full on the 11th. Therefore, on the 6th, about $\frac{5}{18}$ of the moon's disc will remain unilluminated, the eastern limb being the dark one.

The immersion will therefore take place in the dark limb, and the emersion from the bright. For all practical purposes this will be sufficient, but it should be noted that if the path of the star is almost tangential to the moon's disc, it is possible for *both immersion and emersion* to take place in the same limb, bright or dark.

In the field, such an occultation would not be of much value owing to the extent by which the longitude will be affected by errors in the predicted co-ordinates of the moon; however, a method is here given by which the angle that the terminator (*i.e.*, the line joining the ends of the bright limb) makes with the north point can be determined.

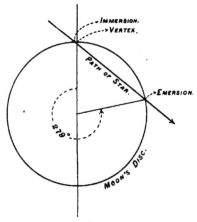

Fig. 65.

In the "Ephemeris for Physical Observations of the Moon," by Mr. Crommelin, published annually in the 'Monthly Notices' of the Royal Astronomical Society, the angle between the

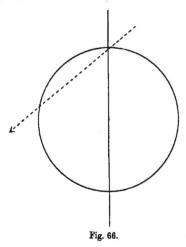

Fig. 66.

north *point* and the north *pole* of the moon is given. If this ephemeris is available we could enter on Plate XXVIII. the line joining the moon's poles, which for all practical purposes could be taken as the 'terminator'; for as the inclination of the moon's axis to the plane of the ecliptic is small, the junctions of the bright limb and dark limb will in general be near the poles, and it may be taken that the angle between them will never exceed 5°, except when the moon is very young or old, or almost full.

An exact solution can be arrived at by the following computation :—

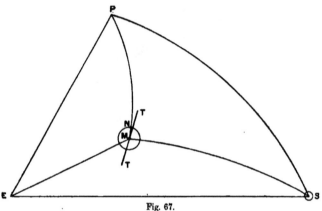

Fig. 67.

M is the moon, E the earth, S the sun, P the pole, and TT the terminator, at right angles to MS.

Required to find the angle PMT :—

Now, in the △ PMS, the two sides PM and PS (the north polar distances of the moon and sun) and the included angle MPS (difference of their R.A.) are known.

Hence the triangle can be solved and PMS, and therefore PMT found. (Note.—N is the moon's north point.)

To insert the amount of the illuminated disc on Plate XXVIII., use must be made of Crommelin's ephemeris, which gives the longitude of the terminator, or its distance from the centre of the disc.

**Moon's Altitude.**—Plate XXVII. can also be utilised to ascertain the *altitude* of the moon, at the instants of immersion and emersion as follows :—

> Measure a distance equal to $L_0$, $G_0$, vertically upwards from C, in Plate XXVII., and note what parallel of latitude is reached (26° in the example).
>
> This gives us the zenith distance of the moon, and the altitude
>
> $= 90° -$ zenith distance
>
> $= 90° - 26° = 64°$ for immersion
>
> and $= 90° - 38\frac{1}{2}° = 51\frac{1}{2}°$ for emersion.

The foregoing explanation is rather a long one, as the intention has been to provide for every example that may occur.

**Summary.**—In brief, the procedure is as follows :—

> On Plate XXVII. draw the "moon line" and the latitude line and set off on the latter the various hour angles.

Draw the perpendiculars for the parallaxes in declination.

On Plate XXVIII. plot $q_0$, $p'$, $q'$, and the parallaxes in declination, and R.A.; and we thus get $L_0$, $L_1$, $L_2$.

The computation for longitude, from the *observed* immersion of the star dealt with in the prediction, is appended. (*See* p. 194.)

## LONGITUDE BY PHOTOGRAPHS OF THE MOON.

(6) **Longitudes by Photographs of the Moon.**—A photographic camera on a stable and rigid stand is set up so that the image of the moon is about in the centre of the plate, and a series of instantaneous exposures is made, allowing such an interval between the exposures that the moon images will not overlap, *i.e.*, from $1\frac{1}{2}$ minutes to $2\frac{1}{2}$ minutes according to the moon's age.

After a set of, say, seven moon exposures, the camera is left untouched until bright stars of approximately the same declination as the moon have arrived at the same point in the heavens. The camera is then opened for periods of 15 to 30 seconds and the stars allowed to impress their trails on the plate. These star exposures should be repeated four or five times.

There should be at least the trails of two stars on the plate. The local time of each moon and star exposure must be known.

From such a negative it is clear that the right ascension of each moon image can be determined, and hence the G.M.T. and longitude. The positions of the centres of the moon images and of the star exposures are measured by a micrometer, and are referred to two co-ordinate axes approximately parallel and perpendicular to the meridian through the centre.

The method in its present form is due to Major E. H. Hills, C.M.G., R.E. Somewhat similar, though less practical, methods had been previously suggested by Dr. Schlichter in England, and Dr. Runge in Germany. For a complete discussion, see the 'Memoirs of the Royal Astronomical Society,' Vol. LIII., 1897, or 'Wilson's Topographic Surveying,' Wiley and Sons, New York.

The accuracy of the method in the field has not been satisfactorily determined; it is perhaps fair to assume that under ordinary conditions the probable error of longitude by one plate would be less than a minute of arc. If time permitted the building of an isolated pillar, this could be reduced. The actual error of the mean of five plates was found at Chatham (in 1894, 1895) to be 0·71 second or about 10 seconds of arc. The measurements and calculations are laborious, and are not readily made during an expedition.

## LONGITUDE BY MOON-CULMINATING STARS.

(7) **Longitude by Moon-Culminating Stars.**—The moon moves so rapidly in right ascension that, if at any instant we can determine its right ascension, we can from this determine the Greenwich time at that instant and hence the longitude. Its right ascension at meridian transit can be most accurately found by the interval in sidereal time between its transit and that of some neighbouring well-known star.

Such stars are called moon-culminating stars and are those which have nearly the same right ascension and declination as the moon. A list of these will be found for each day of the year in the 'Nautical Almanac' (p. 456, *et seq*).

**The observation** must be made with a transit instrument, or transit theodolite, set in the plane of the meridian. The diaphragm of the instrument should be fitted with at least three vertical wires.

To prepare for the observation, select a star from those given in the list of moon-culminating stars and note its R.A. Next determine the moon's R.A. at the moment of transit. Use an approximate value for the longitude and multiply the difference in longitude from Greenwich in hours by the variation in moon's R.A.; if longitude is E., subtract this from R.A. at Greenwich; if W., add. The result will be the approximate R.A. of the bright limb at local transit. Having the right ascensions of star and moon at transit (i.e., the local sidereal times) the mean times can be computed if working with a M.T. chronometer.

Next compute the setting for zenith distance of the star in the usual way ($\zeta = \phi - \delta$).

Having prepared this information beforehand, note carefully the sidereal times to the nearest tenth of a second of the transits of star and bright limb of the moon over each wire. From these deduce the sidereal interval between the star's transit and that of the moon's bright limb.

**Computation.**—In the 'Nautical Almanac,' the R.A. of the bright limb only is given; the R.A. of the other limb may be obtained by adding or subtracting twice the "sidereal time of passage of semi-diameter." It may be necessary to do this, since near full moon the limb changes from I to II (these figures in the 'Nautical Almanac' indicate the limbs which come first or last to the meridian); and the R.A.'s taken from the 'Nautical Almanac' must be those of the limb observed.

Take the R.A.'s of the observed limb from the 'Nautical Almanac,' put them in order and take their differences as far as the fourth difference, thus:—

|  | R.A. ☽'s bright limb. | Differences. | | | |
|---|---|---|---|---|---|
|  |  | 1st. | 2nd. | 3rd. | 4th. |
| Upper passage preceding day ... | $A_{-2}$ |  |  |  |  |
|  |  | $a_{-2}$ |  |  |  |
| Lower  ,,   ,,  ... ... | $A_{-1}$ |  | $b_{-1}$ |  |  |
|  |  | $a_{-1}$ |  | $c_{-1}$ |  |
| Upper passage corresponding day ... | $A_0$ |  | $b_0$ |  | $d_0$ |
|  |  | $a_1$ |  | $c_1$ |  |
| Lower  ,, following ... ... | $A_1$ |  | $b_1$ |  |  |
|  |  | $a_2$ |  |  |  |
| Upper  ,,   ,,  ... ... | $A_2$ |  |  |  |  |

The first difference,
$$a_{-2} = A_{-1} - A_{-2}$$
$$a_{-1} = A_0 - A_{-1}, \&c.$$

The second difference,
$$b_{-1} = a_{-1} - a_{-2}$$
$$b_0 = a_1 - a_{-1}, \&c.$$

The R.A. of the moon's bright limb at transit in W. longitude L will be

$$A_0 + na_1 + \frac{n(n-1)}{1.2}.b_0 + \frac{(n+1)(n)(n-1)}{1.2.3}c_1 + \frac{(n+1)(n)(n-1)(n-2)}{1.2.3.4}d_0,$$

where $n = \frac{L}{12}$, L being expressed in hours.

If the longitude is E., use the same formula, but take the following list of R.A.s of moon's bright limb:—

| Lower passage two days before | ... | ... | $A_{-2}$ |
| Upper passage preceding | ... | ... | $A_{-1}$ |
| Lower    ,,       ,, | ... | ... | $A_0$ |
| Upper    ,,   corresponding day | ... | $A_1$ |
| Lower    ,,   following | ... | ... | $A_2$ |

Then, if the E. longitude is L', $n = \frac{12 - L'}{12}$.

An approximate value for the longitude will be known; assume two values for the longitude, one a few (1 to 5) minutes greater than the approximate value, and one a few minutes less. Work out the R.A. of moon's bright limb for these two values, and find by proportional parts what alteration must be applied to the smaller longitude in order that the right ascension by formula may agree with the right ascension by observation.

The smaller longitude so corrected is the longitude required. *Great care must be taken to apply the signs strictly according to the rules of algebra.*

As regards the accuracy of a longitude by moon culminations, an error of a second of time produces about 30 seconds' error in the longitude. We may assume in the ordinary case an error of $\frac{1}{3}$ second and a resulting error of 10 seconds of longitude. As a rule, only about eight observations can be taken in a month, so that after a month's work the mean of the results would be about 3 or 4 seconds in error, or say nearly 1 minute of arc.

(8) **Longitude by Altitudes of the Moon.**—A fair method, but one not often used. For the details see 'Manual of Surveying for India.' An error of a second of time in an altitude will produce at least 30 seconds of time in longitude, sometimes considerably more. An error of 1 second in zenith distance produces at least 2 seconds of time in longitude. With a small theodolite we may expect an error in time of $\frac{1}{3}$ second and in arc (after four altitudes on two faces) of 5 seconds, so that the resulting error will be about $\sqrt{10^2 + 10^2}$, say 14 seconds of time.

(9) **Lunar Distances.**—A rough method; the observations cannot be taken with a theodolite.

(10) **Eclipses of Jupiter's Satellites.**—A very rough method.

These last two methods should never be used by the topographical surveyor or explorer.

NOTE.—It is of great importance that the explorer and surveyor should have clear ideas on the subject of the relative accuracy of the different methods available for determining longitude.

It should be understood that where longitude can be carried along by a triangulation, this should always be done. Excellent results also can be got by telegraphic signals; the longitude of Mandalay was thus determined by Colonel Hobday, with an error of only 1·2 seconds of arc. (Such accuracy cannot, however, be always expected.)

**Table showing the Terminal Error in Longitude which might be expected after a March of 300 miles in a Hilly Tropical Country.**

| | Method. | Probable error in longitude. |
|---|---|---|
| Relative Methods | 1. Triangulation | 100 yards to $\frac{1}{4}$ mile |
| | 2. Telegraph | $\frac{1}{8}$ to $\frac{1}{4}$ mile |
| | 3. Latitudes and azimuths (N. and S.) | $\frac{1}{2}$ mile |
| | 4. Chronometers (5) | 1 mile |
| Absolute Methods | 5. Occultation | $\frac{1}{4}$ mile |
| | 6. Moon photographs (1 plate) | 1 mile |
| | 7. Moon culminations (8 nights) | 1 mile |
| | 8. Moon altitudes (1 night) | 4 miles |
| | 9. Lunar distance | 10 miles |
| | 10. Eclipses of Jupiter's Satellites | over 10 miles |

## CHAPTER XVII.

## THE THEORY OF ERRORS AS APPLIED TO TOPOGRAPHICAL WORK.

**An error of observation** is the result of subtracting the true value from the measured value of a quantity. Errors are of three kinds:

(1) Mistakes: as for example that which results from reading a micrometer head the wrong way, and booking, say, 38 instead of 22 seconds.

(2) Systematic or constant errors: as for example that which results from using a steel tape without applying the correction for temperature; or the constant personal error of an observer in estimating the moment when a star crosses a wire in a theodolite.

(3) Errors which are neither mistakes nor systematic, but which are the combined result of a number of small causes beyond the control of the observer, and which are subject to the laws of probability. Thus in intersecting a luminous trigonometrical signal, the observer, at each intersection, tries to put the wire through the centre of the light, but the centre is an indefinite spot and no two intersections will be in precisely the same place.

Or, to take another example, suppose a base two miles long to be measured with a steel tape reading to hundredths of a foot, it is clear that it is exceedingly unlikely that a series of measurements will give exactly the same values even when all the proper corrections have been applied.

It is to this third class that the name **error** is strictly to be applied, and the following remarks deal only with such errors.

**Frequency of errors.**—The following will be found on consideration to be the laws which govern the frequency of errors:—

(1) Small errors are more frequent than large ones.

(2) Positive and negative errors are equally likely.

(3) Very large errors do not occur.

Thus in measuring a topographical base with a steel tape, it is to be expected that a measurement will be some hundredths of a foot in error, it is somewhat less likely that it will be tenths of a foot in error, and far less likely that it will be feet in error; also any measurement is just as likely to be in excess as in defect, and it is exceedingly unlikely that any measurement will be, say, ten yards in error.

From the second law it is clear that any expression giving the probability of an error of a given size will involve only even powers of the error.

It is also clear from the first and third laws that if a curve is drawn of which the abscissæ represent the sizes of the errors, and the ordinates the probability of such errors occurring, the form of the curve would be that shown in fig. 68.

If we imagine every finite error to be composed of an infinitely great number of very small equal elementary errors, which are equally likely to be positive as negative, it can be shown that $y$, the probability of the occurrence of an error $x$, can be represented by the expression:—

$$\frac{h}{\sqrt{\pi}} e^{-h^2 x^2} dx \qquad \ldots \ldots \ldots \ldots \text{(i.)},$$

where $h$ is a constant.

This expression gives a small probability for any error however large, and to this extent differs from experience. This probability of a large error derived from this equation is so

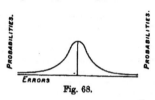

Fig. 68.

small as to be negligible, and results from the hypothesis that finite errors are the result of the algebraic sum of an indefinitely large number of very small elementary errors.

The probability that an error lies between two limits $x_1$, $x_2$, is*

$$\frac{h}{\sqrt{\pi}} \int_{x_1}^{x_2} e^{-h^2 x^2} dx \quad \ldots \ldots \ldots \ldots \text{(ii.).}$$

**Residuals.**—If a number of measurements are made of the same quantity, the sum of the measured values divided by the number of measurements is the **mean**, and a **residual** is the result of subtracting the mean from a measured value.

With a large number of observations the mean tends to become the true value, and the residuals to become errors.

**Probable Error.**—In an indefinitely large series of errors, the **probable error** is that error than which there are as many errors greater as there are smaller; *i.e.*, the chance that any error taken at random is greater than the probable error is the same as that it is smaller.

**The Probable Error of the Arithmetic Mean** of a series of observations

$$= 0 \cdot 674 \sqrt{\frac{\Sigma r^2}{n(n-1)}},$$

when $n$ is the number of observation, and $\Sigma r^2$ is the sum of the squares of the residuals.

The probable error is a valuable means of comparing the accuracy of sets of observations of the same class. Thus suppose two observers, A and B, have measured the same angle with the same instrument and have obtained the following results (seconds only are given):—

| A. | " | $r$ | $r^2$ | B. | " | $r$ | $r^2$ |
|---|---|---|---|---|---|---|---|
| | 47 | 0 | 0 | | 48 | 3 | 9 |
| | 49 | 2 | 4 | | 46 | 1 | 1 |
| | 50 | 3 | 9 | | 49 | 4 | 16 |
| | 42 | 5 | 25 | | 41 | 4 | 16 |
| | 48 | 1 | 1 | | 47 | 2 | 4 |
| | 46 | 1 | 1 | | 39 | 6 | 36 |
| Mean | 47 | $\Sigma r^2 = 40$ | | | 45 | $\Sigma r^2 = 82$ | |

The probable error of the mean of A's set is

$$\cdot 67 \sqrt{\tfrac{40}{30}} = \cdot 78 \text{ and of B's set is } \cdot 67 \sqrt{\tfrac{82}{30}} = 1 \cdot 11.$$

A's set is thus the most reliable.

* Note.—For the investigation of this expression and of the Theory of Errors generally, see 'Text Book on the Method of Least Squares.' Merriman.—J. Wiley & Sons, New York.

The means would be expressed as

$$47'' \pm 0.78 \text{ and } 45'' \pm 1.11.$$

Owing to the fact that it is very difficult to free observations of any kind from constant or systematic errors, the probable error obtained in this way is better considered as a test of the relative value of similar observations than as a statement of the amount by which the mean of a series is probably in error. Thus the probable error of angles in a geodetic triangulation derived from the inter-agreement of observed values of the angles is generally smaller than that derived from a consideration of the triangular errors.

**Law of the Propagation of Error.**—*The probable error of the sum or difference of two quantities is equal to the square root of the sum of the squares of their probable errors*; and so on for any number of quantities. This rule is of frequent use in all kinds of survey work.

As an example, suppose that the probable error of a carefully measured base is $\pm 1.2$ inches, and that subsequently an extension is measured with a probable error of $\pm 0.7$ inch, then the probable error of the whole base will be $\sqrt{(1.2^2 + .7^2)} = 1.4$.

The probable error of the arithmetic mean = probable error of a single observation divided by the square root of the number of observations.

*The probable error of the mean therefore varies inversely as the square root of the number of observations.*

Thus suppose the probable error of a determination of longitude by a single moon culmination is 3 miles; the probable error of the mean of the result of 9 culminations will be 1 mile.

If the probable error of longitude from 1 lunar distance is 10 miles, it will require 100 lunar distances to obtain a result with a probable error of 1 mile.

**Examples :—**

(1) *Latitudes and Azimuths.*—Suppose that it is found impossible to measure a base directly, and it is decided to use one determined by latitudes at two stations and the observed azimuth of the ray between them. It is required that this base shall have a probable error not greater than $\frac{1}{500}$. The azimuth of the ray is about 45°, how long must the ray be?

The probable error of each observed latitude is about 1″, say; the probable error due to local attraction is about 2″; the probable error of the difference of the two latitudes will therefore be

$$\sqrt{(1 + 1 + 4 + 4)} = 3''.16 = 316 \text{ feet (about)},$$

and as the azimuth of the ray is 45°, this will produce a probable error in the length of the ray of $316 \times \sqrt{2}$ feet = 442 feet, and if the probable error of the length of the ray is not to exceed $\frac{1}{500}$, the ray must not be less than $500 \times 442$ feet long, or say 40 miles.

(2) *Long Traverses.*—American experience shows that the longitudinal error of a long traverse under favourable conditions is not greater than $\frac{1}{3000}$. Assuming that this error in the unfavourable conditions of the Gold Coast may amount to $\frac{1}{1000}$, and that a traverse is run approximately north and south, how long should the traverse be in order to obtain a check from latitude observations? The probable error of an observation for latitude with a zenith telescope may be put down as 0″.1.

The probable error due to local attraction is about 2″.0.

Total probable error in difference of latitude

$$= \sqrt{(.01 + .01 + 4.0 + 4.0)} = 2.83 \text{ or } 283 \text{ feet.}$$

If the latitudes are to be any check on the traverse, the probable error of their difference must not exceed that of the traverse, and the traverse must therefore not be shorter than 283,000 feet, or 54 miles.

**Probable Errors of a Triangulation.**—If the probable error of each angle of a triangle is $p$, then the probable error of the sum of the three angles is $p\sqrt{3}$.

If the triangular error is $e$, the probable error of an angle is $\dfrac{e}{\sqrt{3}}$.

In N triangles the probable error of an angle is $\cdot 67 \sqrt{\dfrac{\Sigma e^2}{3N}}$, since the probable value of a triangular error is $\cdot 67 \sqrt{\dfrac{\Sigma e^2}{N}}$.

*Example.*—The following were the triangular errors of a chain of secondary triangles. What is the probable error of an angle?

Triangular errors ($e$):—

| " | $e^2$ |
|---|---|
| 2 | 4 |
| 5 | 25 |
| 3 | 9 |
| 4 | 16 |
| 6 | 36 |
| 1 | 1 |
| 4 | 16 |
| 3 | 9 |
| 2 | 4 |
| 6 | 36 |
| 4 | 16 |
| 0 | 0 |
| 5 | 25 |
| 7 | 49 |
| | $\Sigma e^2 = 246$ |

The probable value of a triangular error $= \cdot 67 \sqrt{\dfrac{246}{14}}$, and the probable error of an angle $= \cdot 67 \sqrt{\dfrac{246}{42}} = 1''\cdot 62$.

**Adjustment of the Errors in a Net-work of Long Traverses.**

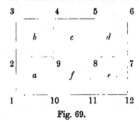

Fig. 69.

Let fig. 69 be a diagrammatic sketch of a net-work of traverses, each line being 50 to 70 miles long.

Each traverse circuit is given a letter, $a$, $b$, $c$, &c.
Each latitude station has a number, 1, 2, 3, &c.
Including the effect of local attraction, each observed latitude has a probable error of about $1'' \cdot 8$.

*First, adjust the traverse errors, leaving out of consideration the observed latitudes.*
Thus:—

> For each traverse circuit such as $a$, we have, by computing round to the starting point, a closing error in the northings and southings, and one in the eastings and westings.
>
> Give each portion of the traverse from one junction point to the next an error $e$ in the northings and southings, thus, $e_{10-2}$, $e_{2-9}$, &c.

Then in each case, such as $a$,

$$e_{10-2} + e_{2-9} + e_{9-10} = m, \text{ a known quantity,}$$

and so on for each traverse circuit, so that there will be $f$ equations of this form, and by the ordinary method of least squares solve so that $e^2_{10-2} + e^2_{2-9} + \ldots$ is a minimum.

Do the same with the eastings and westings.

The result will be the most probable consistent solution of the problem, so far only as concerns the traverse adjustment, leaving out of consideration the observed latitudes.

*Next, with some initial latitude, say of 1, deduce latitudes of all the junction points.* Compare these with the astronomical (observed) latitudes. Now, by inspection, we can probably tell:—

(1) Whether there are any large local attractions, shown by abrupt discordance at a point surrounded by concordances.

(2) Whether a northing in an E. to W. traverse is more or less accurate than a northing in a N. to S. traverse.

It should now be possible to weight the traverse amplitudes as compared with the astronomical amplitudes.

Then, starting afresh, we have for any line, such as 9-10:

> Error of observed latitude 9 − error of observed latitude 10
> − Error in traverse northing 9-10 = a known quantity,

and so on for the other lines, added to the condition that the sum of the "weighted squares" of the errors is to be a minimum.

NOTE.—It is probable that this brief sketch of a particular case of an adjustment of errors by the method of least squares will not be very intelligible to those who have not studied the subject before. It is, however, impossible, even in a book devoted to topographical surveying, to omit all mention of the subject, which enters largely into the highest field of surveying, viz., geodesy. And it is very desirable that at least some of the main principles of the theory of errors should be appreciated by the topographer.

The following books may be studied:—

"Treatise on the Adjustment of Observations." Wright. New York. 1884.
"Method of Least Squares." Merriman. J. Wiley & Sons. New York. 1884.
"Geodesy." Col. A. R. Clarke, C.B., R.E., F.R.S. Clarendon Press. 1880.

Examples of the application of the theory will be found in—

The various volumes of the "G.T. Survey of India."
"Report on the Geodetic Survey of S. Africa," by Sir David Gill, K.C.B., F.R.S., 1896.

# TABLES.

## Summary of Constants for Tables I., II., IX. to XIII., XVII. and XVIII.
### Based on Clarke's First Figure (1858).

$a$ = Equatorial Radius = 20,926,348 feet.
$b$ = Polar Radius = 20,855,233 feet.
$c$ = Compression or Ellipticity = $\dfrac{a-b}{a}$ = $\dfrac{1}{294\cdot 26}$.
$e$ = Eccentricity = $\left\{\dfrac{a^2 - b^2}{a^2}\right\}^{\frac{1}{2}}$.
$\nu$ = Normal to meridian at latitude $\lambda$ terminated by minor axis (or radius of curvature at right angles to meridian).
   = $\dfrac{a}{(1 - e^2 \sin^2 \lambda)^{\frac{1}{2}}}$.
$\rho$ = Radius of curvature to meridian at latitude $\lambda$
   = $\dfrac{a(1 - e^2)}{(1 - e^2 \sin^2 \lambda)^{\frac{3}{2}}}$.

NOTE.—In Clarke's later spheroid (1880) the values of $a$ and $b$ only differ from those given above by 146 feet and 338 feet respectively. These differences are so small that Tables based on Clarke's 1880 figure practically give the same results. The radii employed by the Survey of India differ from Clarke's by 3270 feet and 1520 feet respectively.

---

**TABLE I.**—Logarithms for facilitating the Computation of Terrestrial Latitudes, Longitudes, and Reverse Azimuths.

| LATITUDE | P $\dfrac{1}{\rho \sin 1''}$ | | Q $\dfrac{\rho}{\nu}$ | | R $\dfrac{1}{2\rho}$ | | S $2 \sec \lambda \dfrac{\rho}{\nu}$ | | T $\left(2 \tan^2 \lambda + \dfrac{\nu}{\rho}\right) \times \left(\dfrac{\cot \lambda \cos \lambda}{2}\right)$ | |
|---|---|---|---|---|---|---|---|---|---|---|
| ° ′ | | Diff. | | Diff. | | Diff. | | Diff. | | Diff. |
| 0  0 | − 3·9966885 | − 1 | − 1·9970432 | + 0 | 8·38123 | − 0 | 0·29807 | + 0 | Infinite | − Infin. |
| 10 | 6884 | 1 | 0432 | 1 | 23 | 0 | 807 | 1 | 2·23820 | 30101 |
| 20 | 6883 | 2 | 0433 | 1 | 23 | 0 | 808 | 1 | 1·93719 | 17607 |
| 30 | 6881 | 2 | 0434 | 2 | 23 | 0 | 809 | 1 | 76112 | 12491 |
| 40 | 6879 | 4 | 0436 | 2 | 23 | 0 | 810 | 2 | 63621 | 9687 |
| 50 | 6875 | 4 | 0438 | 3 | 23 | 0 | 812 | 2 | 53934 | 7914 |
| 1  0 | − 3·9966871 | 5 | − 1·9970441 | 3 | 8·38123 | 0 | 0·29814 | 2 | 1·46020 | 6689 |
| 10 | 6866 | 5 | 0444 | 4 | 23 | 0 | 16 | 3 | 39331 | 5793 |
| 20 | 6861 | 6 | 0448 | 4 | 23 | 0 | 19 | 3 | 33538 | 5108 |
| 30 | 6855 | 8 | 0452 | 5 | 23 | 0 | 22 | 4 | 28430 | 4568 |
| 40 | 6847 | 7 | 0457 | 5 | 23 | 0 | 26 | 4 | 23862 | 4130 |
| 50 | 6840 | 9 | 0462 | 6 | 23 | 0 | 30 | 4 | 19732 | 8769 |

## TABLE I.—Logarithms for facilitating the Computation of

| Latitude. | P | Diff. − | Q | Diff. + | R | Diff. − | S | Diff. + | T | Diff. − |
|---|---|---|---|---|---|---|---|---|---|---|
| ° ′ | | | | | | | | | | |
| 2 0 | 3·9966831 | | 1·9970468 | | 8·38123 | | 0·29834 | | 1·15963 | |
| | | 9 | | 6 | | 0 | | 5 | | 3466 |
| 10 | 6822 | 10 | 0474 | 7 | 23 | 0 | 39 | 5 | 12497 | 3207 |
| 20 | 6812 | 11 | 0481 | 7 | 23 | 0 | 44 | 5 | 09290 | 2984 |
| 30 | 6801 | 12 | 0488 | 8 | 23 | 1 | 49 | 6 | 06306 | 2790 |
| 40 | 6789 | 12 | 0496 | 8 | 22 | 0 | 55 | 6 | 03516 | 2619 |
| 50 | 6777 | 13 | 0504 | 9 | 22 | 0 | 61 | 7 | 00897 | 2468 |
| 3 0 | 3·9966764 | 14 | 1·9970513 | 9 | 8·38122 | 0 | 0·29868 | 7 | 0·98429 | 2333 |
| 10 | 6750 | 15 | 0522 | 10 | 22 | 0 | 75 | 7 | 96096 | 2211 |
| 20 | 6735 | 15 | 0532 | 10 | 22 | 0 | 82 | 7 | 93885 | 2102 |
| 30 | 6720 | 16 | 0542 | 10 | 22 | 0 | 89 | 9 | 91783 | 2002 |
| 40 | 6704 | 17 | 0552 | 12 | 22 | 1 | 98 | 8 | 89781 | 1912 |
| 50 | 6687 | 17 | 0564 | 11 | 21 | 0 | 0·29906 | 9 | 87869 | 1828 |
| 4 0 | 3·9966670 | 19 | 1·9970575 | 13 | 8·38121 | 0 | 0·29915 | 9 | 0·86041 | 1752 |
| 10 | 6651 | 19 | 0588 | 12 | 21 | 0 | 24 | 9 | 84289 | 1683 |
| 20 | 6632 | 19 | 0600 | 13 | 21 | 0 | 33 | 10 | 82606 | 1617 |
| 30 | 6613 | 21 | 0613 | 14 | 21 | 1 | 43 | 10 | 80989 | 1556 |
| 40 | 6592 | 21 | 0627 | 14 | 20 | 0 | 53 | 11 | 79433 | 1500 |
| 50 | 6571 | 22 | 0641 | 15 | 20 | 0 | 64 | 11 | 77933 | 1448 |
| 5 0 | 3·9966549 | 23 | 1·9970656 | 15 | 8·38120 | 0 | 0·29975 | 12 | 0·76485 | 1399 |
| 10 | 6526 | 23 | 0671 | 16 | 20 | 0 | 87 | 11 | 75086 | 1352 |
| 20 | 6503 | 24 | 0687 | 16 | 20 | 1 | 98 | 12 | 73734 | 1309 |
| 30 | 6479 | 25 | 0703 | 16 | 19 | 0 | 0·30010 | 13 | 72425 | 1269 |
| 40 | 6454 | 26 | 0719 | 17 | 19 | 0 | 23 | 13 | 71156 | 1230 |
| 50 | 6428 | 26 | 0736 | 18 | 19 | 0 | 36 | 13 | 69926 | 1194 |
| 6 0 | 3·9966402 | 27 | 1·9970754 | 18 | 8·38119 | 1 | 0·30049 | 14 | 0·68732 | 1160 |
| 10 | 6375 | 28 | 0772 | 19 | 18 | 0 | 63 | 14 | 67572 | 1127 |
| 20 | 6347 | 29 | 0791 | 19 | 18 | 0 | 77 | 14 | 66445 | 1096 |
| 30 | 6318 | 29 | 0810 | 19 | 18 | 1 | 91 | 15 | 65349 | 1067 |
| 40 | 6289 | 30 | 0829 | 20 | 17 | 0 | 0·30106 | 15 | 64282 | 1039 |
| 50 | 6259 | 31 | 0849 | 21 | 17 | 0 | 21 | 16 | 63243 | 1013 |

## Terrestrial Latitudes, Longitudes, and Reverse Azimuths.

| LATI-TUDE. | P | Diff. − | Q | Diff. + | R | Diff. − | S | Diff. + | T | Diff. − |
|---|---|---|---|---|---|---|---|---|---|---|
| ° ′ | | | | | | | | | | |
| 7 0 | 3·9966228 | | 1·9970870 | | 8·38117 | | 0·30137 | | 0·62230 | |
| | | 31 | | 21 | | 1 | | 16 | | 986 |
| 10 | 6197 | | 0891 | | 16 | | 53 | | 61244 | |
| | | 32 | | 21 | | 0 | | 16 | | 963 |
| 20 | 6165 | | 0912 | | 16 | | 69 | | 60281 | |
| | | 33 | | 22 | | 0 | | 16 | | 939 |
| 30 | 6132 | | 0934 | | 16 | | 85 | | 59342 | |
| | | 34 | | 23 | | 1 | | 18 | | 917 |
| 40 | 6098 | | 0957 | | 15 | | 0·30203 | | 58425 | |
| | | 34 | | 22 | | 0 | | 17 | | 896 |
| 50 | 6064 | | 0979 | | 15 | | 20 | | 57529 | |
| | | 35 | | 24 | | 0 | | 18 | | 875 |
| 8 0 | 3·9966029 | | 1·9371003 | | 8·38115 | | 0·30238 | | 0·56654 | |
| | | 36 | | 24 | | 1 | | 18 | | 856 |
| 10 | 5993 | | 1027 | | 14 | | 56 | | 55798 | |
| | | 37 | | 24 | | 0 | | 18 | | 837 |
| 20 | 5956 | | 1051 | | 14 | | 74 | | 54961 | |
| | | 37 | | 25 | | 0 | | 19 | | 819 |
| 30 | 5919 | | 1076 | | 14 | | 93 | | 54112 | |
| | | 38 | | 25 | | 1 | | 20 | | 801 |
| 40 | 5881 | | 1101 | | 13 | | 0·30313 | | 53341 | |
| | | 39 | | 26 | | 0 | | 19 | | 784 |
| 50 | 5842 | | 1127 | | 13 | | 32 | | 52557 | |
| | | 39 | | 26 | | 0 | | 21 | | 769 |
| 9 0 | 3·9965803 | | 1·9971153 | | 8·38113 | | 0·30353 | | 0·51788 | |
| | | 40 | | 27 | | 1 | | 20 | | 752 |
| 10 | 5763 | | 1180 | | 12 | | 373 | | 51036 | |
| | | 41 | | 27 | | 0 | | 21 | | 737 |
| 20 | 5722 | | 1207 | | 12 | | 394 | | 50299 | |
| | | 41 | | 28 | | 1 | | 21 | | 723 |
| 30 | 5681 | | 1235 | | 11 | | 415 | | 49576 | |
| | | 43 | | 28 | | 0 | | 22 | | 709 |
| 40 | 5638 | | 1263 | | 11 | | 437 | | 48867 | |
| | | 43 | | 29 | | 1 | | 22 | | 695 |
| 50 | 5595 | | 1292 | | 10 | | 459 | | 48172 | |
| | | 43 | | 29 | | 0 | | 22 | | 682 |
| 10 0 | 3·9965552 | | 1·9971321 | | 8·38110 | | 0·30481 | | 0·47490 | |
| | | 45 | | 29 | | 0 | | 23 | | 669 |
| 10 | 5507 | | 1350 | | 10 | | 504 | | 46821 | |
| | | 45 | | 30 | | 1 | | 23 | | 656 |
| 20 | 5462 | | 1380 | | 09 | | 527 | | 46165 | |
| | | 45 | | 31 | | 0 | | 23 | | 645 |
| 30 | 5417 | | 1411 | | 09 | | 550 | | 45520 | |
| | | 47 | | 31 | | 1 | | 24 | | 633 |
| 40 | 5370 | | 1442 | | 08 | | 574 | | 44887 | |
| | | 47 | | 31 | | 0 | | 25 | | 621 |
| 50 | 5323 | | 1473 | | 08 | | 599 | | 44266 | |
| | | 48 | | 32 | | 1 | | 24 | | 611 |
| 11 0 | 3·9965275 | | 1·9971505 | | 8·38107 | | 0·30623 | | 0·43655 | |
| | | 48 | | 32 | | 0 | | 25 | | 600 |
| 10 | 5227 | | 1537 | | 07 | | 648 | | 43055 | |
| | | 49 | | 33 | | 1 | | 26 | | 589 |
| 20 | 5178 | | 1570 | | 06 | | 674 | | 42466 | |
| | | 50 | | 33 | | 0 | | 26 | | 579 |
| 30 | 5128 | | 1603 | | 06 | | 700 | | 41887 | |
| | | 51 | | 34 | | 1 | | 26 | | 570 |
| 40 | 5077 | | 1637 | | 05 | | 726 | | 41317 | |
| | | 51 | | 34 | | 0 | | 27 | | 560 |
| 50 | 5026 | | 1671 | | 05 | | 753 | | 40757 | |
| | | 52 | | 35 | | 1 | | 27 | | 550 |

(5455)　　　　　　　　　　　　　　　　　　　　　　　　　　　　2 E

## TABLE I.—Logarithms for facilitating the Computation of

| LATI-TUDE. | | P | Diff. − | Q | Diff. + | R | Diff. − | S | Diff. + | T | Diff. − |
|---|---|---|---|---|---|---|---|---|---|---|---|
| ° | ′ | | | | | | | | | | |
| 12 | 0 | 3·9964074 | | 1·9971706 | | 8·38104 | | 0·30780 | | 0·40207 | |
| | | | 53 | | 35 | | 0 | | 27 | | 542 |
| | 10 | 4921 | | 1741 | | 04 | | 807 | | 39665 | |
| | | | 53 | | 36 | | | 1 | | 28 | | 532 |
| | 20 | 4868 | | 1777 | | 03 | | 835 | | 39133 | |
| | | | 54 | | 36 | | | 0 | | 28 | | 524 |
| | 30 | 4814 | | 1813 | | 03 | | 863 | | 38609 | |
| | | | 55 | | 36 | | | 1 | | 29 | | 516 |
| | 40 | 4759 | | 1849 | | 02 | | 892 | | 38093 | |
| | | | 55 | | 37 | | | 0 | | 28 | | 507 |
| | 50 | 4704 | | 1886 | | 02 | | 920 | | 37586 | |
| | | | 56 | | 37 | | | 1 | | 30 | | 500 |
| 13 | 0 | 3·9964648 | | 1·9971923 | | 8·38101 | | 0·30950 | | 0·37086 | |
| | | | 57 | | 38 | | | 1 | | 30 | | 492 |
| | 10 | 4591 | | 1961 | | 00 | | 980 | | 36594 | |
| | | | 57 | | 39 | | | 0 | | 30 | | 484 |
| | 20 | 4534 | | 2000 | | 00 | | 0·31010 | | 36110 | |
| | | | 58 | | 38 | | | 1 | | 30 | | 477 |
| | 30 | 4476 | | 2038 | | 8·38099 | | 040 | | 35633 | |
| | | | 59 | | 39 | | | 0 | | 31 | | 469 |
| | 40 | 4417 | | 2077 | | 99 | | 071 | | 35164 | |
| | | | 60 | | 40 | | | 1 | | 31 | | 462 |
| | 50 | 4357 | | 2117 | | 98 | | 102 | | 34702 | |
| | | | 60 | | 40 | | | 1 | | 32 | | 455 |
| 14 | 0 | 3·9964297 | | 1·9972157 | | 8·38097 | | 0·31134 | | 0·34247 | |
| | | | 60 | | 40 | | | 0 | | 32 | | 448 |
| | 10 | 4237 | | 2197 | | 97 | | 166 | | 33799 | |
| | | | 62 | | 41 | | | 1 | | 33 | | 442 |
| | 20 | 4175 | | 2238 | | 96 | | 199 | | 33357 | |
| | | | 62 | | 42 | | | 0 | | 33 | | 435 |
| | 30 | 4113 | | 2280 | | 96 | | 232 | | 32922 | |
| | | | 62 | | 41 | | | 1 | | 33 | | 429 |
| | 40 | 4051 | | 2321 | | 95 | | 265 | | 32493 | |
| | | | 64 | | 43 | | | 1 | | 34 | | 422 |
| | 50 | 3987 | | 2364 | | 94 | | 299 | | 32071 | |
| | | | 64 | | 42 | | | 0 | | 34 | | 416 |
| 15 | 0 | 3·9963923 | | 1·9972406 | | 8·38094 | | 0·31333 | | 0·31655 | |
| | | | 64 | | 44 | | | 1 | | 34 | | 410 |
| | 10 | 3859 | | 2450 | | 93 | | 367 | | 31245 | |
| | | | 66 | | 43 | | | 1 | | 35 | | 404 |
| | 20 | 3793 | | 2493 | | 92 | | 402 | | 30841 | |
| | | | 66 | | 44 | | | 0 | | 35 | | 399 |
| | 30 | 3727 | | 2537 | | 92 | | 437 | | 30442 | |
| | | | 66 | | 44 | | | 1 | | 36 | | 393 |
| | 40 | 3661 | | 2581 | | 91 | | 473 | | 30049 | |
| | | | 67 | | 45 | | | 1 | | 36 | | 388 |
| | 50 | 3594 | | 2626 | | 90 | | 509 | | 29661 | |
| | | | 68 | | 45 | | | 0 | | 37 | | 383 |
| 16 | 0 | 3·9963526 | | 1·9972671 | | 8·38090 | | 0·31546 | | 0·29278 | |
| | | | 69 | | 46 | | | 1 | | 36 | | 377 |
| | 10 | 3457 | | 2717 | | 89 | | 582 | | 28901 | |
| | | | 39 | | 46 | | | 1 | | 38 | | 371 |
| | 20 | 3388 | | 2763 | | 88 | | 620 | | 28530 | |
| | | | 70 | | 47 | | | 0 | | 37 | | 367 |
| | 30 | 3318 | | 2810 | | 88 | | 657 | | 28163 | |
| | | | 70 | | 46 | | | 1 | | 38 | | 361 |
| | 40 | 3248 | | 2856 | | 87 | | 695 | | 27802 | |
| | | | 71 | | 48 | | | 1 | | 39 | | 357 |
| | 50 | 3177 | | 2904 | | 86 | | 734 | | 27445 | |
| | | | 72 | | 48 | | | 0 | | 39 | | 351 |

Terrestrial Latitudes, Longitudes, and Reverse Azimuths.

| Latitude | P | Diff. − | Q | Diff. + | R | Diff. − | S | Diff. + | T | Diff. − |
|---|---|---|---|---|---|---|---|---|---|---|
| ° ′ | | | | | | | | | | |
| 17 0 | 3·9963105 | | 1·9972952 | | 8·38056 | | 0·31773 | | 0·27094 | |
|  | | 72 | | 48 | | 1 | | 39 | | 347 |
| 10 | 3033 | | 3000 | | 85 | | 812 | | 26747 | |
|  | | 73 | | 48 | | 1 | | 40 | | 342 |
| 20 | 2960 | | 3048 | | 84 | | 852 | | 26405 | |
|  | | 73 | | 49 | | 1 | | 40 | | 338 |
| 30 | 2887 | | 3097 | | 83 | | 892 | | 26067 | |
|  | | 74 | | 50 | | 0 | | 41 | | 333 |
| 40 | 2813 | | 3147 | | 83 | | 933 | | 25734 | |
|  | | 75 | | 50 | | 1 | | 40 | | 329 |
| 50 | 2738 | | 3197 | | 82 | | 973 | | 25405 | |
|  | | 75 | | 50 | | 1 | | 42 | | 324 |
| 18 0 | 3·9962663 | | 1·9973247 | | 8·38081 | | 0·32015 | | 0·25081 | |
|  | | 76 | | 50 | | 1 | | 42 | | 320 |
| 10 | 2587 | | 3297 | | 80 | | 057 | | 24761 | |
|  | | 77 | | 51 | | 0 | | 42 | | 316 |
| 20 | 2510 | | 3348 | | 80 | | 099 | | 24445 | |
|  | | 77 | | 52 | | 1 | | 42 | | 312 |
| 30 | 2433 | | 3400 | | 79 | | 141 | | 24133 | |
|  | | 78 | | 52 | | 1 | | 43 | | 307 |
| 40 | 2355 | | 3452 | | 78 | | 184 | | 23826 | |
|  | | 78 | | 52 | | 1 | | 44 | | 304 |
| 50 | 2277 | | 3504 | | 77 | | 228 | | 23522 | |
|  | | 79 | | 52 | | 1 | | 44 | | 300 |
| 19 0 | 3·9962198 | | 1·9973556 | | 8·38076 | | 0·32272 | | 0·23222 | |
|  | | 80 | | 54 | | 0 | | 44 | | 295 |
| 10 | 2118 | | 3610 | | 76 | | 316 | | 22927 | |
|  | | 80 | | 53 | | 1 | | 44 | | 292 |
| 20 | 2038 | | 3663 | | 75 | | 360 | | 22635 | |
|  | | 80 | | 54 | | 1 | | 45 | | 289 |
| 30 | 1958 | | 3717 | | 74 | | 405 | | 22346 | |
|  | | 82 | | 54 | | 1 | | 46 | | 284 |
| 40 | 1876 | | 3771 | | 73 | | 451 | | 22062 | |
|  | | 81 | | 55 | | 1 | | 46 | | 281 |
| 50 | 1795 | | 3826 | | 72 | | 497 | | 21781 | |
|  | | 83 | | 54 | | 0 | | 46 | | 278 |
| 20 0 | 3·9961712 | | 1·9973880 | | 8·38072 | | 0·32543 | | 0·21503 | |
|  | | 83 | | 56 | | 1 | | 47 | | 273 |
| 10 | 1629 | | 3936 | | 71 | | 590 | | 21230 | |
|  | | 83 | | 56 | | 1 | | 47 | | 271 |
| 20 | 1546 | | 3992 | | 70 | | 637 | | 20959 | |
|  | | 85 | | 56 | | 1 | | 48 | | 267 |
| 30 | 1461 | | 4048 | | 69 | | 685 | | 20692 | |
|  | | 84 | | 56 | | 1 | | 48 | | 263 |
| 40 | 1377 | | 4104 | | 68 | | 733 | | 20429 | |
|  | | 86 | | 57 | | 1 | | 48 | | 261 |
| 50 | 1291 | | 4161 | | 67 | | 781 | | 20168 | |
|  | | 85 | | 57 | | 0 | | 49 | | 257 |
| 21 0 | 3·9961206 | | 1·9974218 | | 8·38067 | | 0·32830 | | 0·19911 | |
|  | | 87 | | 58 | | 1 | | 49 | | 253 |
| 10 | 1119 | | 4276 | | 66 | | 879 | | 19658 | |
|  | | 87 | | 58 | | 1 | | 50 | | 251 |
| 20 | 1032 | | 4334 | | 65 | | 929 | | 19407 | |
|  | | 87 | | 58 | | 1 | | 50 | | 248 |
| 30 | 0945 | | 4392 | | 64 | | 979 | | 19159 | |
|  | | 88 | | 59 | | 1 | | 51 | | 244 |
| 40 | 0857 | | 4451 | | 63 | | 0·33030 | | 18915 | |
|  | | 89 | | 59 | | 1 | | 51 | | 243 |
| 50 | 0768 | | 4510 | | 62 | | 081 | | 18672 | |
|  | | 89 | | 59 | | 1 | | 51 | | 239 |

(5455)          2 z 2

## TABLE I.—Logarithms for facilitating the Computation of

| Lati-tude. | P | Diff − | Q | Diff + | R | Diff − | S | Diff + | T | Diff − |
|---|---|---|---|---|---|---|---|---|---|---|
| ° ′ | | | | | | | | | | |
| 22  0 | 3·9930679 |    | 1·9974569 |    | 8·38061 |   | 0·33132 |    | 0·18434 |     |
|       |           | 90 |           | 60 |         | 1 |         | 52 |         | 235 |
|    10 |    0589   |    |    4629   |    |    60   |   |    184  |    |   18199 |     |
|       |           | 90 |           | 60 |         | 0 |         | 52 |         | 233 |
|    20 |    0499   |    |    4689   |    |    60   |   |    236  |    |   17966 |     |
|       |           | 91 |           | 61 |         | 1 |         | 53 |         | 230 |
|    30 |    0408   |    |    4750   |    |    59   |   |    289  |    |   17736 |     |
|       |           | 91 |           | 60 |         | 1 |         | 53 |         | 226 |
|    40 |    0317   |    |    4810   |    |    58   |   |    342  |    |   17510 |     |
|       |           | 92 |           | 62 |         | 1 |         | 54 |         | 225 |
|    50 |    0225   |    |    4872   |    |    57   |   |    396  |    |   17285 |     |
|       |           | 92 |           | 61 |         | 1 |         | 54 |         | 221 |
| 23  0 | 3·9960133 |    | 1·9974933 |    | 8·38056 |   | 0·33450 |    | 0·17064 |     |
|       |           | 93 |           | 62 |         | 1 |         | 54 |         | 219 |
|    10 |    0040   |    |    4995   |    |    55   |   |    504  |    |   16845 |     |
|       |           | 93 |           | 62 |         | 1 |         | 55 |         | 216 |
|    20 | 3·9959947 |    |    5057   |    |    54   |   |    559  |    |   16629 |     |
|       |           | 94 |           | 63 |         | 1 |         | 55 |         | 213 |
|    30 |    9853   |    |    5120   |    |    53   |   |    614  |    |   16416 |     |
|       |           | 94 |           | 63 |         | 1 |         | 56 |         | 211 |
|    40 |    9759   |    |    5183   |    |    52   |   |    670  |    |   16205 |     |
|       |           | 95 |           | 63 |         | 1 |         | 56 |         | 208 |
|    50 |    9664   |    |    5246   |    |    51   |   |    726  |    |   15997 |     |
|       |           | 96 |           | 64 |         | 1 |         | 57 |         | 205 |
| 24  0 | 3·9959568 |    | 1·9975310 |    | 8·38050 |   | 0·33783 |    | 0·15792 |     |
|       |           | 96 |           | 64 |         | 1 |         | 57 |         | 204 |
|    10 |    9472   |    |    5374   |    |    49   |   |    840  |    |   15588 |     |
|       |           | 96 |           | 64 |         | 1 |         | 58 |         | 200 |
|    20 |    9376   |    |    5438   |    |    48   |   |    898  |    |   15388 |     |
|       |           | 97 |           | 64 |         | 1 |         | 58 |         | 199 |
|    30 |    9279   |    |    5502   |    |    47   |   |    956  |    |   15189 |     |
|       |           | 97 |           | 65 |         | 1 |         | 58 |         | 195 |
|    40 |    9182   |    |    5567   |    |    46   |   | 0·34014 |    |   14994 |     |
|       |           | 98 |           | 66 |         | 1 |         | 59 |         | 194 |
|    50 |    9084   |    |    5633   |    |    45   |   |    073  |    |   14800 |     |
|       |           | 99 |           | 65 |         | 1 |         | 59 |         | 191 |
| 25  0 | 3·9958985 |    | 1·9975698 |    | 8·38044 |   | 0·34132 |    | 0·14609 |     |
|       |           | 98 |           | 66 |         | 1 |         | 60 |         | 189 |
|    10 |    8887   |    |    5764   |    |    43   |   |    192  |    |   14420 |     |
|       |           |100 |           | 66 |         | 1 |         | 60 |         | 187 |
|    20 |    8787   |    |    5830   |    |    42   |   |    252  |    |   14233 |     |
|       |           |100 |           | 67 |         | 1 |         | 61 |         | 184 |
|    30 |    8687   |    |    5897   |    |    41   |   |    313  |    |   14049 |     |
|       |           |100 |           | 67 |         | 1 |         | 61 |         | 182 |
|    40 |    8587   |    |    5964   |    |    40   |   |    374  |    |   13867 |     |
|       |           |101 |           | 67 |         | 1 |         | 62 |         | 180 |
|    50 |    8486   |    |    6031   |    |    39   |   |    436  |    |   13687 |     |
|       |           |101 |           | 67 |         | 1 |         | 62 |         | 178 |
| 26  0 | 3·9958385 |    | 1·9976098 |    | 8·38038 |   | 0·34498 |    | 0·13509 |     |
|       |           |102 |           | 68 |         | 1 |         | 62 |         | 175 |
|    10 |    8283   |    |    6166   |    |    37   |   |    560  |    |   13334 |     |
|       |           |102 |           | 68 |         | 1 |         | 63 |         | 174 |
|    20 |    8181   |    |    6234   |    |    36   |   |    623  |    |   13160 |     |
|       |           |102 |           | 69 |         | 1 |         | 64 |         | 172 |
|    30 |    8079   |    |    6303   |    |    35   |   |    687  |    |   12988 |     |
|       |           |103 |           | 68 |         | 1 |         | 64 |         | 169 |
|    40 |    7976   |    |    6371   |    |    34   |   |    751  |    |   12819 |     |
|       |           |104 |           | 69 |         | 1 |         | 64 |         | 167 |
|    50 |    7872   |    |    6440   |    |    33   |   |    815  |    |   12652 |     |
|       |           |104 |           | 70 |         | 1 |         | 65 |         | 166 |

Terrestrial Latitudes, Longitudes, and Reverse Azimuths.

| Latitude | P | Diff − | Q | Diff + | R | Diff − | S | Diff + | T | Diff − |
|---|---|---|---|---|---|---|---|---|---|---|
| ° ′ | | | | | | | | | | |
| 27 0 | 3·2957768 | | 1·9976510 | | 8·38032 | | 0·34880 | | 0·12486 | |
|  |  | 104 |  | 69 |  | 1 |  | 65 |  | 163 |
| 10 | 7664 |  | 6579 |  | 31 |  | 945 |  | 12323 |  |
|  |  | 105 |  | 70 |  | 1 |  | 66 |  | 162 |
| 20 | 7559 |  | 6648 |  | 30 |  | 0·35011 |  | 12161 |  |
|  |  | 105 |  | 70 |  | 1 |  | 66 |  | 160 |
| 30 | 7454 |  | 6719 |  | 29 |  | 077 |  | 12001 |  |
|  |  | 106 |  | 71 |  | 1 |  | 67 |  | 158 |
| 40 | 7348 |  | 6790 |  | 28 |  | 144 |  | 11843 |  |
|  |  | 106 |  | 71 |  | 1 |  | 67 |  | 155 |
| 50 | 7242 |  | 6861 |  | 27 |  | 211 |  | 11688 |  |
|  |  | 107 |  | 71 |  | 1 |  | 68 |  | 154 |
| 28 0 | 3·9957135 |  | 1·9976932 |  | 8·38026 |  | 0·35279 |  | 0·11534 |  |
|  |  | 107 |  | 71 |  | 1 |  | 68 |  | 152 |
| 10 | 7028 |  | 7003 |  | 25 |  | 347 |  | 11382 |  |
|  |  | 107 |  | 71 |  | 1 |  | 69 |  | 151 |
| 20 | 6921 |  | 7074 |  | 24 |  | 416 |  | 11231 |  |
|  |  | 108 |  | 72 |  | 1 |  | 69 |  | 148 |
| 30 | 6813 |  | 7146 |  | 23 |  | 485 |  | 11083 |  |
|  |  | 108 |  | 72 |  | 1 |  | 69 |  | 147 |
| 40 | 6705 |  | 7218 |  | 22 |  | 554 |  | 10936 |  |
|  |  | 109 |  | 73 |  | 2 |  | 70 |  | 145 |
| 50 | 6596 |  | 7291 |  | 20 |  | 624 |  | 10791 |  |
|  |  | 109 |  | 73 |  | 1 |  | 71 |  | 143 |
| 29 0 | 3·9956487 |  | 1·9977364 |  | 8·38019 |  | 0·35695 |  | 0·10648 |  |
|  |  | 109 |  | 72 |  | 1 |  | 71 |  | 141 |
| 10 | 6378 |  | 7436 |  | 18 |  | 766 |  | 10507 |  |
|  |  | 110 |  | 74 |  | 1 |  | 71 |  | 140 |
| 20 | 6268 |  | 7510 |  | 17 |  | 837 |  | 10367 |  |
|  |  | 110 |  | 73 |  | 1 |  | 72 |  | 138 |
| 30 | 6158 |  | 7583 |  | 16 |  | 909 |  | 10229 |  |
|  |  | 111 |  | 74 |  | 1 |  | 73 |  | 137 |
| 40 | 6047 |  | 7657 |  | 15 |  | 982 |  | 10092 |  |
|  |  | 111 |  | 74 |  | 1 |  | 73 |  | 134 |
| 50 | 5936 |  | 7731 |  | 14 |  | 0·36055 |  | 09958 |  |
|  |  | 111 |  | 74 |  | 1 |  | 73 |  | 134 |
| 30 0 | 3·9955825 |  | 1·9977805 |  | 8·38013 |  | 0·36128 |  | 0·09824 |  |
|  |  | 112 |  | 75 |  | 1 |  | 74 |  | 131 |
| 10 | 5713 |  | 7880 |  | 12 |  | 202 |  | 09693 |  |
|  |  | 112 |  | 74 |  | 1 |  | 74 |  | 130 |
| 20 | 5601 |  | 7954 |  | 11 |  | 276 |  | 09563 |  |
|  |  | 112 |  | 75 |  | 2 |  | 75 |  | 129 |
| 30 | 5489 |  | 8029 |  | 09 |  | 351 |  | 09434 |  |
|  |  | 113 |  | 76 |  | 1 |  | 76 |  | 127 |
| 40 | 5376 |  | 8105 |  | 08 |  | 427 |  | 09307 |  |
|  |  | 113 |  | 75 |  | 1 |  | 76 |  | 125 |
| 50 | 5263 |  | 8180 |  | 07 |  | 503 |  | 09182 |  |
|  |  | 114 |  | 76 |  | 1 |  | 76 |  | 124 |
| 31 0 | 3·9955149 |  | 1·9078256 |  | 8·38006 |  | 0·36579 |  | 0·09058 |  |
|  |  | 114 |  | 76 |  | 1 |  | 77 |  | 123 |
| 10 | 5035 |  | 8332 |  | 05 |  | 656 |  | 08935 |  |
|  |  | 114 |  | 76 |  | 1 |  | 77 |  | 121 |
| 20 | 4921 |  | 8408 |  | 04 |  | 733 |  | 08814 |  |
|  |  | 115 |  | 76 |  | 1 |  | 78 |  | 119 |
| 30 | 4808 |  | 8484 |  | 03 |  | 811 |  | 08695 |  |
|  |  | 115 |  | 77 |  | 2 |  | 79 |  | 118 |
| 40 | 4691 |  | 8561 |  | 01 |  | 890 |  | 08577 |  |
|  |  | 115 |  | 77 |  | 1 |  | 79 |  | 117 |
| 50 | 4576 |  | 8638 |  | 00 |  | 969 |  | 08460 |  |
|  |  | 115 |  | 77 |  | 1 |  | 79 |  | 116 |

## TABLE I.—Logarithms for facilitating the Computation of

| LATI-TUDE. | P | Diff. − | Q | Diff. + | R | Diff. − | S | Diff. + | T | Diff. − |
|---|---|---|---|---|---|---|---|---|---|---|
| ° , ′ | | | | | | | | | | |
| 32  0 | 3·9954461 |     | 1·9978715 |    | 8·37999 |   | 0·37048 |    | 0·08344 |     |
|       |           | 116 |           | 77 |         | 1 |         | 80 |         | 114 |
|    10 |      4345 |     |      8792 |    |      98 |   |     128 |    |   08230 |     |
|       |           | 117 |           | 78 |         | 1 |         | 81 |         | 113 |
|    20 |      4228 |     |      8870 |    |      97 |   |     209 |    |   08117 |     |
|       |           | 116 |           | 77 |         | 1 |         | 81 |         | 111 |
|    30 |      4112 |     |      8947 |    |      96 |   |     290 |    |   08006 |     |
|       |           | 117 |           | 78 |         | 2 |         | 81 |         | 110 |
|    40 |      3995 |     |      9025 |    |      94 |   |     371 |    |   07896 |     |
|       |           | 117 |           | 78 |         | 1 |         | 82 |         | 108 |
|    50 |      3878 |     |      9103 |    |      93 |   |     453 |    |   07788 |     |
|       |           | 118 |           | 79 |         | 1 |         | 83 |         | 107 |
| 33  0 | 3·9953760 |     | 1·9979182 |    | 8·37992 |   | 0·37536 |    | 0·07681 |     |
|       |           | 118 |           | 78 |         | 1 |         | 83 |         | 106 |
|    10 |      3642 |     |      9260 |    |      91 |   |     619 |    |   07575 |     |
|       |           | 118 |           | 79 |         | 1 |         | 83 |         | 105 |
|    20 |      3524 |     |      9339 |    |      90 |   |     702 |    |   07470 |     |
|       |           | 118 |           | 79 |         | 1 |         | 85 |         | 104 |
|    30 |      3406 |     |      9418 |    |      89 |   |     787 |    |   07366 |     |
|       |           | 119 |           | 79 |         | 2 |         | 84 |         | 102 |
|    40 |      3287 |     |      9497 |    |      87 |   |     871 |    |   07264 |     |
|       |           | 119 |           | 80 |         | 1 |         | 85 |         | 101 |
|    50 |      3168 |     |      9577 |    |      86 |   |     956 |    |   07163 |     |
|       |           | 119 |           | 79 |         | 1 |         | 86 |         |  99 |
| 34  0 | 3·9953049 |     | 1·9979656 |    | 8·37985 |   | 0·38042 |    | 0·07064 |     |
|       |           | 120 |           | 80 |         | 1 |         | 86 |         |  99 |
|    10 |      2929 |     |      9736 |    |      84 |   |     128 |    |   06965 |     |
|       |           | 120 |           | 80 |         | 1 |         | 87 |         |  97 |
|    20 |      2809 |     |      9816 |    |      83 |   |     215 |    |   06868 |     |
|       |           | 120 |           | 80 |         | 2 |         | 88 |         |  96 |
|    30 |      2689 |     |      9896 |    |      81 |   |     303 |    |   06772 |     |
|       |           | 121 |           | 80 |         | 1 |         | 87 |         |  95 |
|    40 |      2568 |     |      9976 |    |      80 |   |     390 |    |   06677 |     |
|       |           | 120 |           | 81 |         | 1 |         | 89 |         |  94 |
|    50 |      2448 |     | 1·9980057 |    |      79 |   |     479 |    |   06583 |     |
|       |           | 121 |           | 80 |         | 1 |         | 89 |         |  93 |
| 35  0 | 3·0952327 |     | 1·9980137 |    | 8·37978 |   | 0·38568 |    | 0·06490 |     |
|       |           | 121 |           | 81 |         | 1 |         | 89 |         |  91 |
|    10 |      2206 |     |      0218 |    |      77 |   |     657 |    |   06399 |     |
|       |           | 122 |           | 81 |         | 2 |         | 91 |         |  91 |
|    20 |      2084 |     |      0299 |    |      75 |   |     748 |    |   06308 |     |
|       |           | 122 |           | 81 |         | 1 |         | 90 |         |  89 |
|    30 |      1962 |     |      0380 |    |      74 |   |     838 |    |   06219 |     |
|       |           | 122 |           | 82 |         | 1 |         | 91 |         |  88 |
|    40 |      1840 |     |      0462 |    |      73 |   |     929 |    |   06131 |     |
|       |           | 122 |           | 81 |         | 1 |         | 92 |         |  87 |
|    50 |      1718 |     |      0543 |    |      72 |   | 0·39021 |    |   06044 |     |
|       |           | 122 |           | 82 |         | 2 |         | 92 |         |  86 |
| 36  0 | 3·9951596 |     | 1·9980625 |    | 8·37970 |   | 0·39113 |    | 0·05958 |     |
|       |           | 123 |           | 82 |         | 1 |         | 93 |         |  85 |
|    10 |      1473 |     |      0707 |    |      69 |   |     206 |    |   05873 |     |
|       |           | 123 |           | 82 |         | 1 |         | 94 |         |  84 |
|    20 |      1350 |     |      0789 |    |      68 |   |     300 |    |   05789 |     |
|       |           | 123 |           | 82 |         | 1 |         | 94 |         |  83 |
|    30 |      1227 |     |      0871 |    |      67 |   |     394 |    |   05706 |     |
|       |           | 123 |           | 82 |         | 1 |         | 94 |         |  82 |
|    40 |      1104 |     |      0953 |    |      66 |   |     488 |    |   05624 |     |
|       |           | 124 |           | 82 |         | 2 |         | 96 |         |  81 |
|    50 |      0980 |     |      1035 |    |      64 |   |     584 |    |   05543 |     |
|       |           | 124 |           | 83 |         | 1 |         | 95 |         |  80 |

## 217

Terrestrial Latitudes, Longitudes, and Reverse Azimuths.

| LATI-TUDE. | P | Diff. − | Q | Diff. + | R | Diff. − | S | Diff. + | T | Diff. − |
|---|---|---|---|---|---|---|---|---|---|---|
| ° ′ | | | | | | | | | | |
| 37 0 | 3·9950856 | | 1·9981118 | | 8·37963 | | 0·39679 | | 0·05463 | |
| | | 124 | | 82 | | 1 | | 97 | | 79 |
| 10 | 0732 | | 1200 | | 62 | | 776 | | 05384 | |
| | | 124 | | 83 | | 1 | | 96 | | 78 |
| 20 | 0608 | | 1283 | | 61 | | 872 | | 05306 | |
| | | 124 | | 83 | | 2 | | 98 | | 77 |
| 30 | 0484 | | 1366 | | 59 | | 970 | | 05229 | |
| | | 125 | | 83 | | 1 | | 98 | | 76 |
| 40 | 0359 | | 1449 | | 58 | | 0·40068 | | 05153 | |
| | | 125 | | 83 | | 1 | | 99 | | 75 |
| 50 | 0234 | | 1532 | | 57 | | 167 | | 05078 | |
| | | 125 | | 84 | | 1 | | 99 | | 75 |
| 38 0 | 3·9950109 | | 1·9981616 | | 8·37956 | | 0·40266 | | 0·05003 | |
| | | 125 | | 83 | | 2 | | 100 | | 74 |
| 10 | 3·9949984 | | 1699 | | 54 | | 366 | | 04929 | |
| | | 125 | | 84 | | 1 | | 100 | | 72 |
| 20 | 9859 | | 1783 | | 53 | | 466 | | 04857 | |
| | | 126 | | 83 | | 1 | | 101 | | 71 |
| 30 | 9733 | | 1866 | | 52 | | 567 | | 04786 | |
| | | 126 | | 84 | | 1 | | 102 | | 71 |
| 40 | 9607 | | 1950 | | 51 | | 669 | | 04715 | |
| | | 125 | | 84 | | 2 | | 102 | | 70 |
| 50 | 9482 | | 2034 | | 49 | | 771 | | 04645 | |
| | | 126 | | 84 | | 1 | | 103 | | 68 |
| 39 0 | 3·9949356 | | 1·9982118 | | 8·37948 | | 0·40874 | | 0·04577 | |
| | | 127 | | 84 | | 1 | | 103 | | 68 |
| 10 | 9229 | | 2202 | | 47 | | 977 | | 04509 | |
| | | 126 | | 85 | | 1 | | 104 | | 67 |
| 20 | 9103 | | 2287 | | 46 | | 0·41081 | | 04442 | |
| | | 126 | | 84 | | 2 | | 105 | | 67 |
| 30 | 8977 | | 2371 | | 44 | | 186 | | 04375 | |
| | | 127 | | 84 | | 1 | | 105 | | 65 |
| 40 | 8850 | | 2455 | | 43 | | 291 | | 04310 | |
| | | 127 | | 85 | | 1 | | 106 | | 65 |
| 50 | 8723 | | 2540 | | 42 | | 397 | | 04245 | |
| | | 127 | | 84 | | 2 | | 107 | | 63 |
| 40 0 | 3·9948596 | | 1·9982624 | | 8·37940 | | 0·41504 | | 0·04182 | |
| | | 127 | | 85 | | 1 | | 107 | | 63 |
| 10 | 8469 | | 2709 | | 39 | | 611 | | 04119 | |
| | | 127 | | 85 | | 1 | | 108 | | 62 |
| 20 | 8342 | | 2794 | | 38 | | 719 | | 04057 | |
| | | 127 | | 85 | | 1 | | 108 | | 61 |
| 30 | 8215 | | 2879 | | 37 | | 827 | | 03996 | |
| | | 128 | | 85 | | 2 | | 109 | | 61 |
| 40 | 8087 | | 2964 | | 35 | | 936 | | 03935 | |
| | | 127 | | 85 | | 1 | | 110 | | 60 |
| 50 | 7960 | | 3049 | | 34 | | 0·42046 | | 03875 | |
| | | 128 | | 85 | | 1 | | 110 | | 59 |
| 41 0 | 3·9947832 | | 1·9983134 | | 8·37933 | | 0·42156 | | 0·03816 | |
| | | 128 | | 85 | | 1 | | 111 | | 59 |
| 10 | 7704 | | 3219 | | 32 | | 267 | | 03757 | |
| | | 127 | | 85 | | 2 | | 112 | | 58 |
| 20 | 7577 | | 3304 | | 30 | | 379 | | 03699 | |
| | | 128 | | 86 | | 1 | | 112 | | 57 |
| 30 | 7449 | | 3390 | | 29 | | 491 | | 03642 | |
| | | 128 | | 85 | | 1 | | 113 | | 56 |
| 40 | 7321 | | 3475 | | 28 | | 604 | | 03586 | |
| | | 129 | | 85 | | 2 | | 114 | | 55 |
| 50 | 7192 | | 3560 | | 26 | | 718 | | 03531 | |
| | | 128 | | 86 | | 1 | | 114 | | 54 |

## TABLE I.—Logarithms for facilitating the Computation of

| LATITUDE. | P | Diff. − | Q | Diff. + | R | Diff − | S | Diff. + | T | Diff. − |
|---|---|---|---|---|---|---|---|---|---|---|
| ° ′ | | | | | | | | | | |
| 42  0 | 3·9947064 | | 1·9983646 | | 8·37925 | | 0·42832 | | 0·03477 | |
| | | 128 | | 85 | | 1 | | 115 | | 54 |
| 10 | 6936 | | 3731 | | 24 | | • 947 | | 03423 | |
| | | 129 | | 86 | | 2 | | 116 | | 53 |
| 20 | 6807 | | 3817 | | 22 | | 0·43063 | | 03370 | |
| | | 128 | | 85 | | 1 | | 116 | | 53 |
| 30 | 6679 | | 3902 | | 21 | | 179 | | 03317 | |
| | | 128 | | 86 | | 1 | | 117 | | 52 |
| 40 | 6551 | | 3988 | | 20 | | 296 | | 03265 | |
| | | 129 | | 86 | | 1 | | 118 | | 51 |
| 50 | 6422 | | 4074 | | 19 | | 414 | | 03214 | |
| | | 129 | | 86 | | 2 | | 118 | | 50 |
| 43  0 | 3·9946293 | | 1·9984160 | | 8·37917 | | 0·43532 | | 0·03164 | |
| | | 128 | | 85 | | 1 | | 119 | | 50 |
| 10 | 6165 | | 4245 | | 16 | | 651 | | 03114 | |
| | | 129 | | 86 | | 1 | | 120 | | 49 |
| 20 | 6036 | | 4331 | | 15 | | 771 | | 03065 | |
| | | 129 | | 86 | | 1 | | 120 | | 49 |
| 30 | 5907 | | 4417 | | 14 | | 891 | | 03016 | |
| | | 129 | | 86 | | 2 | | 121 | | 48 |
| 40 | 5778 | | 4503 | | 12 | | 0·44012 | | 02968 | |
| | | 129 | | 86 | | 1 | | 122 | | 47 |
| 50 | 5649 | | 4589 | | 11 | | 134 | | 02921 | |
| | | 129 | | 86 | | 1 | | 122 | | 47 |
| 44  0 | 3·9945520 | | 1·9984675 | | 8·37910 | | 0·44256 | | 0·02874 | |
| | | 128 | | 86 | | 2 | | 124 | | 46 |
| 10 | 5392 | | 4761 | | 08 | | 380 | | 02828 | |
| | | 129 | | 86 | | 1 | | 123 | | 45 |
| 20 | 5263 | | 4847 | | 07 | | 503 | | 02783 | |
| | | 129 | | 86 | | 1 | | 125 | | 45 |
| 30 | 5134 | | 4933 | | 06 | | 628 | | 02738 | |
| | | 129 | | 86 | | 1 | | 125 | | 44 |
| 40 | 5005 | | 5019 | | 05 | | 753 | | 02694 | |
| | | 129 | | 86 | | 2 | | 126 | | 44 |
| 50 | 4876 | | 5105 | | 03 | | 879 | | 02650 | |
| | | 129 | | 86 | | 1 | | 127 | | 44 |
| 45  0 | 3·9944747 | | 1·9985191 | | 8·37902 | | 0·45006 | | 0·02606 | |
| | | 129 | | 86 | | 1 | | 128 | | 43 |
| 10 | 4618 | | 5277 | | 01 | | 134 | | 02563 | |
| | | 129 | | 86 | | 2 | | 128 | | 42 |
| 20 | 4489 | | 5363 | | 8·37899 | | 262 | | 02521 | |
| | | 129 | | 86 | | 1 | | 129 | | 41 |
| 30 | 4360 | | 5449 | | 98 | | 391 | | 02480 | |
| | | 129 | | 86 | | 1 | | 130 | | 40 |
| 40 | 4231 | | 5535 | | 97 | | 521 | | 02440 | |
| | | 129 | | 86 | | 2 | | 131 | | 40 |
| 50 | 4102 | | 5621 | | 95 | | 652 | | 02400 | |
| | | 129 | | 86 | | 1 | | 131 | | 39 |
| 46  0 | 3·9943973 | | 1·9985707 | | 8·37894 | | 0·45783 | | 0·02361 | |
| | | 129 | | 86 | | 1 | | 132 | | 39 |
| 10 | 3844 | | 5793 | | 93 | | 915 | | 02322 | |
| | | 129 | | 86 | | 1 | | 133 | | 39 |
| 20 | 3715 | | 5879 | | 92 | | 0·46048 | | 02283 | |
| | | 129 | | 86 | | 2 | | 133 | | 38 |
| 30 | 3586 | | 5965 | | 90 | | 181 | | 02245 | |
| | | 129 | | 85 | | 1 | | 135 | | 38 |
| 40 | 3457 | | 6050 | | 89 | | 316 | | 02207 | |
| | | 129 | | 86 | | 1 | | 135 | | 37 |
| 50 | 3328 | | 6136 | | 88 | | 451 | | 02170 | |
| | | 129 | | 86 | | 2 | | 136 | | 37 |

## Terrestrial Latitudes, Longitudes, and Reverse Azimuths.

| Latitude | P | Diff. − | Q | Diff. + | R | Diff. − | S | Diff. + | T | Diff. − |
|---|---|---|---|---|---|---|---|---|---|---|
| ° ′ | | | | | | | | | | |
| 47 0 | 3·9943199 | | 1·9986222 | | 8·37886 | | 0·46587 | | 0·02133 | |
|  | | 128 | | 86 | | 1 | | 137 | | 36 |
| 10 | 3071 | | 6308 | | 85 | | 724 | | 02097 | |
|  | | 129 | | 86 | | 1 | | 137 | | 36 |
| 20 | 2942 | | 6394 | | 84 | | 861 | | 02061 | |
|  | | 128 | | 85 | | 1 | | 138 | | 35 |
| 30 | 2814 | | 6479 | | 83 | | 999 | | 02026 | |
|  | | 129 | | 86 | | 2 | | 140 | | 34 |
| 40 | 2685 | | 6565 | | 81 | | 0·47139 | | 01992 | |
|  | | 128 | | 86 | | 1 | | 140 | | 34 |
| 50 | 2557 | | 6651 | | 80 | | 279 | | 01958 | |
|  | | 129 | | 85 | | 1 | | 140 | | 33 |
| 48 0 | 3·9942428 | | 1·9986736 | | 8·37879 | | 0·47419 | | 0·01925 | |
|  | | 128 | | 86 | | 2 | | 142 | | 33 |
| 10 | 2300 | | 6822 | | 77 | | 561 | | 01892 | |
|  | | 128 | | 85 | | 1 | | 142 | | 33 |
| 20 | 2172 | | 6907 | | 76 | | 703 | | 01859 | |
|  | | 129 | | 86 | | 1 | | 143 | | 32 |
| 30 | 2043 | | 6993 | | 75 | | 846 | | 01827 | |
|  | | 128 | | 85 | | 1 | | 145 | | 31 |
| 40 | 1915 | | 7078 | | 74 | | 991 | | 01796 | |
|  | | 128 | | 86 | | 2 | | 144 | | 31 |
| 50 | 1787 | | 7164 | | 72 | | 0·48135 | | 01765 | |
|  | | 127 | | 85 | | 1 | | 146 | | 30 |
| 49 0 | 3·9941660 | | 1·9987249 | | 8·37871 | | 0·48281 | | 0·01735 | |
|  | | 128 | | 85 | | 1 | | 147 | | 31 |
| 10 | 1532 | | 7334 | | 70 | | 428 | | 704 | |
|  | | 128 | | 85 | | 1 | | 147 | | 31 |
| 20 | 1404 | | 7419 | | 69 | | 575 | | 673 | |
|  | | 127 | | 85 | | 2 | | 149 | | 30 |
| 30 | 1277 | | 7504 | | 67 | | 724 | | 643 | |
|  | | 128 | | 85 | | 1 | | 149 | | 29 |
| 40 | 1149 | | 7589 | | 66 | | 873 | | 614 | |
|  | | 127 | | 85 | | 1 | | 150 | | 29 |
| 50 | 1022 | | 7674 | | 65 | | 0·49023 | | 585 | |
|  | | 127 | | 85 | | 2 | | 151 | | 29 |
| 50 0 | 3·9940895 | | 1·9987759 | | 8·37863 | | 0·49174 | | 0·01556 | |
|  | | 127 | | 85 | | 1 | | 152 | | 28 |
| 10 | 0768 | | 7844 | | 62 | | 326 | | 528 | |
|  | | 127 | | 84 | | 1 | | 152 | | 27 |
| 20 | 0641 | | 7928 | | 61 | | 478 | | 501 | |
|  | | 127 | | 85 | | 1 | | 154 | | 27 |
| 30 | 0514 | | 8013 | | 60 | | 632 | | 474 | |
|  | | 127 | | 84 | | 2 | | 155 | | 27 |
| 40 | 0387 | | 8097 | | 58 | | 787 | | 447 | |
|  | | 126 | | 84 | | 1 | | 155 | | 26 |
| 50 | 0261 | | 8181 | | 57 | | 942 | | 421 | |
|  | | 127 | | 85 | | 1 | | 156 | | 26 |
| 51 0 | 3·9940134 | | 3·9988266 | | 8·37856 | | 0·50098 | | 0·01395 | |
|  | | 126 | | 84 | | 1 | | 158 | | 26 |
| 10 | 0008 | | 8350 | | 55 | | 256 | | 369 | |
|  | | 126 | | 84 | | 2 | | 158 | | 25 |
| 20 | 3·9939882 | | 8434 | | 53 | | 414 | | 344 | |
|  | | 126 | | 84 | | 1 | | 159 | | 25 |
| 30 | 9756 | | 8518 | | 52 | | 573 | | 319 | |
|  | | 126 | | 84 | | 1 | | 160 | | 24 |
| 40 | 9630 | | 8602 | | 51 | | 733 | | 295 | |
|  | | 125 | | 83 | | 2 | | 161 | | 24 |
| 50 | 9505 | | 8685 | | 49 | | 894 | | 271 | |
|  | | 125 | | 84 | | 1 | | 162 | | 24 |

(5455)   2 F

## TABLE I.—Logarithms for facilitating the Computation of

| Lati-tude. | P | Diff. − | Q | Diff. + | R | Diff. − | S | Diff. + | T | Diff. − |
|---|---|---|---|---|---|---|---|---|---|---|
| ° ′ | | | | | | | | | | |
| 52 0 | 3·9939380 | | 1·9988769 | | 8·37848 | | 0·51056 | | 0·01247 | |
| | | 126 | | 83 | | 1 | | 163 | | 24 |
| 10 | 9254 | | 8852 | | 47 | | 219 | | 223 | |
| | | 125 | | 84 | | 1 | | 164 | | 23 |
| 20 | 9129 | | 8936 | | 46 | | 383 | | 200 | |
| | | 124 | | 83 | | 1 | | 165 | | 23 |
| 30 | 9005 | | 9019 | | 45 | | 548 | | 177 | |
| | | 125 | | 83 | | 2 | | 166 | | 22 |
| 40 | 8880 | | 9102 | | 43 | | 714 | | 155 | |
| | | 125 | | 83 | | 1 | | 167 | | 22 |
| 50 | 8755 | | 9185 | | 42 | | 881 | | 133 | |
| | | 124 | | 83 | | 1 | | 168 | | 21 |
| 53 0 | 3·9938631 | | 1·9989268 | | 8·37841 | | 0·52049 | | 0·01112 | |
| | | 124 | | 82 | | 1 | | 169 | | 22 |
| 10 | 8507 | | 9350 | | 40 | | 218 | | 090 | |
| | | 124 | | 83 | | 2 | | 170 | | 21 |
| 20 | 8383 | | 9433 | | 38 | | 388 | | 069 | |
| | | 123 | | 82 | | 1 | | 171 | | 21 |
| 30 | 8260 | | 9515 | | 37 | | 559 | | 048 | |
| | | 124 | | 83 | | 1 | | 172 | | 20 |
| 40 | 8136 | | 9598 | | 36 | | 731 | | 028 | |
| | | 123 | | 82 | | 1 | | 174 | | 20 |
| 50 | 8013 | | 9680 | | 35 | | 905 | | 008 | |
| | | 123 | | 82 | | 2 | | 174 | | 19 |
| 54 0 | 3·9937890 | | 1·9989762 | | 8·37833 | | 0·53079 | | 0·00989 | |
| | | 122 | | 82 | | 1 | | 175 | | 20 |
| 10 | 7768 | | 9844 | | 32 | | 254 | | 969 | |
| | | 123 | | 81 | | 1 | | 176 | | 19 |
| 20 | 7645 | | 9925 | | 31 | | 430 | | 950 | |
| | | 122 | | 82 | | 1 | | 178 | | 19 |
| 30 | 7523 | | 1·9990007 | | 30 | | 608 | | 931 | |
| | | 122 | | 81 | | 1 | | 178 | | 18 |
| 40 | 7401 | | 0088 | | 29 | | 786 | | 913 | |
| | | 122 | | 81 | | 2 | | 180 | | 18 |
| 50 | 7279 | | 0169 | | 27 | | 966 | | 895 | |
| | | 121 | | 81 | | 1 | | 180 | | 17 |
| 55 0 | 3·9937158 | | 1·9990250 | | 8·37826 | | 0·54146 | | 0·00878 | |
| | | 122 | | 81 | | 1 | | 182 | | 18 |
| 10 | 7036 | | 0331 | | 25 | | 328 | | 860 | |
| | | 121 | | 81 | | 1 | | 183 | | 18 |
| 20 | 6915 | | 0412 | | 24 | | 511 | | 842 | |
| | | 120 | | 80 | | 2 | | 184 | | 17 |
| 30 | 6795 | | 0492 | | 22 | | 695 | | 825 | |
| | | 121 | | 80 | | 1 | | 185 | | 17 |
| 40 | 6674 | | 0572 | | 21 | | 880 | | 808 | |
| | | 120 | | 81 | | 1 | | 187 | | 16 |
| 50 | 6554 | | 0653 | | 20 | | 0·55067 | | 792 | |
| | | 120 | | 80 | | 1 | | 187 | | 16 |
| 56 0 | 3·9936434 | | 1·9990733 | | 8·37819 | | 0·55254 | | 0·00776 | |
| | | 120 | | 79 | | 1 | | 189 | | 16 |
| 10 | 6314 | | 0812 | | 18 | | 443 | | 760 | |
| | | 119 | | 80 | | 2 | | 190 | | 16 |
| 20 | 6195 | | 0892 | | 16 | | 633 | | 744 | |
| | | 119 | | 79 | | 1 | | 191 | | 16 |
| 30 | 6076 | | 0971 | | 15 | | 824 | | 728 | |
| | | 119 | | 79 | | 1 | | 192 | | 15 |
| 40 | 5957 | | 1050 | | 14 | | 0·56016 | | 713 | |
| | | 118 | | 79 | | 1 | | 193 | | 15 |
| 50 | 5839 | | 1129 | | 13 | | 209 | | 698 | |
| | | 118 | | 79 | | 1 | | 195 | | 14 |

## Terrestrial Latitudes, Longitudes, and Reverse Azimuths.

| LATI-TUDE. | P | Diff. − | Q | Diff. + | R | Diff. − | S | Diff. + | T | Diff. − |
|---|---|---|---|---|---|---|---|---|---|---|
| ° ′ | | | | | | | | | | |
| 57  0 | 3·9935721 | | 1·9991208 | | 8·37812 | | 0·56404 | | 0·00684 | |
|  | | 118 | | 79 | | 1 | | 196 | | 15 |
|    10 | 5603 | | 1287 | | 11 | | 600 | | 669 | |
|  | | 118 | | 78 | | 2 | | 197 | | 14 |
|    20 | 5485 | | 1365 | | 09 | | 797 | | 655 | |
|  | | 117 | | 78 | | 1 | | 199 | | 14 |
|    30 | 5368 | | 1443 | | 08 | | 996 | | 641 | |
|  | | 117 | | 78 | | 1 | | 199 | | 14 |
|    40 | 5251 | | 1521 | | 07 | | 0·57195 | | 627 | |
|  | | 117 | | 78 | | 1 | | 201 | | 13 |
|    50 | 5134 | | 1599 | | 06 | | 396 | | 614 | |
|  | | 116 | | 78 | | 1 | | 203 | | 13 |
| 58  0 | 3·9935018 | | 1·9991677 | | 8·37805 | | 0·57599 | | 0·00601 | |
|  | | 116 | | 77 | | 1 | | 203 | | 13 |
|    10 | 4902 | | 1754 | | 04 | | 802 | | 588 | |
|  | | 116 | | 77 | | 2 | | 205 | | 13 |
|    20 | 4786 | | 1831 | | 02 | | 0·58007 | | 575 | |
|  | | 115 | | 77 | | 1 | | 207 | | 13 |
|    30 | 4671 | | 1908 | | 01 | | 214 | | 562 | |
|  | | 115 | | 76 | | 1 | | 207 | | 12 |
|    40 | 4556 | | 1984 | | 00 | | 421 | | 550 | |
|  | | 114 | | 77 | | 1 | | 209 | | 12 |
|    50 | 4442 | | 2061 | | 8·37799 | | 630 | | 538 | |
|  | | 115 | | 76 | | 1 | | 210 | | 12 |
| 59  0 | 3·9934327 | | 1·9992137 | | 8·37798 | | 0·58840 | | 0·00526 | |
|  | | 114 | | 76 | | 1 | | 212 | | 12 |
|    10 | 4213 | | 2213 | | 97 | | 0·59052 | | 514 | |
|  | | 113 | | 76 | | 1 | | 213 | | 12 |
|    20 | 4100 | | 2289 | | 96 | | 265 | | 502 | |
|  | | 113 | | 75 | | 2 | | 215 | | 12 |
|    30 | 3987 | | 2364 | | 94 | | 480 | | 490 | |
|  | | 113 | | 75 | | 1 | | 216 | | 11 |
|    40 | 3874 | | 2439 | | 93 | | 696 | | 479 | |
|  | | 113 | | 75 | | 1 | | 217 | | 11 |
|    50 | 3761 | | 2514 | | 92 | | 913 | | 468 | |
|  | | 112 | | 75 | | 1 | | 219 | | 10 |
| 60  0 | 3·9933649 | | 1·9992589 | | 8·37791 | | 0·60132 | | 0·00458 | |

## TABLE II. (A).

**Directions for applying the signs to the terms of the Latitude, Longitude, and Azimuth formulæ, when the fixed station A lies North of the Equator.**

The Azimuth is reckoned from South by West. The Latitude North, and Longitude East, should always be considered *positive*.

| Terms of the Formulæ. | Magnitude of the given Azimuth A. | | | |
|---|---|---|---|---|
| | 0° to 90° | 90° to 180° | 180° to 270° | 270° to 360° |
| $\delta_1 \lambda$ | − | + | + | − |
| $\delta_1 L$ | − | − | + | + |
| $\delta_1 A$ | − | − | + | + |
| $\delta_2 \lambda$ | − | − | − | − |
| $\delta_2 L$ | + | − | + | − |
| $\delta_2 A$ | + | − | + | − |

For use in Distance and Azimuth Computation:—

$A + \dfrac{\Delta A}{2}$ is in    1st    2nd    3rd    4th Quadrant

when $\dfrac{\Delta \lambda}{\Delta L}$ are    $\dfrac{-}{-}$    $\dfrac{+}{-}$    $\dfrac{+}{+}$    $\dfrac{-}{+}$

## TABLE II. (B).

**Directions for applying the signs to the terms of the Latitude, Longitude, and Azimuth formulæ, when the fixed station A lies to the South of the Equator.**

The Latitude South, and the Longitude East, should be considered *positive*.

| Terms of the Formulæ. | Magnitude of the given Azimuth A. | | | |
|---|---|---|---|---|
| | 0° to 90° | 90° to 180° | 180° to 270° | 270° to 360° |
| $\delta_1 \lambda$ | + | − | − | + |
| $\delta_1 L$ | − | − | + | + |
| $\delta_1 A$ | + | + | − | − |
| $\delta_2 \lambda$ | − | − | − | − |
| $\delta_2 L$ | − | + | − | + |
| $\delta_2 A$ | + | − | + | − |

For use in Distance and Azimuth Computation:—

$A + \dfrac{\Delta A}{2}$ is in    1st    2nd    3rd    4th Quadrant

when $\dfrac{\Delta \lambda}{\Delta L}$ are    $\dfrac{+}{-}$    $\dfrac{-}{-}$    $\dfrac{-}{+}$    $\dfrac{+}{+}$

## TABLE III.

### Computation of Heights from Vertical Angles.

*Angle K, for Use in Computing Heights when only One Angle has been Observed.*

| Distance between Stations in feet. | K |
|---|---|
| 1,000   | 0′   4″ |
| 2,000   | 0     9 |
| 3,000   | 0    13 |
| 4,000   | 0    17 |
| 5,000   | 0    21 |
| 6,000   | 0    26 |
| 7,000   | 0    30 |
| 8,000   | 0    34 |
| 9,000   | 0    38 |
| 10,000  | 0    43 |
| 20,000  | 1    25 |
| 30,000  | 2     8 |
| 40,000  | 2    50 |
| 50,000  | 3    33 |
| 60,000  | 4    15 |
| 70,000  | 4    58 |
| 80,000  | 5    40 |
| 90,000  | 6    23 |
| 100,000 | 7     5 |

The value of the co-efficient of refraction has been found on the average to be about ·07. It is with this value that the above table has been computed.

## TABLE IV.

### Computation of Heights.

To facilitate finding the Subtended Angle when only one Angle has been observed.

| LATITUDE. | ADOPTED REFRACTION EXPRESSED IN DECIMALS OF CONTAINED ARC. | | | | | | | | |
|---|---|---|---|---|---|---|---|---|---|
|  | 0·20 | 0·10 | 0·09 | 0·08 | 0·07 | 0·06 | 0·05 | 0·04 | 0·03 |
| 0°  | 2·5276 | 2·4027 | 2·3920 | 2·3815 | 2·3713 | 2·3613 | 2·3515 | 2·3420 | 2·3327 |
| 5   | 5277 | 4027 | 3920 | 3815 | 3713 | 3613 | 3516 | 3420 | 3327 |
| 10  | 5277 | 4028 | 3921 | 3816 | 3714 | 3614 | 3516 | 3421 | 3327 |
| 15  | 5278 | 4029 | 3922 | 3817 | 3715 | 3615 | 3517 | 3422 | 3328 |
| 20  | 5280 | 4030 | 3923 | 3818 | 3716 | 3616 | 3519 | 3423 | 3330 |
| 25  | 5281 | 4032 | 3925 | 3820 | 3718 | 3618 | 3521 | 3425 | 3332 |
| 30  | 5284 | 4034 | 3927 | 3822 | 3720 | 3620 | 3523 | 3427 | 3334 |
| 35  | 5286 | 4036 | 3929 | 3825 | 3722 | 3623 | 3525 | 3429 | 3336 |
| 40  | 5288 | 4039 | 3932 | 3827 | 3725 | 3625 | 3527 | 3432 | 3338 |
| 45  | 5291 | 4042 | 3935 | 3830 | 3728 | 3628 | 3530 | 3435 | 3342 |
| 50  | 5294 | 4045 | 3937 | 3833 | 3731 | 3631 | 3533 | 3438 | 3344 |
| 55  | 5297 | 4047 | 3940 | 3835 | 3733 | 3633 | 3536 | 3440 | 3347 |
| 60  | 5299 | 4049 | 3942 | 3838 | 3736 | 3636 | 3538 | 3442 | 3349 |

NOTE.—If nothing is known of the Refraction, adopt 0·07 as the value of it.

## TABLES V. and VI.

### For Determining Differences of Height with the Barometer.

(BAILY.)

TABLE V.—PART I. Thermometers in the Open Air in Degrees Fahrenheit.

| $t + t'$ | A | $t + t'$ | A | $t + t'$ | A |
|---|---|---|---|---|---|
| ° | | ° | | ° | |
| 1 | 4·74913 | 65 | 4·78113 | 130 | 4·81138 |
| 5 | 4·75120 | 70 | 353 | 135 | 362 |
| 10 | 377 | 75 | 592 | 140 | 585 |
| 15 | 633 | 80 | 830 | 145 | 807 |
| 20 | 888 | 85 | 4·79066 | 150 | 4·82028 |
| 25 | 4·76141 | 90 | 301 | 155 | 248 |
| 30 | 392 | 95 | 535 | 160 | 466 |
| 35 | 642 | 100 | 768 | 165 | 684 |
| 40 | 891 | 105 | 999 | 170 | 900 |
| 45 | 4·77138 | 110 | 4·80229 | 175 | 4·83115 |
| 50 | 384 | 115 | 458 | 180 | 330 |
| 55 | 628 | 120 | 686 | | |
| 60 | 871 | 125 | 913 | | |

PART II. Attached Thermometers (not required with aneroid).

| $r - r'$ | B | $r - r'$ | B | $r - r'$ | B |
|---|---|---|---|---|---|
| ° | | ° | | ° | |
| 0 | 0·00000 | 25 | 0·00109 | 50 | 0·00217 |
| 5 | 22 | 30 | 130 | 55 | 238 |
| 10 | 43 | 35 | 152 | 60 | 259 |
| 15 | 65 | 40 | 174 | | |
| 20 | 87 | 45 | 195 | | |

TABLE VI.—Latitude of the place.

| Lat. | C | Lat. | C | Lat. | C |
|---|---|---|---|---|---|
| ° | | ° | | ° | |
| 0 | 0·00117 | 35 | 0·00040 | 70 | 1·99910 |
| 5 | 115 | 40 | 20 | 75 | 00 |
| 10 | 110 | 45 | 0·00000 | 80 | 1·99890 |
| 15 | 101 | 50 | 1·99980 | 85 | 85 |
| 20 | 90 | 55 | 60 | 90 | 83 |
| 25 | 75 | 60 | 42 | | |
| 30 | 58 | 65 | 25 | | |

TABLES V. and VI. (*continued*).

If $\beta$ be the reading of barometer at lower station.
$\beta'$ „ „ „ upper „
$t$ „ detached thermometer in air at lower station
$t'$ „ „ „ „ upper „
$r$ „ attached „ „ lower „
$r'$ „ „ „ „ upper „
$A$, the number in the table corresponding to $t + t'$.
$B$, „ „ „ $r - r'$.

Make $D = \log \beta - (\log \beta' + B)$:
then
$\log D + A + C = \log$: of the difference in altitude.

*Example.*—In lat. 51° 30′ the barometer at upper station read 25·55 in. and 29·81 at lower station. The readings of the detached thermometers were 50°·5 and 63°, and those of the attached thermometers 52° and 62° at the upper and lower stations respectively. What was the difference of altitude?

$$
\begin{array}{rr}
\text{Log } 29·81 = & 1·47436 \\
-\text{Log } 25·55 = & -1·40739 \\
-\quad B = & -0·00043 \\ \hline
D = & 0·06654 \\
\text{Log } D = & \overline{2}·82308 \\
A = & 4·80389 \\
C = & 1·99973 \\ \hline
\text{Log } h = & 3·62670
\end{array}
$$

$h = 4234$ feet, the required difference.

NOTE.—In the case of aneroids the term B is 0, as they are compensated for change of temperature of the *instrument*. The term A, however, which depends on the *air* temperatures, must always be included in the computation.

## TABLES VII. and VIII.

### For Determining Heights with the Boiling-Point Thermometer.

#### TABLE VII.

Approximate Height* in feet Corresponding to the Observed Boiling Point in Degrees Fahrenheit and Corresponding Barometer Readings.

| Boiling Point. | Barometer Reading. | Approximate Height. | Boiling Point. | Barometer Reading. | Approximate Height. |
|---|---|---|---|---|---|
| ° | | | ° | | |
| 212 | 29·921 | — | 195 | 21·133 | 9,084 |
| 211 | 29·329 | 522 | 194 | 20·690 | 9,638 |
| 210 | 28·746 | 1,046 | 193 | 20·255 | 10,193 |
| 209 | 28·174 | 1,571 | 192 | 19·828 | 10,750 |
| 208 | 27·613 | 2,097 | 191 | 19·409 | 11,307 |
| 207 | 27·062 | 2,624 | 190 | 18·998 | 11,867 |
| 206 | 26·521 | 3,151 | 189 | 18·595 | 12,426 |
| 205 | 25·989 | 3,680 | 188 | 18·199 | 12,968 |
| 204 | 25·466 | 4,210 | 187 | 17·809 | 13,556 |
| 203 | 24·952 | 4,741 | 186 | 17·426 | 14,124 |
| 202 | 24·447 | 5,278 | 185 | 17·050 | 14,694 |
| 201 | 23·950 | 5,815 | 184 | 16·681 | 15,266 |
| 200 | 23·461 | 6,354 | 183 | 16·319 | 15,839 |
| 199 | 22·980 | 6,895 | 182 | 15·964 | 16,412 |
| 198 | 22·507 | 7,439 | 181 | 15·616 | 16,987 |
| 197 | 22·042 | 7,984 | 180 | 15·275 | 17,564 |
| 196 | 21·584 | 8,533 | | | |

\* Accurately—this table gives the altitude above level at which water boils at 212°, temperature of intermediate air being 32° F.

#### TABLE VIII.

Correction for Temperature of Intermediate Air.

| Mean Temperature. | Multiplier. | Mean Temperature. | Multiplier. |
|---|---|---|---|
| ° | | ° | |
| 20 | 0·975 | 55 | 1·048 |
| 25 | 0·985 | 60 | 1·058 |
| 30 | 0·996 | 65 | 1·069 |
| 35 | 1·006 | 70 | 1·079 |
| 40 | 1·017 | 75 | 1·090 |
| 45 | 1·027 | 80 | 1·100 |
| 50 | 1·038 | 85 | 1·110 |

*Example.*—The boiling point at upper station was 204°·2, and at lower station 208°·7, and the respective air temperatures were 75° and 83°.

$$\text{Tabular number for } 204·2 = 4107$$
$$208·7 = 1726$$
$$\overline{\phantom{0000}2381}$$

Temperature at upper station, 75
„ lower „ 83

Mean temperature ... 79; multiplier, 1·1044.
And difference of height = 2381 × 1·044 = 2628 feet.

227

**TABLE IX.—Graticules of Maps. Sides and Diagonals of Squares of a quarter of a degree of Lat. and Long. on the scale of 1 inch to 2 miles.**

| Latitude. | m on meridian. $\frac{1}{4} \cdot \frac{\pi}{180} \times \frac{\rho}{5280}$ | n on lower parallel. $\frac{1}{4} \times \frac{\pi}{180 \times 4}$ | p on upper parallel. $\times \frac{\nu \cos \lambda}{5280}$ | q on diagonal. $\sqrt{m^2 + np}$ |
|---|---|---|---|---|
| From 0 0 to 0 15 | 8·588 | 8·647 | 8·647 | 12·187 |
| 0 15 ,, 0 30 | 588 | 647 | 646 | 187 |
| 0 30 ,, 0 45 | 588 | 646 | 646 | 186 |
| 0 45 ,, 1 0 | 588 | 646 | 645 | 186 |
| 1 0 ,, 1 15 | 588 | 645 | 645 | 186 |
| 1 15 ,, 1 30 | 588 | 645 | 644 | 185 |
| 1 30 ,, 1 45 | 588 | 644 | 643 | 184 |
| 1 45 ,, 2 0 | 588 | 643 | 641 | 183 |
| 2 0 ,, 2 15 | 8·588 | 8·641 | 8·640 | 12·182 |
| 2 15 ,, 2 30 | 588 | 640 | 638 | 181 |
| 2 30 ,, 2 45 | 588 | 638 | 637 | 180 |
| 2 45 ,, 3 0 | 588 | 637 | 635 | 179 |
| 3 0 ,, 3 15 | 588 | 635 | 633 | 178 |
| 3 15 ,, 3 30 | 588 | 633 | 631 | 176 |
| 3 30 ,, 3 45 | 588 | 631 | 628 | 175 |
| 3 45 ,, 4 0 | 588 | 628 | 626 | 173 |
| 4 0 ,, 4 15 | 8·588 | 8·626 | 8·623 | 12·171 |
| 4 15 ,, 4 30 | 588 | 623 | 620 | 169 |
| 4 30 ,, 4 45 | 589 | 620 | 617 | 167 |
| 4 45 ,, 5 0 | 589 | 617 | 614 | 165 |
| 5 0 ,, 5 15 | 589 | 614 | 611 | 163 |
| 5 15 ,, 5 30 | 589 | 611 | 607 | 161 |
| 5 30 ,, 5 45 | 589 | 607 | 603 | 158 |
| 5 45 ,, 6 0 | 589 | 603 | 600 | 155 |
| 6 0 ,, 6 15 | 8·589 | 8·600 | 8·596 | 12·153 |
| 6 15 ,, 6 30 | 589 | 596 | 591 | 150 |
| 6 30 ,, 6 45 | 589 | 591 | 587 | 147 |
| 6 45 ,, 7 0 | 589 | 587 | 583 | 144 |
| 7 0 ,, 7 15 | 589 | 583 | 578 | 141 |
| 7 15 ,, 7 30 | 589 | 578 | 573 | 137 |
| 7 30 ,, 7 45 | 589 | 573 | 568 | 134 |
| 7 45 ,, 8 0 | 590 | 568 | 563 | 131 |
| 8 0 ,, 8 15 | 8·590 | 8·563 | 8·558 | 12 127 |
| 8 15 ,, 8 30 | 590 | 558 | 552 | 123 |
| 8 30 ,, 8 45 | 590 | 552 | 547 | 119 |
| 8 45 ,, 9 0 | 590 | 547 | 541 | 116 |
| 9 0 ,, 9 15 | 590 | 541 | 535 | 111 |
| 9 15 ,, 9 30 | 590 | 535 | 529 | 107 |
| 9 30 ,, 9 45 | 590 | 529 | 523 | 103 |
| 9 45 ,, 10 0 | 591 | 523 | 516 | 099 |
| 10 0 ,, 10 15 | 8·591 | 8·516 | 8·510 | 12·094 |
| 10 15 ,, 10 30 | 591 | 510 | 503 | 089 |
| 10 30 ,, 10 45 | 591 | 503 | 496 | 085 |
| 10 45 ,, 11 0 | 591 | 496 | 489 | 081 |
| 11 0 ,, 11 15 | 591 | 489 | 482 | 075 |
| 11 15 ,, 11 30 | 591 | 482 | 474 | 070 |
| 11 30 ,, 11 45 | 592 | 474 | 467 | 065 |
| 11 45 ,, 12 0 | 592 | 467 | 459 | 060 |

(5455)           2 G

## TABLE IX.—Graticules of Maps. Sides and Diagonals of Squares of

| LATITUDE. | Length in Inches. | | | |
|---|---|---|---|---|
| | m<br>on meridian. | n<br>on lower parallel. | p<br>on upper parallel. | q<br>on diagonal. |
| ° ′   ° ′ | | | | |
| From 12  0 to 12 15 | 8·592 | 8·459 | 8·451 | 12·054 |
| 12 15 ,, 12 30 | 592 | 451 | 443 | 049 |
| 12 30 ,, 12 45 | 592 | 443 | 435 | 043 |
| 12 45 ,, 13  0 | 592 | 435 | 426 | 037 |
| 13  0 ,, 13 15 | 592 | 426 | 418 | 031 |
| 13 15 ,, 13 30 | 593 | 418 | 409 | 026 |
| 13 30 ,, 13 45 | 593 | 409 | 400 | 020 |
| 13 45 ,, 14  0 | 593 | 400 | 391 | 014 |
| | | | | |
| 14  0 ,, 14 15 | 8·593 | 8·391 | 8·382 | 12·007 |
| 14 15 ,, 14 30 | 593 | 382 | 373 | 001 |
| 14 30 ,, 14 45 | 594 | 373 | 364 | 11·995 |
| 14 45 ,, 15  0 | 594 | 364 | 354 | 989 |
| 15  0 ,, 15 15 | 594 | 354 | 344 | 982 |
| 15 15 ,, 15 30 | 594 | 344 | 334 | 974 |
| 15 30 ,, 15 45 | 594 | 334 | 324 | 968 |
| 15 45 ,, 16  0 | 595 | 324 | 314 | 962 |
| | | | | |
| 16  0 ,, 16 15 | 8·595 | 8·314 | 8·303 | 11·954 |
| 16 15 ,, 16 30 | 595 | 303 | 293 | 947 |
| 16 30 ,, 16 45 | 595 | 293 | 282 | 940 |
| 16 45 ,, 17  0 | 595 | 282 | 271 | 932 |
| 17  0 ,, 17 15 | 596 | 271 | 260 | 925 |
| 17 15 ,, 17 30 | 596 | 260 | 249 | 918 |
| 17 30 ,, 17 45 | 596 | 249 | 238 | 910 |
| 17 45 ,, 18  0 | 596 | 238 | 226 | 902 |
| | | | | |
| 18  0 ,, 18 15 | 8·596 | 8·226 | 8·214 | 11·894 |
| 18 15 ,, 18 30 | 597 | 214 | 203 | 886 |
| 18 30 ,, 18 45 | 597 | 203 | 191 | 878 |
| 18 45 ,, 19  0 | 597 | 191 | 179 | 870 |
| 19  0 ,, 19 15 | 597 | 178 | 166 | 861 |
| 19 15 ,, 19 30 | 598 | 166 | 154 | 853 |
| 19 30 ,, 19 45 | 598 | 154 | 141 | 845 |
| 19 45 ,, 20  0 | 598 | 141 | 128 | 836 |
| | | | | |
| 20  0 ,, 20 15 | 8·598 | 8·128 | 8·115 | 11·827 |
| 20 15 ,, 20 30 | 599 | 115 | 102 | 819 |
| 20 30 ,, 20 45 | 599 | 102 | 089 | 810 |
| 20 45 ,, 21  0 | 599 | 089 | 076 | 801 |
| 21  0 ,, 21 15 | 599 | 076 | 062 | 792 |
| 21 15 ,, 21 30 | 600 | 062 | 049 | 783 |
| 21 30 ,, 21 45 | 600 | 049 | 035 | 774 |
| 21 45 ,, 22  0 | 600 | 035 | 021 | 765 |
| | | | | |
| 22  0 ,, 22 15 | 8·600 | 8·021 | 8·007 | 11·755 |
| 22 15 ,, 22 30 | 601 | 007 | 7·992 | 746 |
| 22 30 ,, 22 45 | 601 | 7·992 | 978 | 736 |
| 22 45 ,, 23  0 | 601 | 978 | 963 | 726 |
| 23  0 ,, 23 15 | 601 | 963 | 949 | 716 |
| 23 15 ,, 23 30 | 602 | 949 | 934 | 707 |
| 23 30 ,, 23 45 | 602 | 934 | 919 | 697 |
| 23 45 ,, 24  0 | 602 | 919 | 904 | 687 |

a quarter of a degree of Lat. and Long. on the scale of 1 inch to 2 miles.

| Latitude. | | | | | Length in Inches. | | | |
|---|---|---|---|---|---|---|---|---|
| | | | | | m<br>on meridian. | n<br>on lower parallel. | p<br>on upper parallel. | q<br>on diagonal. |
| ° | ′ | | ° | ′ | | | | |
| From 24 | 0 | to | 24 | 15 | 8·603 | 7·904 | 7·888 | 11·677 |
| 24 | 15 | ,, | 24 | 30 | 603 | 888 | 873 | 667 |
| 24 | 30 | ,, | 24 | 45 | 603 | 873 | 857 | 656 |
| 24 | 45 | ,, | 25 | 0 | 604 | 857 | 841 | 646 |
| 25 | 0 | ,, | 25 | 15 | 604 | 841 | 825 | 636 |
| 25 | 15 | ,, | 25 | 30 | 604 | 825 | 809 | 626 |
| 25 | 30 | ,, | 25 | 45 | 604 | 809 | 793 | 615 |
| 25 | 45 | ,, | 26 | 0 | 605 | 793 | 777 | 604 |
| 26 | 0 | ,, | 26 | 15 | 8·605 | 7·777 | 7·760 | 11·593 |
| 26 | 15 | ,, | 26 | 30 | 605 | 760 | 743 | 582 |
| 26 | 30 | ,, | 26 | 45 | 606 | 743 | 727 | 571 |
| 26 | 45 | ,, | 27 | 0 | 606 | 727 | 709 | 559 |
| 27 | 0 | ,, | 27 | 15 | 606 | 709 | 692 | 548 |
| 27 | 15 | ,, | 27 | 30 | 606 | 692 | 675 | 537 |
| 27 | 30 | ,, | 27 | 45 | 607 | 675 | 658 | 526 |
| 27 | 45 | ,, | 28 | 0 | 607 | 658 | 640 | 515 |
| 28 | 0 | ,, | 28 | 15 | 8·607 | 7·640 | 7·623 | 11·503 |
| 28 | 15 | ,, | 28 | 30 | 608 | 623 | 605 | 492 |
| 28 | 30 | ,, | 28 | 45 | 608 | 605 | 587 | 480 |
| 28 | 45 | ,, | 29 | 0 | 608 | 587 | 569 | 468 |
| 29 | 0 | ,, | 29 | 15 | 609 | 569 | 550 | 457 |
| 29 | 15 | ,, | 29 | 30 | 609 | 551 | 532 | 445 |
| 29 | 30 | ,, | 29 | 45 | 609 | 532 | 513 | 433 |
| 29 | 45 | ,, | 30 | 0 | 610 | 513 | 495 | 421 |
| 30 | 0 | ,, | 30 | 15 | 8·610 | 7·495 | 7·476 | 11·409 |
| 30 | 15 | ,, | 30 | 30 | 610 | 476 | 457 | 397 |
| 30 | 30 | ,, | 30 | 45 | 611 | 457 | 438 | 385 |
| 30 | 45 | ,, | 31 | 0 | 611 | 438 | 419 | 372 |
| 31 | 0 | ,, | 31 | 15 | 611 | 419 | 399 | 360 |
| 31 | 15 | ,, | 31 | 30 | 612 | 399 | 379 | 347 |
| 31 | 30 | ,, | 31 | 45 | 612 | 379 | 360 | 335 |
| 31 | 45 | ,, | 32 | 0 | 612 | 360 | 340 | 322 |
| 32 | 0 | ,, | 32 | 15 | 8·613 | 7·340 | 7·320 | 11·310 |
| 32 | 15 | ,, | 32 | 30 | 613 | 320 | 300 | 297 |
| 32 | 30 | ,, | 32 | 45 | 613 | 300 | 279 | 284 |
| 32 | 45 | ,, | 33 | 0 | 614 | 279 | 259 | 271 |
| 33 | 0 | ,, | 33 | 15 | 614 | 259 | 238 | 258 |
| 33 | 15 | ,, | 33 | 30 | 614 | 238 | 218 | 244 |
| 33 | 30 | ,, | 33 | 45 | 615 | 218 | 197 | 232 |
| 33 | 45 | ,, | 34 | 0 | 615 | 197 | 175 | 219 |
| 34 | 0 | ,, | 34 | 15 | 8·616 | 7·175 | 7·155 | 11·206 |
| 34 | 15 | ,, | 34 | 30 | 616 | 155 | 134 | 193 |
| 34 | 30 | ,, | 34 | 45 | 616 | 134 | 112 | 179 |
| 34 | 45 | ,, | 35 | 0 | 617 | 112 | 091 | 166 |
| 35 | 0 | ,, | 35 | 15 | 617 | 091 | 069 | 153 |
| 35 | 15 | ,, | 35 | 30 | 617 | 069 | 047 | 139 |
| 35 | 30 | ,, | 35 | 45 | 618 | 047 | 026 | 126 |
| 35 | 45 | ,, | 36 | 0 | 618 | 026 | 003 | 112 |

230

## TABLE IX.—Graticules of Maps. Sides and Diagonals of Squares of

| Latitude. | | | | m<br>on meridian. | n<br>on lower parallel. | p<br>on upper parallel. | q<br>on diagonal. |
|---|---|---|---|---|---|---|---|
| ° | ′ | | ° | ′ | | | |
| From 36 | 0 | to | 36 | 15 | 8·618 | 7·003 | 6·981 | 11·098 |
| 36 | 15 | ,, | 36 | 30 | 619 | 6·981 | 959 | 085 |
| 36 | 30 | ,, | 36 | 45 | 619 | 959 | 937 | 071 |
| 36 | 45 | ,, | 37 | 0 | 620 | 937 | 914 | 057 |
| 37 | 0 | ,, | 37 | 15 | 620 | 914 | 891 | 043 |
| 37 | 15 | ,, | 37 | 30 | 620 | 891 | 868 | 029 |
| 37 | 30 | ,, | 37 | 45 | 621 | 868 | 846 | 015 |
| 37 | 45 | ,, | 38 | 0 | 621 | 846 | 822 | 001 |
| 38 | 0 | ,, | 38 | 15 | 8·621 | 6·822 | 6·799 | 10·987 |
| 38 | 15 | ,, | 38 | 30 | 622 | 799 | 776 | 973 |
| 38 | 30 | ,, | 38 | 45 | 622 | 776 | 752 | 959 |
| 38 | 45 | ,, | 39 | 0 | 623 | 752 | 729 | 945 |
| 39 | 0 | ,, | 39 | 15 | 623 | 729 | 705 | 930 |
| 39 | 15 | ,, | 39 | 30 | 623 | 705 | 681 | 916 |
| 39 | 30 | ,, | 39 | 45 | 624 | 681 | 657 | 902 |
| 39 | 45 | ,, | 40 | 0 | 624 | 657 | 633 | 887 |
| 40 | 0 | ,, | 40 | 15 | 8·625 | 6·633 | 6·309 | 10·873 |
| 40 | 15 | ,, | 40 | 30 | 625 | 609 | 584 | 858 |
| 40 | 30 | ,, | 40 | 45 | 625 | 584 | 560 | 843 |
| 40 | 45 | ,, | 41 | 0 | 626 | 560 | 535 | 829 |
| 41 | 0 | ,, | 41 | 15 | 626 | 535 | 510 | 814 |
| 41 | 15 | ,, | 41 | 30 | 626 | 510 | 486 | 800 |
| 41 | 30 | ,, | 41 | 45 | 627 | 486 | 461 | 786 |
| 41 | 45 | ,, | 42 | 0 | 627 | 461 | 435 | 770 |
| 42 | 0 | ,, | 42 | 15 | 8·627 | 6·435 | 6·410 | 10·755 |
| 42 | 15 | ,, | 42 | 30 | 628 | 410 | 385 | 741 |
| 42 | 30 | ,, | 42 | 45 | 628 | 385 | 359 | 726 |
| 42 | 45 | ,, | 43 | 0 | 629 | 359 | 334 | 712 |
| 43 | 0 | ,, | 43 | 15 | 629 | 334 | 308 | 696 |
| 43 | 15 | ,, | 43 | 30 | 629 | 308 | 282 | 681 |
| 43 | 30 | ,, | 43 | 45 | 630 | 282 | 256 | 667 |
| 43 | 45 | ,, | 44 | 0 | 630 | 256 | 230 | 651 |
| 44 | 0 | ,, | 44 | 15 | 8·631 | 6·230 | 6·204 | 10·637 |
| 44 | 15 | ,, | 44 | 30 | 631 | 204 | 177 | 621 |
| 44 | 30 | ,, | 44 | 45 | 631 | 177 | 151 | 606 |
| 44 | 45 | ,, | 45 | 0 | 632 | 151 | 124 | 591 |
| 45 | 0 | ,, | 45 | 15 | 632 | 124 | 098 | 576 |
| 45 | 15 | ,, | 45 | 30 | 632 | 098 | 071 | 561 |
| 45 | 30 | ,, | 45 | 45 | 633 | 071 | 044 | 546 |
| 45 | 45 | ,, | 46 | 0 | 633 | 044 | 017 | 531 |
| 46 | 0 | ,, | 46 | 15 | 8·634 | 6·017 | 5·990 | 10·516 |
| 46 | 15 | ,, | 46 | 30 | 634 | 5·990 | 963 | 501 |
| 46 | 30 | ,, | 46 | 45 | 634 | 963 | 935 | 485 |
| 46 | 45 | ,, | 47 | 0 | 635 | 935 | 908 | 470 |
| 47 | 0 | ,, | 47 | 15 | 635 | 908 | 880 | 455 |
| 47 | 15 | ,, | 47 | 30 | 635 | 880 | 852 | 439 |
| 47 | 30 | ,, | 47 | 45 | 636 | 852 | 824 | 424 |
| 47 | 45 | ,, | 48 | 0 | 636 | 824 | 797 | 409 |

a quarter of a degree of Lat. and Long. on the scale of 1 inch to 2 miles.

| LATITUDE. | Length in Inches. | | | |
|---|---|---|---|---|
| | m on meridian. | n on lower parallel. | p on upper parallel. | q on diagonal. |
| From 48 0 to 48 15 | 8·637 | 5·797 | 5·769 | 10·394 |
| 48 15 ,, 48 30 | 637 | 769 | 740 | 378 |
| 48 30 ,, 48 45 | 637 | 740 | 712 | 363 |
| 48 45 ,, 49 0 | 638 | 712 | 684 | 348 |
| 49 0 ,, 49 15 | 638 | 684 | 655 | 333 |
| 49 15 ,, 49 30 | 639 | 655 | 627 | 318 |
| 49 30 ,, 49 45 | 639 | 627 | 598 | 302 |
| 49 45 ,, 50 0 | 639 | 598 | 569 | 286 |
| 50 0 ,, 50 15 | 8·640 | 5·569 | 5·540 | 10·271 |
| 50 15 ,, 50 30 | 640 | 540 | 511 | 256 |
| 50 30 ,, 50 45 | 640 | 511 | 482 | 240 |
| 50 45 ,, 51 0 | 641 | 482 | 453 | 225 |
| 51 0 ,, 51 15 | 641 | 453 | 423 | 210 |
| 51 15 ,, 51 30 | 642 | 423 | 394 | 195 |
| 51 30 ,, 51 45 | 642 | 394 | 364 | 179 |
| 51 45 ,, 52 0 | 642 | 364 | 335 | 164 |
| 52 0 ,, 52 15 | 8·643 | 5·335 | 5·305 | 10·149 |
| 52 15 ,, 52 30 | 643 | 305 | 275 | 133 |
| 52 30 ,, 52 45 | 643 | 275 | 245 | 118 |
| 52 45 ,, 53 0 | 644 | 245 | 215 | 103 |
| 53 0 ,, 53 15 | 644 | 215 | 185 | 088 |
| 53 15 ,, 53 30 | 645 | 185 | 155 | 073 |
| 53 30 ,, 53 45 | 645 | 155 | 124 | 057 |
| 53 45 ,, 54 0 | 645 | 124 | 094 | 042 |
| 54 0 ,, 54 15 | 8·646 | 5·094 | 5·063 | 10·027 |
| 54 15 ,, 54 30 | 646 | 063 | 032 | 012 |
| 54 30 ,, 54 45 | 646 | 032 | 002 | 9·996 |
| 54 45 ,, 55 0 | 647 | 002 | 4·971 | 981 |
| 55 0 ,, 55 15 | 647 | 4·971 | 940 | 966 |
| 55 15 ,, 55 30 | 647 | 940 | 909 | 951 |
| 55 30 ,, 55 45 | 648 | 909 | 878 | 936 |
| 55 45 ,, 56 0 | 648 | 878 | 846 | 921 |
| 56 0 ,, 56 15 | 8·649 | 4·846 | 4·815 | 9·906 |
| 56 15 ,, 56 30 | 649 | 815 | 784 | 891 |
| 56 30 ,, 56 45 | 649 | 784 | 752 | 876 |
| 56 45 ,, 57 0 | 650 | 752 | 721 | 861 |
| 57 0 ,, 57 15 | 650 | 721 | 689 | 847 |
| 57 15 ,, 57 30 | 650 | 689 | 657 | 832 |
| 57 30 ,, 57 45 | 651 | 657 | 625 | 817 |
| 57 45 ,, 58 0 | 651 | 625 | 593 | 802 |
| 58 0 ,, 58 15 | 8·651 | 4·593 | 4·561 | 9·787 |
| 58 15 ,, 58 30 | 652 | 561 | 529 | 773 |
| 58 30 ,, 58 45 | 652 | 529 | 497 | 758 |
| 58 45 ,, 59 0 | 652 | 497 | 464 | 743 |
| 59 0 ,, 59 15 | 653 | 464 | 432 | 729 |
| 59 15 ,, 59 30 | 653 | 432 | 400 | 715 |
| 59 30 ,, 59 45 | 653 | 400 | 367 | 700 |
| 59 45 ,, 60 0 | 654 | 367 | 334 | 686 |

## TABLE X.—Length in Miles of

| Latitude. | Length of 15' of Latitude. | Length of 15' of Longitude. | Latitude. | Length of 15' of Latitude. | Length of 15' of Longitude. |
|---|---|---|---|---|---|
| ° ' <br> 0 0 to 0 15 | 17·176 | 17·294 | ° ' <br> 16 0 to 16 15 | 17·190 | 16·628 |
| 1 0 to 1 15 | ·176 | ·290 | 17 0 to 17 15 | ·191 | ·542 |
| 2 0 to 2 15 | ·176 | ·282 | 18 0 to 18 15 | ·192 | ·452 |
| 3 0 to 3 15 | ·176 | ·270 | 19 0 to 19 15 | ·194 | ·356 |
| 4 0 to 4 15 | ·176 | ·252 | 20 0 to 20 15 | ·196 | ·256 |
| 5 0 to 5 15 | ·178 | ·228 | 21 0 to 21 15 | ·198 | ·152 |
| 6 0 to 6 15 | ·178 | ·200 | 22 0 to 22 15 | ·200 | ·042 |
| 7 0 to 7 15 | ·178 | ·166 | 23 0 to 23 15 | ·202 | 15·926 |
| 8 0 to 8 15 | ·180 | ·126 | 24 0 to 24 15 | ·206 | ·806 |
| 9 0 to 9 15 | ·180 | ·082 | 25 0 to 25 15 | ·208 | ·682 |
| 10 0 to 10 15 | ·182 | ·032 | 26 0 to 26 15 | ·210 | ·554 |
| 11 0 to 11 15 | ·182 | 16·978 | 27 0 to 27 15 | ·212 | ·418 |
| 12 0 to 12 15 | ·183 | ·918 | 28 0 to 28 15 | ·214 | ·280 |
| 13 0 to 13 15 | ·184 | ·852 | 29 0 to 29 15 | ·218 | ·138 |
| 14 0 to 14 15 | ·186 | ·782 | 30 0 to 30 15 | ·220 | 14·990 |
| 15 0 to 15 15 | ·188 | ·708 | 31 0 to 31 15 | ·222 | ·838 |

a quarter of a degree (15′) of Latitude and Longitude.

| Latitude. | Length of 15′ of Latitude. | Length of 15′ of Longitude. | Latitude. | Length of 15′ of Latitude. | Length of 15′ of Longitude. |
|---|---|---|---|---|---|
| ° ′ | | | ° ′ | | |
| 32  0 to 32 15 | 17·226 | 14·680 | 47  0 to 47 15 | 17·270 | 11·816 |
| 33  0 to 33 15 | ·228 | ·518 | 48  0 to 48 15 | ·274 | ·594 |
| 34  0 to 34 15 | ·232 | ·350 | 49  0 to 49 15 | ·276 | ·368 |
| 35  0 to 35 15 | ·234 | ·182 | 50  0 to 50 15 | ·280 | ·138 |
| 36  0 to 36 15 | ·236 | ·006 | 51  0 to 51 15 | ·282 | 10·906 |
| 37  0 to 37 15 | ·240 | 13·828 | 52  0 to 52 15 | ·285 | ·670 |
| 38  0 to 38 15 | ·242 | ·644 | 53  0 to 53 15 | ·288 | ·430 |
| 39  0 to 39 15 | ·246 | ·458 | 54  0 to 54 15 | ·292 | ·188 |
| 40  0 to 40 15 | ·250 | ·266 | 55  0 to 55 15 | ·294 | 9·942 |
| 41  0 to 41 15 | ·252 | ·070 | 56  0 to 56 15 | ·298 | ·692 |
| 42  0 to 42 15 | ·254 | 12·870 | 57  0 to 57 15 | ·300 | ·442 |
| 43  0 to 43 15 | ·258 | ·668 | 58  0 to 58 15 | ·303 | ·186 |
| 44  0 to 44 15 | ·262 | ·460 | 59  0 to 59 15 | ·306 | 8·928 |
| 45  0 to 45 15 | ·264 | ·248 | 60  0 to 60 15 | ·308 | ·669 |
| 46  0 to 46 15 | ·268 | ·034 | | | |

The length of the diagonal, $q$, of a space bounded by meridians and parallels $m$, $n$, $p$,
$$= \sqrt{m^2 + np}$$

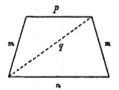

## TABLE XI.—Linear value in feet of one second of Arc

Formula: Length in feet = $\rho \sin 1''$.

| Latitude. | Length in feet. | Logarithm. | Diff. | Latitude. | Length in feet. | Logarithm. | Diff. |
|---|---|---|---|---|---|---|---|
| ° ′ | | | + | ° ′ | | | + |
| 0  0 | 100·7654 | 2·0033115 | 0 | 2  30 | 100·7674 | 2·0033199 | 6 |
| 5 | 7654 | 115 | 1 | 35 | 7675 | 205 | 6 |
| 10 | 7654 | 116 | 0 | 40 | 7676 | 211 | 6 |
| 15 | 7654 | 116 | 1 | 45 | 7678 | 217 | 6 |
| 20 | 7654 | 117 | 1 | 50 | 7679 | 223 | 6 |
| 25 | 7654 | 118 | 1 | 55 | 7681 | 229 | 7 |
| 30 | 7655 | 119 | 1 | 3  0 | 100·7682 | 2·0033236 | 7 |
| 35 | 7655 | 120 | 1 | 5 | 7684 | 243 | 7 |
| 40 | 7656 | 121 | 1 | 10 | 7685 | 250 | 7 |
| 45 | 7656 | 122 | 2 | 15 | 7687 | 257 | 8 |
| 50 | 7656 | 124 | 2 | 20 | 7689 | 265 | 7 |
| 55 | 7657 | 126 | 3 | 25 | 7690 | 272 | 8 |
| 1  0 | 100·7657 | 2·0033129 | 2 | 30 | 7692 | 280 | 8 |
| 5 | 7658 | 131 | 3 | 35 | 7694 | 288 | 8 |
| 10 | 7658 | 134 | 2 | 40 | 7696 | 296 | 8 |
| 15 | 7659 | 136 | 3 | 45 | 7698 | 304 | 9 |
| 20 | 7660 | 139 | 3 | 50 | 7700 | 313 | 8 |
| 25 | 7660 | 142 | 3 | 55 | 7702 | 321 | 9 |
| 30 | 7661 | 145 | 4 | 4  0 | 100·7704 | 2·0033330 | 9 |
| 35 | 7662 | 149 | 4 | 5 | 7706 | 339 | 9 |
| 40 | 7663 | 153 | 3 | 10 | 7708 | 348 | 9 |
| 45 | 7664 | 156 | 4 | 15 | 7710 | 357 | 10 |
| 50 | 7665 | 160 | 4 | 20 | 7713 | 367 | 10 |
| 55 | 7666 | 164 | 5 | 25 | 7715 | 377 | 10 |
| 2  0 | 100·7667 | 2·0033169 | 4 | 30 | 7717 | 387 | 10 |
| 5 | 7668 | 173 | 5 | 35 | 7719 | 397 | 11 |
| 10 | 7669 | 178 | 5 | 40 | 7722 | 408 | 10 |
| 15 | 7670 | 183 | 5 | 45 | 7724 | 418 | 11 |
| 20 | 7671 | 188 | 5 | 50 | 7727 | 429 | 11 |
| 25 | 7672 | 193 | 6 | 55 | 7729 | 440 | 11 |

## and its Logarithm, measured along the Meridian.

| Latitude. | Length in feet. | Logarithm. | Diff. | Latitude. | Length in feet. | Logarithm. | Diff. |
|---|---|---|---|---|---|---|---|
| ° ′ | | | + | ° ′ | | | + |
| 5  0 | 100·7732 | 2·0033451 | | 7 30 | 100·7829 | 2·0033868 | |
| | | | 11 | | | | 17 |
| 5 | 7735 | 462 | | 35 | 7833 | 885 | |
| | | | 12 | | | | 17 |
| 10 | 7737 | 474 | | 40 | 7837 | 902 | |
| | | | 11 | | | | 17 |
| 15 | 7740 | 485 | | 45 | 7841 | 919 | |
| | | | 12 | | | | 17 |
| 20 | 7743 | 497 | | 50 | 7845 | 936 | |
| | | | 12 | | | | 17 |
| 25 | 7745 | 509 | | 55 | 7849 | 953 | |
| | | | 12 | | | | 18 |
| 30 | 7748 | 521 | | 8  0 | 100·7853 | 2·0033971 | |
| | | | 12 | | | | 18 |
| 35 | 7751 | 533 | | 5 | 7857 | 989 | |
| | | | 13 | | | | 18 |
| 40 | 7754 | 546 | | 10 | 7861 | 2·0034007 | |
| | | | 13 | | | | 18 |
| 45 | 7757 | 559 | | 15 | 7865 | 025 | |
| | | | 13 | | | | 19 |
| 50 | 7760 | 572 | | 20 | 7870 | 044 | |
| | | | 13 | | | | 18 |
| 55 | 7763 | 585 | | 25 | 7874 | 062 | |
| | | | 13 | | | | 19 |
| 6  0 | 100·7766 | 2·0033598 | | 30 | 7878 | 081 | |
| | | | 13 | | | | 19 |
| 5 | 7769 | 611 | | 35 | 7883 | 100 | |
| | | | 14 | | | | 19 |
| 10 | 7772 | 625 | | 40 | 7887 | 119 | |
| | | | 14 | | | | 19 |
| 15 | 7776 | 639 | | 45 | 7892 | 138 | |
| | | | 14 | | | | 19 |
| 20 | 7779 | 653 | | 50 | 7896 | 157 | |
| | | | 14 | | | | 20 |
| 25 | 7782 | 667 | | 55 | 7901 | 177 | |
| | | | 15 | | | | 20 |
| 30 | 7786 | 682 | | 9  0 | 100·7905 | 2·0034197 | |
| | | | 14 | | | | 20 |
| 35 | 7789 | 696 | | 5 | 7910 | 217 | |
| | | | 15 | | | | 20 |
| 40 | 7792 | 711 | | 10 | 7914 | 237 | |
| | | | 15 | | | | 20 |
| 45 | 7796 | 726 | | 15 | 7919 | 257 | |
| | | | 15 | | | | 21 |
| 50 | 7799 | 741 | | 20 | 7924 | 278 | |
| | | | 15 | | | | 20 |
| 55 | 7803 | 756 | | 25 | 7929 | 298 | |
| | | | 16 | | | | 21 |
| 7  0 | 100·7806 | 2·0033772 | | 30 | 7934 | 319 | |
| | | | 15 | | | | 21 |
| 5 | 7810 | 787 | | 35 | 7939 | 340 | |
| | | | 16 | | | | 22 |
| 10 | 7814 | 803 | | 40 | 7944 | 362 | |
| | | | 16 | | | | 21 |
| 15 | 7817 | 819 | | 45 | 7949 | 383 | |
| | | | 16 | | | | 22 |
| 20 | 7821 | 835 | | 50 | 7953 | 405 | |
| | | | 16 | | | | 21 |
| 25 | 7825 | 851 | | 55 | 7959 | 426 | |
| | | | 17 | | | | 22 |

## TABLE XI.—Linear value in feet of one second of Arc

| Latitude. | Length in feet. | Logarithm. | Diff. | Latitude. | Length in feet. | Logarithm. | Diff. |
|---|---|---|---|---|---|---|---|
| ° ′ | | | + | ° ′ | | | + |
| 10  0 | 100·7964 | 2·0034448 | 22 | 12 30 | 100·8135 | 2·0035186 | 27 |
| 5 | 7969 | 470 | 22 | 35 | 8141 | 213 | 28 |
| 10 | 7974 | 492 | 22 | 40 | 8148 | 241 | 27 |
| 15 | 7979 | 514 | 33 | 45 | 8154 | 268 | 28 |
| 20 | 7984 | 537 | 23 | 50 | 8160 | 296 | 28 |
| 25 | 7990 | 560 | 23 | 55 | 8167 | 324 | 28 |
| 30 | 7995 | 583 | 23 | 13  0 | 100·8173 | 2·0035352 | 28 |
| 35 | 8000 | 606 | 24 | 5 | 8180 | 380 | 29 |
| 40 | 8006 | 630 | 23 | 10 | 8187 | 409 | 28 |
| 45 | 8011 | 653 | 24 | 15 | 8193 | 437 | 29 |
| 50 | 8017 | 677 | 24 | 20 | 8200 | 466 | 29 |
| 55 | 8022 | 701 | 24 | 25 | 8206 | 495 | 29 |
| 11  0 | 100·8028 | 2·0034725 | 24 | 30 | 8213 | 524 | 29 |
| 5 | 8034 | 749 | 24 | 35 | 8220 | 553 | 30 |
| 10 | 8039 | 773 | 24 | 40 | 8227 | 583 | 30 |
| 15 | 8045 | 797 | 25 | 45 | 8234 | 613 | 30 |
| 20 | 8050 | 822 | 25 | 50 | 8241 | 643 | 30 |
| 25 | 8056 | 847 | 25 | 55 | 8248 | 673 | 30 |
| 30 | 8062 | 872 | 25 | 14  0 | 100·8255 | 2·0035703 | 30 |
| 35 | 8068 | 897 | 26 | 5 | 8262 | 733 | 30 |
| 40 | 8074 | 923 | 25 | 10 | 8269 | 763 | 31 |
| 45 | 8080 | 948 | 26 | 15 | 8276 | 794 | 31 |
| 50 | 8086 | 974 | 26 | 20 | 8283 | 825 | 31 |
| 55 | 8092 | 2·0035000 | 26 | 25 | 8290 | 856 | 31 |
| 12  0 | 100·8098 | 2·0035026 | 26 | 30 | 8297 | 887 | 31 |
| 5 | 8104 | 052 | 27 | 35 | 8305 | 918 | 31 |
| 10 | 8110 | 079 | 26 | 40 | 8312 | 949 | 32 |
| 15 | 8116 | 105 | 27 | 45 | 8319 | 981 | 32 |
| 20 | 8122 | 132 | 27 | 50 | 8327 | 2·0036013 | 32 |
| 25 | 8129 | 159 | 27 | 55 | 8334 | 045 | 32 |

and its Logarithm, measured along the Meridian.

| LATITUDE. | | Length in feet. | Logarithm. | Diff. | LATITUDE. | | Length in feet. | Logarithm. | Diff. |
|---|---|---|---|---|---|---|---|---|---|
| °  | ′ | | | + | ° | ′ | | | + |
| 15 | 0 | 100·8342 | 2·0036077 | 32 | 17 | 30 | 100·8582 | 2·0037113 | 37 |
|  | 5 | 8349 | 109 | 32 |  | 35 | 8591 | 150 | 37 |
|  | 10 | 8357 | 141 | 33 |  | 40 | 8599 | 187 | 37 |
|  | 15 | 8364 | 174 | 33 |  | 45 | 8608 | 224 | 38 |
|  | 20 | 8372 | 207 | 33 |  | 50 | 8617 | 262 | 37 |
|  | 25 | 8379 | 240 | 33 |  | 55 | 8626 | 299 | 38 |
|  | 30 | 8387 | 273 | 33 | 18 | 0 | 100·8634 | 2·0037337 | 38 |
|  | 35 | 8395 | 306 | 33 |  | 5 | 8643 | 375 | 38 |
|  | 40 | 8403 | 339 | 33 |  | 10 | 8652 | 413 | 38 |
|  | 45 | 8410 | 372 | 34 |  | 15 | 8661 | 451 | 39 |
|  | 50 | 8418 | 406 | 34 |  | 20 | 8670 | 490 | 38 |
|  | 55 | 8426 | 440 | 34 |  | 25 | 8679 | 528 | 39 |
| 16 | 0 | 100·8434 | 2·0036474 | 34 |  | 30 | 8688 | 567 | 39 |
|  | 5 | 8442 | 508 | 35 |  | 35 | 8697 | 606 | 39 |
|  | 10 | 8450 | 543 | 34 |  | 40 | 8706 | 645 | 39 |
|  | 15 | 8458 | 577 | 35 |  | 45 | 8715 | 684 | 39 |
|  | 20 | 8466 | 612 | 35 |  | 50 | 8724 | 723 | 39 |
|  | 25 | 8474 | 647 | 35 |  | 55 | 8733 | 762 | 40 |
|  | 30 | 8482 | 682 | 35 | 19 | 0 | 100·8742 | 2·0037802 | 40 |
|  | 35 | 8490 | 717 | 35 |  | 5 | 8752 | 842 | 40 |
|  | 40 | 8498 | 752 | 35 |  | 10 | 8761 | 882 | 40 |
|  | 45 | 8506 | 787 | 36 |  | 15 | 8770 | 922 | 40 |
|  | 50 | 8515 | 823 | 36 |  | 20 | 8779 | 962 | 40 |
|  | 55 | 8523 | 859 | 36 |  | 25 | 8789 | 2·0038002 | 40 |
| 17 | 0 | 100·8531 | 2·0036895 | 36 |  | 30 | 8798 | 042 | 40 |
|  | 5 | 8540 | 931 | 36 |  | 35 | 8807 | 082 | 41 |
|  | 10 | 8548 | 967 | 36 |  | 40 | 8817 | 123 | 41 |
|  | 15 | 8557 | 2·0037003 | 37 |  | 45 | 8826 | 164 | 41 |
|  | 20 | 8565 | 040 | 36 |  | 50 | 8836 | 205 | 41 |
|  | 25 | 8574 | 076 | 37 |  | 55 | 8845 | 246 | 42 |

## TABLE XI.—Linear value in feet of one second of Arc

| Latitude. | | Length in feet. | Logarithm. | Diff. | Latitude. | | Length in feet. | Logarithm. | Diff. |
|---|---|---|---|---|---|---|---|---|---|
| ° | ′ | | | + | ° | ′ | | | + |
| 20 | 0 | 100·8855 | 2·0038288 | | 22 | 30 | 100·9158 | 2·0039592 | |
| | | | | 41 | | | | | 45 |
| | 5 | 8865 | 329 | | | 35 | 9169 | 637 | |
| | | | | 42 | | | | | 46 |
| | 10 | 8874 | 371 | | | 40 | 9179 | 683 | |
| | | | | 41 | | | | | 46 |
| | 15 | 8884 | 412 | | | 45 | 9190 | 729 | |
| | | | | 42 | | | | | 46 |
| | 20 | 8894 | 454 | | | 50 | 9200 | 775 | |
| | | | | 42 | | | | | 46 |
| | 25 | 8903 | 496 | | | 55 | 9211 | 821 | |
| | | | | 42 | | | | | 46 |
| | 30 | 8913 | 538 | | 23 | 0 | 100·9222 | 2·0039867 | |
| | | | | 42 | | | | | 46 |
| | 35 | 8923 | 580 | | | 5 | 9233 | 913 | |
| | | | | 43 | | | | | 47 |
| | 40 | 8933 | 623 | | | 10 | 9244 | 960 | |
| | | | | 42 | | | | | 46 |
| | 45 | 8943 | 665 | | | 15 | 9254 | 2·0040006 | |
| | | | | 43 | | | | | ·47 |
| | 50 | 8953 | 708 | | | 20 | 9265 | 053 | |
| | | | | 43 | | | | | 47 |
| | 55 | 8963 | 751 | | | 25 | 9276 | 100 | |
| | | | | 43 | | | | | 47 |
| 21 | 0 | 100·8973 | 2·0038794 | | | 30 | 9287 | 147 | |
| | | | | 43 | | | | | 47 |
| | 5 | 8983 | 837 | | | 35 | 9298 | 194 | |
| | | | | 44 | | | | | 47 |
| | 10 | 8993 | 881 | | | 40 | 9309 | 241 | |
| | | | | 43 | | | | | 47 |
| | 15 | 9003 | 924 | | | 45 | 9320 | 288 | |
| | | | | 44 | | | | | 48 |
| | 20 | 9013 | 968 | | | 50 | 9331 | 336 | |
| | | | | 43 | | | | | 48 |
| | 25 | 9023 | 2·0039011 | | | 55 | 9342 | 384 | |
| | | | | 44 | | | | | 48 |
| | 30 | 9033 | 055 | | 24 | 0 | 100·9353 | 2·0040432 | |
| | | | | 44 | | | | | 48 |
| | 35 | 9043 | 099 | | | 5 | 9364 | 480 | |
| | | | | 44 | | | | | 48 |
| | 40 | 9054 | 143 | | | 10 | 9375 | 528 | |
| | | | | 44 | | | | | 48 |
| | 45 | 9064 | 187 | | | 15 | 9387 | 576 | |
| | | | | 45 | | | | | 48 |
| | 50 | 9074 | 232 | | | 20 | 9398 | 624 | |
| | | | | 44 | | | | | 48 |
| | 55 | 9085 | 276 | | | 25 | 9409 | 672 | |
| | | | | 45 | | | | | 49 |
| 22 | 0 | 100·9095 | 2·0039321 | | | 30 | 9420 | 721 | |
| | | | | 45 | | | | | 48 |
| | 5 | 9105 | 366 | | | 35 | 9432 | 769 | |
| | | | | 45 | | | | | 49 |
| | 10 | 9116 | 411 | | | 40 | 9443 | 818 | |
| | | | | 45 | | | | | 49 |
| | 15 | 9127 | 456 | | | 45 | 9454 | 867 | |
| | | | | 45 | | | | | 49 |
| | 20 | 9137 | 501 | | | 50 | 9466 | 916 | |
| | | | | 45 | | | | | 49 |
| | 25 | 9147 | 546 | | | 55 | 9477 | 965 | |
| | | | | 46 | | | | | 50 |

and its Logarithm, measured along the Meridian.

| Latitude | Length in feet. | Logarithm. | Diff. | Latitude | Length in feet. | Logarithm. | Diff. |
|---|---|---|---|---|---|---|---|
| ° ′ | | | + | ° ′ | | | + |
| 25  0 | 100·9489 | 2·0041015 | 49 | 27 30 | 100 9845 | 2·0042546 | 53 |
| 5 | 9500 | 064 | 49 | 35 | 9857 | 599 | 53 |
| 10 | 9512 | 113 | 50 | 40 | 9869 | 652 | 53 |
| 15 | 9523 | 163 | 50 | 45 | 9882 | 705 | 53 |
| 20 | 9535 | 213 | 50 | 50 | 9894 | 758 | 53 |
| 25 | 9547 | 263 | 50 | 55 | 9906 | 811 | 54 |
| 30 | 9558 | 313 | 50 | 28  0 | 100·9919 | 2·0042865 | 53 |
| 35 | 9570 | 363 | 50 | 5 | 9931 | 918 | 54 |
| 40 | 9581 | 413 | 50 | 10 | 9944 | 972 | 53 |
| 45 | 9593 | 463 | 51 | 15 | 9956 | 2·0043025 | 54 |
| 50 | 9605 | 514 | 50 | 20 | 9969 | 079 | 54 |
| 55 | 9616 | 564 | 51 | 25 | 9981 | 133 | 54 |
| 26  0 | 100·9628 | 2·0041615 | 51 | 30 | 9994 | 187 | 54 |
| 5 | 9640 | 666 | 51 | 35 | 101·0006 | 241 | 54 |
| 10 | 9652 | 717 | 51 | 40 | 0019 | 295 | 54 |
| 15 | 9664 | 768 | 51 | 45 | 0031 | 349 | 54 |
| 20 | 9676 | 819 | 51 | 50 | 0044 | 403 | 54 |
| 25 | 9687 | 870 | 51 | 55 | 0056 | 457 | 55 |
| 30 | 9699 | 921 | 51 | 29  0 | 101·0069 | 2·0043512 | 55 |
| 35 | 9711 | 972 | 52 | 5 | 0082 | 567 | 55 |
| 40 | 9723 | 2·0042024 | 52 | 10 | 0095 | 622 | 55 |
| 45 | 9735 | 076 | 52 | 15 | 0107 | 677 | 55 |
| 50 | 9747 | 128 | 52 | 20 | 0120 | 732 | 55 |
| 55 | 9760 | 180 | 52 | 25 | 0133 | 787 | 55 |
| 27  0 | 100·9772 | 2·0042232 | 52 | 30 | 0146 | 812 | 55 |
| 5 | 9784 | 284 | 52 | 35 | 0159 | 897 | 55 |
| 10 | 9796 | 336 | 52 | 40 | 0172 | 952 | 55 |
| 15 | 9808 | 388 | 53 | 45 | 0185 | 2·0044007 | 56 |
| 20 | 9820 | 441 | 52 | 50 | 0.98 | 063 | 56 |
| 25 | 9832 | 493 | 53 | 55 | 0210 | 119 | 56 |

## TABLE XI.—Linear value in feet of one second of Arc

| Latitude. | Length in feet. | Logarithm. | Diff. | Latitude. | Length in feet. | Logarithm. | Diff. |
|---|---|---|---|---|---|---|---|
| ° ′ | | | + | ° ′ | | | + |
| 30  0 | 101·0223 | 2·0044175 | 56 | 32  30 | 101·0622 | 2·0045888 | 58 |
| 5 | 0236 | 231 | 56 | 35 | 0635 | 946 | 59 |
| 10 | 0249 | 287 | 56 | 40 | 0649 | 2·0046005 | 58 |
| 15 | 0262 | 343 | 56 | 45 | 0663 | 063 | 59 |
| 20 | 0275 | 399 | 56 | 50 | 0677 | 122 | 59 |
| 25 | 0289 | 455 | 56 | 55 | 0690 | 181 | 59 |
| 30 | 0302 | 511 | 56 | 33  0 | 101·0704 | 2·0046240 | 59 |
| 35 | 0315 | 567 | 57 | 5 | 0718 | 299 | 59 |
| 40 | 0328 | 624 | 56 | 10 | 0731 | 358 | 59 |
| 45 | 0341 | 680 | 57 | 15 | 0745 | 417 | 59 |
| 50 | 0354 | 737 | 57 | 20 | 0759 | 476 | 59 |
| 55 | 0367 | 794 | 57 | 25 | 0773 | 535 | 59 |
| 31  0 | 101·0381 | 2·0044851 | 57 | 30 | 0787 | 594 | 59 |
| 5 | 0394 | 908 | 57 | 35 | 0800 | 653 | 60 |
| 10 | 0407 | 965 | 57 | 40 | 0814 | 713 | 59 |
| 15 | 0421 | 2·0045022 | 57 | 45 | 0828 | 772 | 60 |
| 20 | 0434 | 079 | 57 | 50 | 0842 | 832 | 59 |
| 25 | 0447 | 136 | 57 | 55 | 0856 | 891 | 60 |
| 30 | 0461 | 193 | 57 | 34  0 | 101·0870 | 2·0046951 | 60 |
| 35 | 0474 | 250 | 58 | 5 | 0883 | 2·0047011 | 60 |
| 0 | 0487 | 308 | 58 | 10 | 0897 | 071 | 60 |
| 45 | 0501 | 366 | 58 | 15 | 0911 | 131 | 60 |
| 50 | 0514 | 424 | 57 | 20 | 0925 | 191 | 60 |
| 55 | 0527 | 481 | 58 | 25 | 0939 | 251 | 60 |
| 32  0 | 101·0541 | 2·0045539 | 58 | 30 | 0953 | 311 | 60 |
| 5 | 0554 | 597 | 58 | 35 | 0967 | 371 | 60 |
| 10 | 0568 | 655 | 58 | 40 | 0981 | 431 | 60 |
| 15 | 0581 | 713 | 59 | 45 | 0995 | 491 | 61 |
| 20 | 0595 | 772 | 58 | 50 | 1009 | 552 | 60 |
| 25 | 0608 | 830 | 58 | 55 | 1023 | 612 | 61 |

and its Logarithm, measured along the Meridian.

| Latitude. | Length in feet. | Logarithm. | Diff. | Latitude. | Length in feet. | Logarithm. | Diff. |
|---|---|---|---|---|---|---|---|
| ° ′ | | | + | ° ′ | | | + |
| 35 0 | 101·1038 | 2·0047673 | | 37 30 | 101·1467 | 2·0049516 | |
| | | | 60 | | | | 62 |
| 5 | 1052 | 733 | | 35 | 1481 | 578 | |
| | | | 61 | | | | 63 |
| 10 | 1066 | 794 | | 40 | 1496 | 641 | |
| | | | 61 | | | | 62 |
| 15 | 1080 | 855 | | 45 | 1510 | 703 | |
| | | | 61 | | | | 63 |
| 20 | 1094 | 916 | | 50 | 1525 | 766 | |
| | | | 61 | | | | 62 |
| 25 | 1108 | 977 | | 55 | 1539 | 828 | |
| | | | 61 | | | | 63 |
| 30 | 1122 | 2·0048038 | | 38 0 | 101·1554 | 2·0049891 | |
| | | | 61 | | | | 62 |
| 35 | 1137 | 099 | | 5 | 1569 | 953 | |
| | | | 61 | | | | 63 |
| 40 | 1151 | 160 | | 10 | 1583 | 2·0050016 | |
| | | | 61 | | | | 62 |
| 45 | 1165 | 221 | | 15 | 1598 | 078 | |
| | | | 61 | | | | 63 |
| 50 | 1179 | 282 | | 20 | 1612 | 141 | |
| | | | 61 | | | | 63 |
| 55 | 1193 | 343 | | 25 | 1627 | 204 | |
| | | | 61 | | | | 63 |
| 36 0 | 101·1208 | 2·0048404 | | 30 | 1642 | 267 | |
| | | | 61 | | | | 62 |
| 5 | 1222 | 465 | | 35 | 1656 | 329 | |
| | | | 62 | | | | 63 |
| 10 | 1236 | 527 | | 40 | 1671 | 392 | |
| | | | 61 | | | | 63 |
| 15 | 1250 | 588 | | 45 | 1686 | 455 | |
| | | | 62 | | | | 63 |
| 20 | 1265 | 650 | | 50 | 1700 | 518 | |
| | | | 61 | | | | 63 |
| 25 | 1279 | 711 | | 55 | 1715 | 581 | |
| | | | 62 | | | | 63 |
| 30 | 1294 | 773 | | 39 0 | 101·1730 | 2·0050644 | |
| | | | 61 | | | | 63 |
| 35 | 1308 | 834 | | 5 | 1744 | 707 | |
| | | | 62 | | | | 64 |
| 40 | 1322 | 896 | | 10 | 1759 | 771 | |
| | | | 62 | | | | 63 |
| 45 | 1337 | 958 | | 15 | 1774 | 834 | |
| | | | 62 | | | | 63 |
| 50 | 1351 | 2·0049020 | | 20 | 1788 | 897 | |
| | | | 62 | | | | 63 |
| 55 | 1366 | 082 | | 25 | 1803 | 960 | |
| | | | 62 | | | | 63 |
| 37 0 | 101·1380 | 2·0049144 | | 30 | 1818 | 2·0051023 | |
| | | | 62 | | | | 63 |
| 5 | 1395 | 206 | | 35 | 1832 | 086 | |
| | | | 62 | | | | 64 |
| 10 | 1409 | 268 | | 40 | 1847 | 150 | |
| | | | 62 | | | | 63 |
| 15 | 1423 | 330 | | 45 | 1862 | 213 | |
| | | | 62 | | | | 64 |
| 20 | 1438 | 392 | | 50 | 1877 | 277 | |
| | | | 62 | | | | 63 |
| 25 | 1452 | 454 | | 55 | 1891 | 340 | |
| | | | 62 | | | | 64 |

## TABLE XI.—Linear value in feet of one second of Arc

| Latitude. | | Length in feet. | Logarithm. | Diff. | Latitude. | | Length in feet. | Logarithm. | Diff. |
|---|---|---|---|---|---|---|---|---|---|
| ° | ′ | | | + | ° | ′ | | | + |
| 40 | 0 | 101·1905 | 2·0051404 | 63 | 42 | 30 | 101·2353 | 2·0053321 | 64 |
| | 5 | 1921 | 467 | 64 | | 35 | 2368 | 385 | 64 |
| | 10 | 1936 | 531 | 63 | | 40 | 2383 | 449 | 64 |
| | 15 | 1951 | 594 | 64 | | 45 | 2398 | 513 | 65 |
| | 20 | 1966 | 658 | 63 | | 50 | 2413 | 578 | 64 |
| | 25 | 1980 | 721 | 64 | | 55 | 2428 | 642 | 65 |
| | 30 | 1995 | 785 | 64 | 43 | 0 | 101·2443 | 2·0053707 | 64 |
| | 35 | 2010 | 849 | 64 | | 5 | 2458 | 771 | 64 |
| | 40 | 2025 | 913 | 63 | | 10 | 2473 | 835 | 64 |
| | 45 | 2040 | 976 | 64 | | 15 | 2488 | 899 | 65 |
| | 50 | 2055 | 2·0052040 | 64 | | 20 | 2503 | 964 | 64 |
| | 55 | 2070 | 104 | 64 | | 25 | 2518 | 2·0054028 | 65 |
| 41 | 0 | 101·2085 | 2·0052168 | 64 | | 30 | 2533 | 093 | 64 |
| | 5 | 2100 | 232 | 64 | | 35 | 2548 | 157 | 65 |
| | 10 | 2114 | 296 | 63 | | 40 | 2563 | 222 | 64 |
| | 15 | 2129 | 359 | 64 | | 45 | 2578 | 286 | 65 |
| | 20 | 2144 | 423 | 64 | | 50 | 2593 | 351 | 64 |
| | 25 | 2159 | 487 | 64 | | 55 | 2608 | 415 | 64 |
| | 30 | 2174 | 551 | 64 | 44 | 0 | 101·2623 | 2·0054479 | 64 |
| | 35 | 2189 | 615 | 64 | | 5 | 2638 | 543 | 65 |
| | 40 | 2204 | 679 | 64 | | 10 | 2653 | 608 | 64 |
| | 45 | 2219 | 743 | 65 | | 15 | 2668 | 672 | 65 |
| | 50 | 2234 | 808 | 64 | | 20 | 2684 | 737 | 64 |
| | 55 | 2249 | 872 | 64 | | 25 | 2699 | 801 | 65 |
| 42 | 0 | 101·2264 | 2·0052936 | 64 | | 30 | 2714 | 866 | 64 |
| | 5 | 2279 | 2·0053000 | 64 | | 35 | 2729 | 930 | 65 |
| | 10 | 2294 | 064 | 64 | | 40 | 2744 | 995 | 64 |
| | 15 | 2308 | 128 | 64 | | 45 | 2759 | 2·0055059 | 65 |
| | 20 | 2323 | 192 | 64 | | 50 | 2774 | 124 | 64 |
| | 25 | 2338 | 256 | 65 | | 55 | 2789 | 188 | 65 |

and its Logarithm, measured along the Meridian.

| Latitude. | | Length in feet. | Logarithm. | Diff. | Latitude. | | Length in feet. | Logarithm. | Diff. |
|---|---|---|---|---|---|---|---|---|---|
| ° | ′ | | | + | ° | ′ | | | + |
| 45 | 0 | 101·2804 | 2·0055253 | 64 | 47 | 30 | 101·3255 | 2·0057186 | 64 |
| | 5 | 2819 | 317 | 65 | | 35 | 3270 | 250 | 65 |
| | 10 | 2834 | 382 | 64 | | 40 | 3285 | 315 | 64 |
| | 15 | 2849 | 446 | 65 | | 45 | 3300 | 379 | 64 |
| | 20 | 2864 | 511 | 64 | | 50 | 3315 | 443 | 64 |
| | 25 | 2879 | 575 | 65 | | 55 | 3330 | 507 | 65 |
| | 30 | 2894 | 640 | 64 | 48 | 0 | 101·3345 | 2·0057572 | 64 |
| | 35 | 2909 | 704 | 65 | | 5 | 3360 | 636 | 64 |
| | 40 | 2924 | 769 | 64 | | 10 | 3375 | 700 | 64 |
| | 45 | 2939 | 833 | 65 | | 15 | 3390 | 764 | 64 |
| | 50 | 2954 | 898 | 64 | | 20 | 3405 | 828 | 64 |
| | 55 | 2969 | 962 | 65 | | 25 | 3419 | 892 | 64 |
| 46 | 0 | 101·2984 | 2·0056027 | 64 | | 30 | 3434 | 956 | 64 |
| | 5 | 2999 | 091 | 65 | | 35 | 3449 | 2·0058020 | 65 |
| | 10 | 3015 | 156 | 64 | | 40 | 3464 | 085 | 64 |
| | 15 | 3030 | 220 | 65 | | 45 | 3479 | 149 | 64 |
| | 20 | 3045 | 285 | 64 | | 50 | 3494 | 213 | 63 |
| | 25 | 3060 | 349 | 65 | | 55 | 3509 | 276 | 64 |
| | 30 | 3075 | 414 | 64 | 49 | 0 | 101·3524 | 2·0058340 | 64 |
| | 35 | 3090 | 478 | 65 | | 5 | 3539 | 404 | 64 |
| | 40 | 3105 | 543 | 64 | | 10 | 3554 | 468 | 64 |
| | 45 | 3120 | 607 | 65 | | 15 | 3569 | 532 | 64 |
| | 50 | 3135 | 672 | 64 | | 20 | 3584 | 596 | 63 |
| | 55 | 3150 | 736 | 64 | | 25 | 3598 | 659 | 64 |
| 47 | 0 | 101·3165 | 2·0056800 | 64 | | 30 | 3613 | 723 | 64 |
| | 5 | 3180 | 864 | 65 | | 35 | 3628 | 787 | 64 |
| | 10 | 3195 | 929 | 64 | | 40 | 3643 | 851 | 63 |
| | 15 | 3210 | 993 | 65 | | 45 | 3658 | 914 | 64 |
| | 20 | 3225 | 2·0057058 | 64 | | 50 | 3673 | 978 | 63 |
| | 25 | 3240 | 122 | 64 | | 55 | 3688 | 2·0059041 | 64 |

(5455)

## TABLE XI.—Linear value in feet of one second of Arc

| Latitude. | Length in feet. | Logarithm. | Diff. | Latitude. | Length in feet. | Logarithm. | Diff. |
|---|---|---|---|---|---|---|---|
| ° ′ | | | + | ° ′ | | | + |
| 50  0 | 101·3703 | 2·0059105 | 63 | 52 30 | 101·4144 | 2·0060995 | 62 |
|     5 | 3717 | 168 | 64 |     35 | 4158 | 2·0061057 | 63 |
|    10 | 3732 | 232 | 63 |     40 | 4173 | 120 | 62 |
|    15 | 3747 | 295 | 64 |     45 | 4187 | 182 | 63 |
|    20 | 3762 | 359 | 63 |     50 | 4202 | 245 | 62 |
|    25 | 3776 | 422 | 64 |     55 | 4217 | 307 | 62 |
|    30 | 3791 | 486 | 63 | 53  0 | 101·4231 | 2·0061369 | 62 |
|    35 | 3806 | 549 | 64 |      5 | 4246 | 431 | 62 |
|    40 | 3821 | 613 | 63 |     10 | 4260 | 493 | 62 |
|    45 | 3836 | 676 | 63 |     15 | 4275 | 555 | 62 |
|    50 | 3851 | 739 | 63 |     20 | 4289 | 617 | 61 |
|    55 | 3865 | 802 | 64 |     25 | 4303 | 678 | 62 |
| 51  0 | 101·3880 | 2·0059866 | 63 |     30 | 4318 | 740 | 62 |
|     5 | 3895 | 929 | 63 |     35 | 4332 | 802 | 62 |
|    10 | 3909 | 992 | 63 |     40 | 4347 | 864 | 61 |
|    15 | 3924 | 2·0060055 | 63 |     45 | 4361 | 925 | 62 |
|    20 | 3939 | 118 | 63 |     50 | 4375 | 987 | 61 |
|    25 | 3954 | 181 | 63 |     55 | 4390 | 2·0062048 | 62 |
|    30 | 3968 | 244 | 62 | 54  0 | 101·4404 | 2·0062110 | 61 |
|    35 | 3983 | 306 | 63 |      5 | 4418 | 171 | 61 |
|    40 | 3998 | 369 | 63 |     10 | 4433 | 232 | 61 |
|    45 | 4012 | 432 | 63 |     15 | 4447 | 293 | 62 |
|    50 | 4027 | 495 | 62 |     20 | 4461 | 355 | 61 |
|    55 | 4041 | 557 | 63 |     25 | 4476 | 416 | 61 |
| 52  0 | 101·4056 | 2·0060620 | 63 |     30 | 4490 | 477 | 61 |
|     5 | 4071 | 683 | 63 |     35 | 4504 | 538 | 61 |
|    10 | 4085 | 746 | 62 |     40 | 4518 | 599 | 61 |
|    15 | 4100 | 808 | 63 |     45 | 4532 | 660 | 61 |
|    20 | 4115 | 871 | 62 |     50 | 4547 | 721 | 60 |
|    25 | 4129 | 933 | 62 |     55 | 4561 | 781 | 61 |

245

and its Logarithm, measured along the Meridian.

| Latitude. | | Length in feet. | Logarithm. | Diff. | Latitude. | | Length in feet. | Logarithm. | Diff. |
|---|---|---|---|---|---|---|---|---|---|
| ° | ′ | | | + | ° | ′ | | | + |
| 55 | 0 | 101·4575 | 2·0062842 | 61 | 57 | 30 | 101·4993 | 2·0064632 | 58 |
| | 5 | 4589 | 903 | 61 | | 35 | 5007 | 690 | 59 |
| | 10 | 4603 | 964 | 60 | | 40 | 5021 | 749 | 58 |
| | 15 | 4618 | 2·0063024 | 61 | | 45 | 5034 | 807 | 59 |
| | 20 | 4632 | 085 | 60 | | 50 | 5048 | 866 | 58 |
| | 25 | 4646 | 145 | 60 | | 55 | 5062 | 924 | 58 |
| | 30 | 4660 | 205 | 60 | 58 | 0 | 101·5075 | 2·0064982 | 58 |
| | 35 | 4674 | 265 | 61 | | 5 | 5089 | 2·0065040 | 58 |
| | 40 | 4688 | 326 | 60 | | 10 | 5102 | 098 | 58 |
| | 45 | 4702 | 386 | 60 | | 15 | 5116 | 156 | 58 |
| | 50 | 4716 | 446 | 60 | | 20 | 5129 | 214 | 57 |
| | 55 | 4730 | 506 | 60 | | 25 | 5143 | 271 | 58 |
| | | | | | | 30 | 5156 | 329 | 57 |
| 56 | 0 | 101·4744 | 2·0063566 | 60 | | 35 | 5170 | 386 | 58 |
| | 5 | 4758 | 626 | 60 | | 40 | 5183 | 444 | 57 |
| | 10 | 4772 | 686 | 59 | | 45 | 5197 | 501 | 57 |
| | 15 | 4786 | 745 | 60 | | 50 | 5201 | 558 | 57 |
| | 20 | 4800 | 805 | 59 | | 55 | 5223 | 615 | 58 |
| | 25 | 4814 | 864 | 60 | 59 | 0 | 101·5237 | 2·0065673 | 57 |
| | 30 | 4828 | 924 | 59 | | 5 | 5250 | 730 | 57 |
| | 35 | 4842 | 963 | 60 | | 10 | 5263 | 787 | 56 |
| | 40 | 4856 | 2·0064043 | 59 | | 15 | 5276 | 843 | 57 |
| | 45 | 4869 | 102 | 59 | | 20 | 5289 | 900 | 56 |
| | 50 | 4883 | 161 | 59 | | 25 | 5303 | 956 | 57 |
| | 55 | 4897 | 220 | 59 | | 30 | 5316 | 2·0066013 | 56 |
| 57 | 0 | 101·4911 | 2·0064279 | 59 | | 35 | 5329 | 069 | 57 |
| | 5 | 4925 | 338 | 59 | | 40 | 5343 | 126 | 56 |
| | 10 | 4938 | 397 | 59 | | 45 | 5356 | 182 | 57 |
| | 15 | 4952 | 456 | 59 | | 50 | 5369 | 239 | 56 |
| | 20 | 4966 | 515 | 58 | | 55 | 5382 | 295 | 56 |
| | 25 | 4980 | 573 | 59 | 60 | 0 | 101·5395 | 2·0066351 | |

## TABLE XII.—Linear value in feet of one second of Arc

Formula: Length in feet $= \dfrac{\pi}{180 \times 3600} \nu \cos \lambda$.

| Latitude | Length in feet. | Diff. | Logarithm. | Diff. | Latitude | Length in feet. | Diff. | Logarithm. | Diff. |
|---|---|---|---|---|---|---|---|---|---|
| ° ′ | | | | | ° ′ | | | | |
| 0 0 | 101·4538 | — | 2·0062683 | — | 2 30 | 101·3579 | — | 2·0058576 | — |
| 5 | 4537 | 1 | 2678 | 5 | 35 | 3514 | 65 | 8297 | 279 |
| 10 | 4533 | 4 | 2665 | 13 | 40 | 3447 | 67 | 8010 | 287 |
| 15 | 4528 | 5 | 2642 | 23 | 45 | 3378 | 69 | 7713 | 297 |
| 20 | 4521 | 7 | 2611 | 31 | 50 | 3306 | 72 | 7407 | 306 |
| 25 | 4511 | 10 | 2569 | 42 | 55 | 3232 | 74 | 7091 | 316 |
| 30 | 4500 | 11 | 2519 | 50 | 3 0 | 101·3157 | 75 | 2·0056767 | 324 |
| 35 | 4486 | 14 | 2460 | 59 | 5 | 3079 | 78 | 6434 | 333 |
| 40 | 4470 | 16 | 2391 | 69 | 10 | 3000 | 79 | 6092 | 342 |
| 45 | 4452 | 18 | 2314 | 77 | 15 | 2917 | 83 | 5739 | 353 |
| 50 | 4431 | 21 | 2227 | 87 | 20 | 2833 | 84 | 5379 | 360 |
| 55 | 4409 | 22 | 2131 | 96 | 25 | 2747 | 86 | 5009 | 370 |
| 1 0 | 101·4385 | 24 | 2·0062026 | 105 | 30 | 2659 | 88 | 4630 | 379 |
| 5 | 4358 | 27 | 1912 | 114 | 35 | 2568 | 91 | 4242 | 388 |
| 10 | 4329 | 29 | 1789 | 123 | 40 | 2475 | 93 | 3844 | 398 |
| 15 | 4298 | 31 | 1656 | 133 | 45 | 2380 | 95 | 3437 | 407 |
| 20 | 4265 | 33 | 1515 | 141 | 50 | 2284 | 96 | 3022 | 415 |
| 25 | 4230 | 35 | 1364 | 151 | 55 | 2185 | 99 | 2597 | 425 |
| 30 | 4193 | 37 | 1205 | 159 | 4 0 | 101·2083 | 102 | 2·0052163 | 434 |
| 35 | 4153 | 40 | 1036 | 169 | 5 | 1980 | 103 | 1720 | 443 |
| 40 | 4112 | 41 | 0858 | 178 | 10 | 1875 | 105 | 1267 | 453 |
| 45 | 4068 | 44 | 0671 | 187 | 15 | 1767 | 108 | 0805 | 462 |
| 50 | 4022 | 46 | 0474 | 197 | 20 | 1657 | 110 | 0334 | 471 |
| 55 | 3974 | 48 | 0270 | 204 | 25 | 1545 | 112 | 2·0049854 | 480 |
| 2 0 | 101·3924 | 50 | 2·0060055 | 215 | 30 | 1432 | 113 | 9365 | 489 |
| 5 | 3872 | 52 | 2·0059831 | 224 | 35 | 1316 | 116 | 8867 | 498 |
| 10 | 3818 | 54 | 9598 | 233 | 40 | 1197 | 119 | 8360 | 507 |
| 15 | 3761 | 57 | 9356 | 242 | 45 | 1077 | 120 | 7842 | 518 |
| 20 | 3703 | 58 | 9105 | 251 | 50 | 0955 | 122 | 7317 | 525 |
| 25 | 3642 | 61 | 8845 | 260 | 55 | 0830 | 125 | 6781 | 536 |
| | | 63 | | 269 | | | 127 | | 544 |

and its Logarithm, measured along Parallels of Latitude.

| Lati-tude. | Length in feet. | Diff. | Logarithm. | Diff. | Lati-tude. | Length in feet. | Diff. | Logarithm. | Diff. |
|---|---|---|---|---|---|---|---|---|---|
| ° ′ | | — | | — | ° ′ | | — | | — |
| 5 0 | 101·0703 | | 2·0046237 | | 7 30 | 100·5917 | | 2·0025620 | |
| | | 128 | | 544 | | | 193 | | 831 |
| 5 | 0575 | | 5684 | | 35 | 5724 | | 4789 | |
| | | 131 | | 563 | | | 194 | | 840 |
| 10 | 0444 | | 5121 | | 40 | 5530 | | 3949 | |
| | | 134 | | 572 | | | 197 | | 849 |
| 15 | 0310 | | 4549 | | 45 | 5333 | | 3100 | |
| | | 135 | | 581 | | | 199 | | 859 |
| 20 | 0175 | | 3968 | | 50 | 5134 | | 2241 | |
| | | 137 | | 591 | | | 200 | | 867 |
| 25 | 0038 | | 3377 | | 55 | 4934 | | 1374 | |
| | | 139 | | 599 | | | 203 | | 878 |
| 30 | 100·9899 | | 2778 | | 8 0 | 100·4731 | | 2·0020496 | |
| | | 142 | | 608 | | | 206 | | 887 |
| 35 | 9757 | | 2170 | | 5 | 4525 | | 2·0019609 | |
| | | 143 | | 618 | | | 207 | | 895 |
| 40 | 9614 | | 1552 | | 10 | 4318 | | 8714 | |
| | | 146 | | 628 | | | 209 | | 906 |
| 45 | 9468 | | 0924 | | 15 | 4109 | | 7808 | |
| | | 148 | | 636 | | | 211 | | 913 |
| 50 | 9320 | | 0288 | | 20 | 3898 | | 6895 | |
| | | 150 | | 645 | | | 214 | | 921 |
| 55 | 9170 | | 2·0039643 | | 25 | 3684 | | 5971 | |
| | | 153 | | 656 | | | 215 | | 933 |
| 6 0 | 100·9017 | | 2·0038987 | | 30 | 3469 | | 5038 | |
| | | 154 | | 663 | | | 218 | | 943 |
| 5 | 8863 | | 8324 | | 35 | 3251 | | 4095 | |
| | | 156 | | 674 | | | 220 | | 951 |
| 10 | 8707 | | 7650 | | 40 | 3031 | | 3144 | |
| | | 159 | | 682 | | | 222 | | 962 |
| 15 | 8548 | | 6968 | | 45 | 2809 | | 2182 | |
| | | 160 | | 692 | | | 224 | | 970 |
| 20 | 8388 | | 6276 | | 50 | 2585 | | 1212 | |
| | | 163 | | 701 | | | 226 | | 980 |
| 25 | 8225 | | 5575 | | 55 | 2359 | | 0232 | |
| | | 165 | | 710 | | | 229 | | 989 |
| 30 | 8060 | | 4865 | | 9 0 | 100·2130 | | 2·0009243 | |
| | | 167 | | 720 | | | 230 | | 999 |
| 35 | 7893 | | 4145 | | 5 | 1900 | | 8244 | |
| | | 169 | | 728 | | | 232 | | 1007 |
| 40 | 7724 | | 3417 | | 10 | 1668 | | 7237 | |
| | | 171 | | 738 | | | 235 | | 1017 |
| 45 | 7553 | | 2679 | | 15 | 1433 | | 6220 | |
| | | 173 | | 747 | | | 237 | | 1027 |
| 50 | 7380 | | 1932 | | 20 | 1196 | | 5193 | |
| | | 176 | | 757 | | | 238 | | 1036 |
| 55 | 7204 | | 1175 | | 25 | 0958 | | 4157 | |
| | | 178 | | 766 | | | 241 | | 1046 |
| 7 0 | 100·7026 | | 2·0030409 | | 30 | 0717 | | 3111 | |
| | | 179 | | 775 | | | 243 | | 1055 |
| 5 | 6847 | | 2·0029634 | | 35 | 0474 | | 2056 | |
| | | 182 | | 785 | | | 245 | | 1063 |
| 10 | 6665 | | 8849 | | 40 | 0229 | | 0993 | |
| | | 184 | | 793 | | | 248 | | 1074 |
| 15 | 6481 | | 8056 | | 45 | 99·9981 | | 1·9999919 | |
| | | 186 | | 803 | | | 249 | | 1083 |
| 20 | 6295 | | 7253 | | 50 | 9732 | | 8836 | |
| | | 188 | | 811 | | | 251 | | 1092 |
| 25 | 6107 | | 6442 | | 55 | 9481 | | 7744 | |
| | | 190 | | 822 | | | 254 | | 1102 |

## TABLE XII.—Linear value in feet of one second of Arc

| Lati-tude. | Length in feet. | Diff. | Logarithm. | Diff. | Lati-tude. | Length in feet. | Diff. | Logarithm. | Diff. |
|---|---|---|---|---|---|---|---|---|---|
| ° ′ | | | | | ° ′ | | | | |
| 10 0 | 99·9227 | — | 1·9996642 | — | 12 30 | 99·0647 | — | 1·9959118 | — |
| 5 | 8972 | 255 | 5531 | 1111 | 35 | 0328 | 319 | 57792 | 1396 |
| 10 | 8714 | 258 | 4410 | 1121 | 40 | 0008 | 320 | 56387 | 1405 |
| 15 | 8454 | 260 | 3281 | 1129 | 45 | 98·9685 | 323 | 54972 | 1415 |
| 20 | 8192 | 262 | 2141 | 1140 | 50 | 9361 | 324 | 53547 | 1425 |
| 25 | 7928 | 264 | 0992 | 1149 | 55 | 9034 | 327 | 52123 | 1435 |
| | | 266 | | 1159 | | | 329 | | 1444 |
| 30 | 7662 | 268 | 1·9989833 | 1167 | 13 0 | 98·8705 | 331 | 1·9950668 | 1454 |
| 35 | 7394 | 271 | 8666 | 1177 | 5 | 8374 | 333 | 49214 | 1463 |
| 40 | 7123 | 272 | 7489 | 1187 | 10 | 8041 | 335 | 47751 | 1473 |
| 45 | 6851 | 275 | 6302 | 1196 | 15 | 7706 | 337 | 46278 | 1482 |
| 50 | 6576 | 276 | 5106 | 2205 | 20 | 7369 | 339 | 44796 | 1493 |
| 55 | 6300 | 279 | 3901 | 1215 | 25 | 7030 | 341 | 43303 | 1502 |
| 11 0 | 99·6021 | 281 | 1·9982686 | 1225 | 30 | 6689 | 344 | 41801 | 1511 |
| 5 | 5740 | 283 | 81461 | 1234 | 35 | 6345 | 345 | 40290 | 1521 |
| 10 | 5457 | 285 | 80227 | 1244 | 40 | 6000 | 347 | 38769 | 1531 |
| 15 | 5172 | 287 | 78983 | 1253 | 45 | 5653 | 350 | 37238 | 1541 |
| 20 | 4885 | 289 | 77730 | 1263 | 50 | 5303 | 352 | 35697 | 1550 |
| 25 | 4596 | 291 | 76467 | 1271 | 55 | 4951 | 354 | 34147 | 1560 |
| 30 | 4305 | 294 | 75196 | 1282 | 14 0 | 98·4597 | 355 | 1·9932587 | 1570 |
| 35 | 4011 | 295 | 73914 | 1290 | 5 | 4242 | 358 | 31017 | 1579 |
| 40 | 3716 | 297 | 72624 | 1301 | 10 | 3884 | 360 | 29438 | 1589 |
| 45 | 3419 | 300 | 71323 | 1310 | 15 | 3524 | 362 | 27849 | 1600 |
| 50 | 3119 | 302 | 70013 | 1320 | 20 | 3162 | 364 | 26249 | 1607 |
| 55 | 2817 | 304 | 68693 | 1329 | 25 | 2798 | 366 | 24642 | 1619 |
| 12 0 | 99·2513 | 305 | 1·9967364 | 1338 | 30 | 2432 | 369 | 23023 | 1629 |
| 5 | 2208 | 306 | 66026 | 1349 | 35 | 2063 | 370 | 21394 | 1637 |
| 10 | 1900 | 311 | 64677 | 1358 | 40 | 1693 | 373 | 19757 | 1648 |
| 15 | 1589 | 312 | 63319 | 1367 | 45 | 1320 | 374 | 18109 | 1657 |
| 20 | 1277 | 314 | 61952 | 1377 | 50 | 0946 | 376 | 16452 | 1667 |
| 25 | 0963 | 316 | 60575 | 1387 | 55 | 0570 | 379 | 14785 | 1677 |

249

and its Logarithm, measured along Parallels of Latitude.

| Lati-tude. | Length in feet. | Diff. | Logarithm. | Diff. | Lati-tude. | Length in feet. | Diff. | Logarithm. | Diff. |
|---|---|---|---|---|---|---|---|---|---|
| ° ′ | | | | | ° ′ | | | | |
| 15 0 | 98·0191 | — | 1·9913108 | — | 17 30 | 96·7879 | — | 1·9858211 | — |
| 5 | 97·9810 | 381 | 11421 | 1687 | 35 | 7437 | 442 | 56226 | 1985 |
| 10 | 9428 | 382 | 09725 | 1696 | 40 | 6993 | 444 | 54232 | 1994 |
| 15 | 9043 | 385 | 08019 | 1706 | 45 | 6547 | 446 | 52228 | 2004 |
| 20 | 8656 | 387 | 06303 | 1716 | 50 | 6098 | 449 | 50213 | 2015 |
| 25 | 8268 | 388 | 04577 | 1726 | 55 | 5648 | 450 | 48189 | 2024 |
| 30 | 7877 | 391 | 02841 | 1736 | 18 0 | 96·5196 | 452 | 1·9846153 | 2036 |
| 35 | 7484 | 393 | 01095 | 1746 | 5 | 4741 | 455 | 44109 | 2044 |
| 40 | 7089 | 395 | 1·9899340 | 1755 | 10 | 4285 | 456 | 42054 | 2055 |
| 45 | 6692 | 397 | 97574 | 1766 | 15 | 3826 | 459 | 39988 | 2066 |
| 50 | 6292 | 400 | 95799 | 1775 | 20 | 3366 | 460 | 37913 | 2075 |
| 55 | 5891 | 401 | 94014 | 1785 | 25 | 2904 | 462 | 35828 | 2085 |
| 16 0 | 97·5488 | 403 | 1·9892219 | 1795 | 30 | 2439 | 465 | 33733 | 2095 |
| 5 | 5083 | 405 | 90414 | 1805 | 35 | 1973 | 466 | 31627 | 2106 |
| 10 | 4675 | 408 | 88600 | 1814 | 40 | 1504 | 469 | 29511 | 2116 |
| 15 | 4266 | 409 | 86775 | 1825 | 45 | 1033 | 471 | 27385 | 2126 |
| 20 | 3854 | 412 | 84941 | 1834 | 50 | 0561 | 472 | 25249 | 2136 |
| 25 | 3441 | 413 | 83096 | 1845 | 55 | 0086 | 475 | 23102 | 2147 |
| 30 | 3025 | 416 | 81242 | 1854 | 19 0 | 95·9610 | 476 | 1·9820946 | 2156 |
| 35 | 2608 | 417 | 79378 | 1864 | 5 | 9131 | 479 | 18780 | 2166 |
| 40 | 2188 | 420 | 77503 | 1875 | 10 | 8650 | 481 | 16602 | 2178 |
| 45 | 1766 | 422 | 75618 | 1885 | 15 | 8167 | 483 | 14414 | 2188 |
| 50 | 1343 | 423 | 73724 | 1894 | 20 | 7683 | 484 | 12217 | 2197 |
| 55 | 0917 | 426 | 71820 | 1904 | 25 | 7196 | 487 | 10009 | 2208 |
| 17 0 | 97·0489 | 428 | 1·9869906 | 1914 | 30 | 6707 | 489 | 07791 | 2218 |
| 5 | 0059 | 430 | 67982 | 1924 | 35 | 6217 | 490 | 05563 | 2228 |
| 10 | 96·9627 | 432 | 66048 | 1934 | 40 | 5724 | 493 | 03324 | 2239 |
| 15 | 9193 | 434 | 64103 | 1945 | 45 | 5229 | 495 | 01075 | 2249 |
| 20 | 8757 | 436 | 62149 | 1954 | 50 | 4732 | 497 | 1·9798815 | 2260 |
| 25 | 8319 | 438 | 60185 | 1964 | 55 | 4233 | 499 | 96545 | 2270 |
| | | 440 | | 1974 | | | 501 | | 2280 |

## TABLE XII.—Linear value in feet of one second of Arc

| LATI-TUDE. | Length in feet. | Diff. | Logarithm. | Diff. | LATI-TUDE. | Length in feet. | Diff. | Logarithm. | Diff. |
|---|---|---|---|---|---|---|---|---|---|
| ° ′ | | | | | ° ′ | | | | |
| 20 0 | 95·8732 | — | 1·9794265 | — | 22 30 | 93·7777 | — | 1·9720995 | — |
| 5 | 8230 | 502 | 91975 | 2290 | 35 | 7214 | 563 | 18389 | 2606 |
| 10 | 7725 | 505 | 89674 | 2301 | 40 | 6650 | 564 | 15771 | 2618 |
| 15 | 7218 | 507 | 87363 | 2311 | 45 | 6083 | 567 | 13144 | 2627 |
| 20 | 6709 | 509 | 85042 | 2321 | 50 | 5514 | 569 | 10505 | 2639 |
| 25 | 6198 | 511 | 82710 | 2332 | 55 | 4944 | 570 | 07855 | 2650 |
| | | 513 | | 2343 | | | 573 | | 2660 |
| 30 | 5685 | | 80367 | | 23 0 | 93·4371 | | 1·9705195 | |
| 35 | 5170 | 515 | 78014 | 2353 | 5 | 3797 | 574 | 02523 | 2672 |
| 40 | 94·9653 | 517 | 75651 | 2363 | 10 | 3220 | 577 | 1·9699842 | 2681 |
| 45 | 9134 | 519 | 73277 | 2374 | 15 | 2642 | 578 | 97148 | 2694 |
| 50 | 8614 | 520 | 70893 | 2384 | 20 | 2061 | 581 | 94445 | 2703 |
| 55 | 8091 | 523 | 68499 | 2394 | 25 | 1479 | 582 | 91730 | 2715 |
| | | 525 | | 2406 | | | 584 | | 2725 |
| 21 0 | 94·7566 | | 1·9766093 | | 30 | 0895 | | 89005 | |
| 5 | 7039 | 527 | 63677 | 2416 | 35 | 0308 | 587 | 86269 | 2736 |
| 10 | 6510 | 529 | 61252 | 2425 | 40 | 92·9720 | 588 | 83521 | 2748 |
| 15 | 5979 | 531 | 58815 | 2437 | 45 | 9130 | 590 | 80763 | 2758 |
| 20 | 5446 | 533 | 56368 | 2447 | 50 | 8537 | 593 | 77994 | 2769 |
| 25 | 4911 | 535 | 53910 | 2458 | 55 | 7943 | 594 | 75214 | 2780 |
| | | 537 | | 2468 | | | 596 | | 2790 |
| 30 | 4374 | | 51442 | | 24 0 | 92·7347 | | 1·9672424 | |
| 35 | 3836 | 538 | 48964 | 2478 | 5 | 6749 | 596 | 69622 | 2802 |
| 40 | 3295 | 541 | 46473 | 2491 | 10 | 6149 | 600 | 66809 | 2813 |
| 45 | 2752 | 543 | 43974 | 2499 | 15 | 5547 | 602 | 63985 | 2824 |
| 50 | 2207 | 545 | 41463 | 2511 | 20 | 4943 | 604 | 61150 | 2835 |
| 55 | 1660 | 547 | 38942 | 2521 | 25 | 4337 | 606 | 58304 | 2846 |
| | | 548 | | 2531 | | | 608 | | 2857 |
| 22 0 | 94·1112 | | 1·9736411 | | 30 | 3729 | | 55447 | |
| 5 | 0561 | 551 | 33868 | 2543 | 35 | 3120 | 609 | 52580 | 2867 |
| 10 | 0008 | 553 | 31315 | 2553 | 40 | 2508 | 612 | 49701 | 2879 |
| 15 | 93·9453 | 555 | 28751 | 2564 | 45 | 1894 | 614 | 46810 | 2891 |
| 20 | 8897 | 556 | 26177 | 2574 | 50 | 1279 | 615 | 43909 | 2901 |
| 25 | 8338 | 559 | 23591 | 2586 | 55 | 0661 | 618 | 40997 | 2912 |
| | | 561 | | 2596 | | | 620 | | 2924 |

and its Logarithm, measured along Parallels of Latitude.

| Lati-tude. | Length in feet. | Diff. | Logarithm. | Diff. | Lati-tude. | Length in feet. | Diff. | Logarithm. | Diff. |
|---|---|---|---|---|---|---|---|---|---|
| ° ′ | | | | | ° ′ | | | | |
| 25 0 | 92·0041 | — | 1·9638078 | — | 27 30 | 90·0558 | — | 1·9545116 | — |
|       |         | 621 |           | 2934 |       |         | 679 |           | 3277 |
| 5  | 91·9420 |     | 35139     |      | 35    | 89·9879 |     | 41839     |      |
|    |         | 624 |           | 2946 |       |         | 681 |           | 3288 |
| 10 | 8796    |     | 32193     |      | 40    | 9198    |     | 38551     |      |
|    |         | 625 |           | 2957 |       |         | 683 |           | 3299 |
| 15 | 8171    |     | 29236     |      | 45    | 8515    |     | 35252     |      |
|    |         | 627 |           | 2968 |       |         | 685 |           | 3312 |
| 20 | 7544    |     | 26268     |      | 50    | 7830    |     | 31940     |      |
|    |         | 629 |           | 2979 |       |         | 687 |           | 3323 |
| 25 | 6915    |     | 23289     |      | 55    | 7143    |     | 28617     |      |
|    |         | 632 |           | 2991 |       |         | 689 |           | 3335 |
| 30 | 6283    |     | 20298     |      | 28 0  | 89·6454 |     | 1·9525282 |      |
|    |         | 633 |           | 3002 |       |         | 690 |           | 3346 |
| 35 | 5650    |     | 17296     |      | 5     | 5764    |     | 21936     |      |
|    |         | 635 |           | 3013 |       |         | 692 |           | 3358 |
| 40 | 5015    |     | 14283     |      | 10    | 5072    |     | 18578     |      |
|    |         | 637 |           | 3024 |       |         | 695 |           | 3372 |
| 45 | 4378    |     | 11259     |      | 15    | 4377    |     | 15206     |      |
|    |         | 638 |           | 3035 |       |         | 696 |           | 3381 |
| 50 | 3740    |     | 08224     |      | 20    | 3681    |     | 11825     |      |
|    |         | 641 |           | 3048 |       |         | 698 |           | 3394 |
| 55 | 3099    |     | 05176     |      | 25    | 2983    |     | 08431     |      |
|    |         | 643 |           | 3058 |       |         | 700 |           | 3406 |
| 26 0 | 91·2456 |   | 1·9602118 |      | 30    | 2283    |     | 05025     |      |
|      |         | 645 |           | 3070 |       |         | 702 |           | 3418 |
| 5  | 1811    |     | 1·9599048 |      | 35    | 1581    |     | 01607     |      |
|    |         | 646 |           | 3080 |       |         | 704 |           | 3429 |
| 10 | 1165    |     | 95968     |      | 40    | 0877    |     | 1·9498178 |      |
|    |         | 649 |           | 3093 |       |         | 706 |           | 3441 |
| 15 | 0516    |     | 92875     |      | 45    | 0171    |     | 94737     |      |
|    |         | 650 |           | 3103 |       |         | 707 |           | 3453 |
| 20 | 90·9866 |     | 89772     |      | 50    | 88·9464 |     | 91284     |      |
|    |         | 653 |           | 3116 |       |         | 709 |           | 3465 |
| 25 | 9213    |     | 86656     |      | 55    | 8755    |     | 87819     |      |
|    |         | 654 |           | 3126 |       |         | 712 |           | 3477 |
| 30 | 8559    |     | 83530     |      | 29 0  | 88·8043 |     | 1·9484342 |      |
|    |         | 656 |           | 3137 |       |         | 713 |           | 3490 |
| 35 | 7903    |     | 80393     |      | 5     | 7330    |     | 80852     |      |
|    |         | 659 |           | 3150 |       |         | 715 |           | 3501 |
| 40 | 7244    |     | 77243     |      | 10    | 6615    |     | 77351     |      |
|    |         | 660 |           | 3161 |       |         | 717 |           | 3513 |
| 45 | 6584    |     | 74082     |      | 15    | 5898    |     | 73838     |      |
|    |         | 662 |           | 3172 |       |         | 719 |           | 3525 |
| 50 | 5922    |     | 70910     |      | 20    | 5179    |     | 70313     |      |
|    |         | 663 |           | 3183 |       |         | 720 |           | 3538 |
| 55 | 5259    |     | 67727     |      | 25    | 4459    |     | 66775     |      |
|    |         | 666 |           | 3196 |       |         | 723 |           | 3548 |
| 27 0 | 90·4593 |   | 1·9564531 |      | 30    | 3736    |     | 63227     |      |
|      |         | 668 |           | 3207 |       |         | 724 |           | 3562 |
| 5  | 3925    |     | 61324     |      | 35    | 3012    |     | 59665     |      |
|    |         | 670 |           | 3218 |       |         | 726 |           | 3573 |
| 10 | 3255    |     | 58106     |      | 40    | 2286    |     | 56092     |      |
|    |         | 671 |           | 3231 |       |         | 728 |           | 3585 |
| 15 | 2584    |     | 54875     |      | 45    | 1558    |     | 52507     |      |
|    |         | 674 |           | 3241 |       |         | 730 |           | 3598 |
| 20 | 1910    |     | 51634     |      | 50    | 0828    |     | 48909     |      |
|    |         | 675 |           | 3253 |       |         | 732 |           | 3611 |
| 25 | 1235    |     | 48381     |      | 55    | 0096    |     | 45298     |      |
|    |         | 677 |           | 3265 |       |         | 734 |           | 3622 |

## TABLE XII.—Linear value in feet of one second of Arc

| Lati-tude. | Length in feet. | Diff. | Logarithm. | Diff. | Lati-tude. | Length in feet. | Diff. | Logarithm. | D'ff. |
|---|---|---|---|---|---|---|---|---|---|
| ° ′ | | | | | ° ′ | | | | |
| 30 0 | 87·9362 | — | 1·9441676 | — | 32 30 | 85·6492 | — | 1·9327233 | — |
| | | 736 | | 3635 | | | 791 | | 4012 |
| 5 | 8626 | | 38041 | | 35 | 5701 | | 23221 | |
| | | 737 | | 3646 | | | 792 | | 4023 |
| 10 | 7889 | | 34396 | | 40 | 4909 | | 19198 | |
| | | 739 | | 3658 | | | 795 | | 4038 |
| 15 | 7150 | | 30737 | | 45 | 4114 | | 15160 | |
| | | 742 | | 3672 | | | 796 | | 4049 |
| 20 | 6408 | | 27065 | | 50 | 3318 | | 11111 | |
| | | 743 | | 3684 | | | 798 | | 4063 |
| 25 | 5665 | | 23381 | | 55 | 2520 | | 07048 | |
| | | 744 | | 3695 | | | 799 | | 4076 |
| 30 | 4921 | | 19686 | | 33 0 | 85·1721 | | 1·9302972 | |
| | | 747 | | 3708 | | | 802 | | 4089 |
| 35 | 4174 | | 15978 | | 5 | 0919 | | 1·9298883 | |
| | | 749 | | 3721 | | | 803 | | 4102 |
| 40 | 3425 | | 12257 | | 10 | 0116 | | 94781 | |
| | | 750 | | 3733 | | | 805 | | 4115 |
| 45 | 2675 | | 08524 | | 15 | 84·9311 | | 90666 | |
| | | 752 | | 3745 | | | 807 | | 4128 |
| 50 | 1923 | | 04779 | | 20 | 8504 | | 86538 | |
| | | 755 | | 3758 | | | 809 | | 4142 |
| 55 | 1168 | | 01021 | | 25 | 7695 | | 82396 | |
| | | 756 | | 3770 | | | 811 | | 4154 |
| 31 0 | 87·0412 | | 1·9397251 | | 30 | 6884 | | 78242 | |
| | | 757 | | 3783 | | | 812 | | 4168 |
| 5 | 86·9655 | | 93468 | | 35 | 6072 | | 74074 | |
| | | 760 | | 3795 | | | 814 | | 4180 |
| 10 | 8895 | | 89673 | | 40 | 5258 | | 69894 | |
| | | 762 | | 3808 | | | 816 | | 4195 |
| 15 | 8133 | | 85865 | | 45 | 4442 | | 65699 | |
| | | 763 | | 3820 | | | 817 | | 4207 |
| 20 | 7370 | | 82045 | | 50 | 3625 | | 61492 | |
| | | 765 | | 3833 | | | 820 | | 4221 |
| 25 | 6605 | | 78212 | | 55 | 2805 | | 57271 | |
| | | 767 | | 3845 | | | 821 | | 4234 |
| 30 | 5838 | | 74367 | | 34 0 | 84·1984 | | 1·9253037 | |
| | | 769 | | 3858 | | | 823 | | 4247 |
| 35 | 5069 | | 70509 | | 5 | 1161 | | 48790 | |
| | | 770 | | 3871 | | | 825 | | 4261 |
| 40 | 4299 | | 66638 | | 10 | 0336 | | 44529 | |
| | | 773 | | 3882 | | | 827 | | 4274 |
| 45 | 3526 | | 62756 | | 15 | 83·9509 | | 40255 | |
| | | 774 | | 3897 | | | 828 | | 4287 |
| 50 | 2752 | | 58859 | | 20 | 8681 | | 35968 | |
| | | 776 | | 3909 | | | 830 | | 4301 |
| 55 | 1976 | | 54950 | | 25 | 7851 | | 31667 | |
| | | 778 | | 3921 | | | 832 | | 4315 |
| 32 0 | 86·1198 | | 1·9351029 | | 30 | 7019 | | 27352 | |
| | | 780 | | 3934 | | | 834 | | 4328 |
| 5 | 0418 | | 47095 | | 35 | 6185 | | 23024 | |
| | | 782 | | 3947 | | | 835 | | 4341 |
| 10 | 85·9636 | | 43148 | | 40 | 5350 | | 18683 | |
| | | 783 | | 3960 | | | 838 | | 4356 |
| 15 | 8853 | | 39188 | | 45 | 4512 | | 14327 | |
| | | 785 | | 3972 | | | 839 | | 4368 |
| 20 | 8068 | | 35216 | | 50 | 3673 | | 09959 | |
| | | 787 | | 3985 | | | 841 | | 4382 |
| 25 | 7281 | | 31231 | | 55 | 2832 | | 05577 | |
| | | 789 | | 3998 | | | 842 | | 4396 |

and its Logarithm, measured along Parallels of Latitude.

| Latitude. | Length in feet. | Diff. | Logarithm. | Diff. | Latitude. | Length in feet. | Diff. | Logarithm. | Diff. |
|---|---|---|---|---|---|---|---|---|---|
| ° ′ | | | | | ° ′ | | | | |
| 35 0 | 83·1990 | — | 1·9201181 | — | 37 30 | 80·5901 | — | 1·9062817 | — |
|  |  | 844 |  | 4410 |  |  | 897 |  | 4834 |
| 5 | ·1146 |  | 1·9196771 |  | 35 | 5004 |  | 57983 |  |
|  |  | 846 |  | 4423 |  |  | 898 |  | 4847 |
| 10 | 0300 |  | 92348 |  | 40 | 4106 |  | 53136 |  |
|  |  | 848 |  | 4437 |  |  | 900 |  | 4864 |
| 15 | 82·9452 |  | 87911 |  | 45 | 3206 |  | 48272 |  |
|  |  | 850 |  | 4450 |  |  | 901 |  | 4877 |
| 20 | 8602 |  | 83461 |  | 50 | 2305 |  | 43395 |  |
|  |  | 851 |  | 4465 |  |  | 903 |  | 4892 |
| 25 | 7751 |  | 78996 |  | 55 | 1402 |  | 38503 |  |
|  |  | 853 |  | 4479 |  |  | 905 |  | 4907 |
| 30 | 6898 |  | 74517 |  | 38 0 | 80·0497 |  | 1·9033596 |  |
|  |  | 855 |  | 4491 |  |  | 907 |  | 4921 |
| 35 | 6043 |  | 70026 |  | 5 | 79·9590 |  | 28675 |  |
|  |  | 857 |  | 4507 |  |  | 908 |  | 4936 |
| 40 | 5186 |  | 65519 |  | 10 | 8682 |  | 23739 |  |
|  |  | 858 |  | 4520 |  |  | 910 |  | 4951 |
| 45 | 4328 |  | 60999 |  | 15 | 7772 |  | 18788 |  |
|  |  | 860 |  | 4533 |  |  | 912 |  | 4966 |
| 50 | 3468 |  | 56466 |  | 20 | 6860 |  | 13822 |  |
|  |  | 862 |  | 4548 |  |  | 913 |  | 4982 |
| 55 | 2606 |  | 51918 |  | 25 | 5947 |  | 08840 |  |
|  |  | 864 |  | 4562 |  |  | 915 |  | 4996 |
| 36 0 | 82·1742 |  | 1·9147356 |  | 30 | 5032 |  | 03844 |  |
|  |  | 865 |  | 4576 |  |  | 917 |  | 5011 |
| 5 | 0877 |  | 42780 |  | 35 | 4115 |  | 1·8998833 |  |
|  |  | 867 |  | 4590 |  |  | 919 |  | 5026 |
| 10 | 0010 |  | 38190 |  | 40 | 3196 |  | 93807 |  |
|  |  | 869 |  | 4604 |  |  | 920 |  | 5041 |
| 15 | 81·9141 |  | 33586 |  | 45 | 2276 |  | 88766 |  |
|  |  | 871 |  | 4618 |  |  | 922 |  | 5056 |
| 20 | 8270 |  | 28968 |  | 50 | 1354 |  | 83710 |  |
|  |  | 872 |  | 4632 |  |  | 923 |  | 5072 |
| 25 | 7398 |  | 24336 |  | 55 | 0431 |  | 78638 |  |
|  |  | 874 |  | 4647 |  |  | 925 |  | 5086 |
| 30 | 6524 |  | 19689 |  | 39 0 | 78·9506 |  | 1·8973552 |  |
|  |  | 876 |  | 4660 |  |  | 927 |  | 5102 |
| 35 | 5648 |  | 15029 |  | 5 | 8579 |  | 68450 |  |
|  |  | 878 |  | 4675 |  |  | 929 |  | 5117 |
| 40 | 4770 |  | 10354 |  | 10 | 7650 |  | 63333 |  |
|  |  | 879 |  | 4689 |  |  | 930 |  | 5132 |
| 45 | 3891 |  | 05665 |  | 15 | 6720 |  | 58201 |  |
|  |  | 880 |  | 4703 |  |  | 932 |  | 5147 |
| 50 | 3011 |  | 00962 |  | 20 | 5788 |  | 53054 |  |
|  |  | 883 |  | 4718 |  |  | 934 |  | 5163 |
| 55 | 2128 |  | 1·9096244 |  | 25 | 4854 |  | 47891 |  |
|  |  | 885 |  | 4732 |  |  | 935 |  | 5178 |
| 37 0 | 81·1243 |  | 1·9091512 |  | 30 | 3919 |  | 42713 |  |
|  |  | 886 |  | 4746 |  |  | 937 |  | 5193 |
| 5 | 0357 |  | 86766 |  | 35 | 2882 |  | 37520 |  |
|  |  | 888 |  | 4761 |  |  | 938 |  | 5209 |
| 10 | 80·9469 |  | 82005 |  | 40 | 2044 |  | 32311 |  |
|  |  | 889 |  | 4775 |  |  | 940 |  | 5225 |
| 15 | 8580 |  | 77230 |  | 45 | 1104 |  | 27086 |  |
|  |  | 891 |  | 4790 |  |  | 942 |  | 5240 |
| 20 | 7689 |  | 72440 |  | 50 | 0162 |  | 21846 |  |
|  |  | 893 |  | 4805 |  |  | 944 |  | 5256 |
| 25 | 6796 |  | 67635 |  | 55 | 77·9218 |  | 16590 |  |
|  |  | 895 |  | 4818 |  |  | 945 |  | 5271 |

## TABLE XII.—Linear value in feet of one second of Arc

| Latitude | Length in feet. | Diff. | Logarithm. | Diff. | Latitude | Length in feet. | Diff. | Logarithm. | Diff. |
|---|---|---|---|---|---|---|---|---|---|
| ° ′ | | | | | ° ′ | | | | |
| 40 0 | 77·8273 | — | 1·8911319 | — | 42 30 | 74·0157 | — | 1·8745727 | — |
| 5 | 7326 | 947 | 06032 | 5287 | 35 | 8161 | 996 | 39952 | 5775 |
| 10 | 6378 | 948 | 00730 | 5302 | 40 | 7164 | 997 | 34160 | 5792 |
| 15 | 5427 | 951 | 1·8895411 | 5319 | 45 | 6165 | 999 | 28351 | 5809 |
| 20 | 4475 | 952 | 90077 | 5334 | 50 | 5165 | 1000 | 22525 | 5826 |
| 25 | 3522 | 953 | 84727 | 5350 | 55 | 4163 | 1002 | 16681 | 5844 |
| 30 | 2567 | 955 | 79361 | 5366 | 43 0 | 74·3160 | 1003 | 1·8710822 | 5859 |
| 35 | 1610 | 957 | 73980 | 5381 | 5 | 2155 | 1005 | 04944 | 5878 |
| 40 | 0652 | 958 | 68583 | 5397 | 10 | 1148 | 1007 | 1·8699050 | 5894 |
| 45 | 76·9692 | 960 | 63169 | 5414 | 15 | 0140 | 1008 | 93137 | 5913 |
| 50 | 8730 | 962 | 57739 | 5430 | 20 | 73·9130 | 1010 | 87209 | 5928 |
| 55 | 7767 | 963 | 52294 | 5445 | 25 | 8119 | 1011 | 81262 | 5947 |
| 41 0 | 76·6803 | 965 | 1·8846833 | 5461 | 30 | 7100 | 1013 | 75298 | 5964 |
| 5 | 5836 | 966 | 41355 | 5478 | 35 | 6091 | 1015 | 69316 | 5982 |
| 10 | 4868 | 968 | 35862 | 5493 | 40 | 5075 | 1016 | 63318 | 5898 |
| 15 | 3898 | 970 | 30351 | 5511 | 45 | 4057 | 1018 | 57301 | 6017 |
| 20 | 2026 | 972 | 24825 | 5526 | 50 | 3038 | 1019 | 51267 | 6034 |
| 25 | 1953 | 973 | 19282 | 5543 | 55 | 2018 | 1020 | 45215 | 6052 |
| 30 | 0978 | 975 | 13723 | 5559 | 44 0 | 73·0995 | 1023 | 1·8639145 | 6070 |
| 35 | 0002 | 976 | 08148 | 5575 | 5 | 72·9971 | 1024 | 33058 | 6087 |
| 40 | 75·9024 | 978 | 02556 | 5592 | 10 | 8946 | 1025 | 26953 | 6105 |
| 45 | 8045 | 979 | 1·8796948 | 5608 | 15 | 7919 | 1027 | 20830 | 6123 |
| 50 | 7064 | 981 | 91323 | 5625 | 20 | 6890 | 1029 | 14689 | 6141 |
| 55 | 6081 | 983 | 85683 | 5640 | 25 | 5860 | 1030 | 08531 | 6158 |
| 42 0 | 75·5097 | 984 | 1·8780025 | 5658 | 30 | 4829 | 1031 | 02354 | 6177 |
| 5 | 4111 | 986 | 74350 | 5675 | 35 | 3796 | 1033 | 1·8596159 | 6195 |
| 10 | 3123 | 988 | 68659 | 5691 | 40 | 2761 | 1035 | 89946 | 6213 |
| 15 | 2134 | 989 | 62951 | 5708 | 45 | 1725 | 1036 | 83715 | 6231 |
| 20 | 1143 | 991 | 57226 | 5725 | 50 | 0687 | 1038 | 77465 | 6250 |
| 25 | 0151 | 992 | 51483 | 5741 | 55 | 71·9648 | 1039 | 71198 | 6267 |
| | | 994 | | 5758 | | | 1041 | | 6286 |

## and its Logarithm, measured along Parallels of Latitude.

| Latitude. | Length in feet. | Diff. | Logarithm. | Diff. | Latitude. | Length in feet. | Diff. | Logarithm. | Diff. |
|---|---|---|---|---|---|---|---|---|---|
| ° ′ | | — | | — | ° ′ | | — | | — |
| 45 0 | 71 8607 | | 1·8564912 | | 47 30 | 68·6679 | | 1·8367540 | |
| | | 1043 | | 6304 | | | 1087 | | 6382 |
| 5 | 7564 | | 58608 | | 35 | 5592 | | 60653 | |
| | | 1044 | | 6323 | | | 1089 | | 8902 |
| 10 | 6520 | | 52285 | | 40 | 4503 | | 53756 | |
| | | 1045 | | 6341 | | | 1090 | | 6922 |
| 15 | 5475 | | 45944 | | 45 | 3413 | | 46834 | |
| | | 1047 | | 6360 | | | 1091 | | 6943 |
| 20 | 4428 | | 39584 | | 50 | 2322 | | 39891 | |
| | | 1048 | | 6378 | | | 1093 | | 6962 |
| 25 | 3380 | | 33206 | | 55 | 1229 | | 32929 | |
| | | 1050 | | 6397 | | | 1095 | | 6985 |
| 30 | 2330 | | 26809 | | 48 0 | 68·0134 | | 1·8325944 | |
| | | 1052 | | 6415 | | | 1096 | | 7004 |
| 35 | 1278 | | 20394 | | 5 | 67·9038 | | 18940 | |
| | | 1053 | | 6435 | | | 1098 | | 7025 |
| 40 | 0225 | | 13959 | | 10 | 7940 | | 11915 | |
| | | 1055 | | 6453 | | | 1099 | | 7046 |
| 45 | 70·9170 | | 07506 | | 15 | 6841 | | 04869 | |
| | | 1056 | | 6472 | | | 1100 | | 7065 |
| 50 | 8114 | | 01034 | | 20 | 5741 | | 1·8297804 | |
| | | 1057 | | 6491 | | | 1102 | | 7087 |
| 55 | 7057 | | 1·8494543 | | 25 | 4639 | | 90717 | |
| | | 1059 | | 6510 | | | 1103 | | 7108 |
| 46 0 | 70·5998 | | 1·8488033 | | 30 | 3536 | | 83609 | |
| | | 1061 | | 6529 | | | 1105 | | 7129 |
| 5 | 4937 | | 81504 | | 35 | 2431 | | 76480 | |
| | | 1062 | | 6548 | | | 1106 | | 7149 |
| 10 | 3875 | | 74956 | | 40 | 1325 | | 69331 | |
| | | 1064 | | 6568 | | | 1107 | | 7171 |
| 15 | 2811 | | 68388 | | 45 | 0218 | | 62160 | |
| | | 1065 | | 6586 | | | 1109 | | 7192 |
| 20 | 1746 | | 61802 | | 50 | 66·9109 | | 54968 | |
| | | 1066 | | 6605 | | | 1110 | | 7212 |
| 25 | 0680 | | 55197 | | 55 | 7999 | | 47756 | |
| | | 1068 | | 6626 | | | 1112 | | 7236 |
| 30 | 69·9612 | | 48571 | | 49 0 | 66·6887 | | 1·8240520 | |
| | | 1070 | | 6644 | | | 1113 | | 7255 |
| 35 | 8542 | | 41927 | | 5 | 5774 | | 33265 | |
| | | 1071 | | 6664 | | | 1115 | | 7277 |
| 40 | 7471 | | 35263 | | 10 | 4659 | | 25988 | |
| | | 1072 | | 6683 | | | 1116 | | 7299 |
| 45 | 6399 | | 28580 | | 15 | 3543 | | 18689 | |
| | | 1074 | | 6704 | | | 1118 | | 7320 |
| 50 | 5325 | | 21876 | | 20 | 2425 | | 11369 | |
| | | 1076 | | 6722 | | | 1119 | | 7342 |
| 55 | 4249 | | 15154 | | 25 | 1306 | | 04027 | |
| | | 1077 | | 6743 | | | 1120 | | 7364 |
| 47 0 | 69·3172 | | 1·8408411 | | 30 | 0186 | | 1·8196663 | |
| | | 1078 | | 6761 | | | 1122 | | 7385 |
| 5 | 2094 | | 01650 | | 35 | 65·9064 | | 89278 | |
| | | 1080 | | 6783 | | | 1123 | | 7407 |
| 10 | 1014 | | 1·8394867 | | 40 | 7941 | | 81871 | |
| | | 1081 | | 6801 | | | 1124 | | 7429 |
| 15 | 68·9933 | | 88066 | | 45 | 6817 | | 74442 | |
| | | 1083 | | 6822 | | | 1126 | | 7452 |
| 20 | 8850 | | 81214 | | 50 | 5691 | | 66990 | |
| | | 1085 | | 6842 | | | 1127 | | 7472 |
| 25 | 7765 | | 74402 | | 55 | 4564 | | 59518 | |
| | | 1086 | | 6862 | | | 1129 | | 7496 |

## TABLE XII.—Linear value in feet of one second of Arc

| Latitude. | Length in feet. | Diff. | Logarithm. | Diff. | Latitude. | Length in feet. | Diff. | Logarithm. | Diff. |
|---|---|---|---|---|---|---|---|---|---|
| ° ′ | | | | | ° ′ | | | | |
| 50  0 | 65 ·3435 | — | 1 ·8152022 | — | 52 30 | 61 ·8935 | — | 1 ·7916447 | — |
|       |          | 1130 |          | 7518 |       |          | 1171 |          | 8223 |
|  5    | 2305     |      | 44504    |      | 35    | 7764     |      | 08224    |      |
|       |          | 1132 |          | 7540 |       |          | 1172 |          | 8248 |
| 10    | 1173     |      | 36964    |      | 40    | 6592     |      | 1 ·7899976 |    |
|       |          | 1133 |          | 7563 |       |          | 1174 |          | 8273 |
| 15    | 0040     |      | 29401    |      | 45    | 5418     |      | 91703    |      |
|       |          | 1134 |          | 7585 |       |          | 1175 |          | 8299 |
| 20    | 64 8906  |      | 21816    |      | 50    | 4243     |      | 83404    |      |
|       |          | 1136 |          | 7607 |       |          | 1176 |          | 8324 |
| 25    | 7770     |      | 14209    |      | 55    | 3067     |      | 75080    |      |
|       |          | 1137 |          | 7631 |       |          | 1177 |          | 8340 |
| 30    | 6633     |      | 06578    |      | 53  0 | 61 ·1890 |      | 1 ·7866731 |    |
|       |          | 1138 |          | 7652 |       |          | 1179 |          | 8374 |
| 35    | 5495     |      | 1 ·8098926 |    |  5    | 0711     |      | 58357    |      |
|       |          | 1140 |          | 7675 |       |          | 1180 |          | 8400 |
| 40    | 4355     |      | 91251    |      | 10    | 60 ·9531 |      | 49957    |      |
|       |          | 1141 |          | 7699 |       |          | 1181 |          | 8425 |
| 45    | 3214     |      | 83552    |      | 15    | 8350     |      | 41532    |      |
|       |          | 1143 |          | 7722 |       |          | 1183 |          | 8451 |
| 50    | 2071     |      | 75830    |      | 20    | 7167     |      | 33081    |      |
|       |          | 1144 |          | 7744 |       |          | 1184 |          | 8478 |
| 55    | 0927     |      | 68086    |      | 25    | 5983     |      | 24603    |      |
|       |          | 1145 |          | 7768 |       |          | 1185 |          | 8502 |
| 51  0 | 63 ·9782 |      | 1 ·8060318 |    | 30    | 4798     |      | 16101    |      |
|       |          | 1147 |          | 7791 |       |          | 1187 |          | 8529 |
|  5    | 8635     |      | 52527    |      | 35    | 3611     |      | 07572    |      |
|       |          | 1148 |          | 7814 |       |          | 1188 |          | 8555 |
| 10    | 7487     |      | 44713    |      | 40    | 2423     |      | 1 ·7799017 |    |
|       |          | 1149 |          | 7838 |       |          | 1189 |          | 8581 |
| 15    | 6338     |      | 36875    |      | 45    | 1234     |      | 90436    |      |
|       |          | 1151 |          | 7861 |       |          | 1190 |          | 8607 |
| 20    | 5187     |      | 29014    |      | 50    | 0044     |      | 81829    |      |
|       |          | 1152 |          | 7884 |       |          | 1192 |          | 8634 |
| 25    | 4035     |      | 21130    |      | 55    | 59 ·8852 |      | 73195    |      |
|       |          | 1154 |          | 7908 |       |          | 1193 |          | 8660 |
| 30    | 2881     |      | 13222    |      | 54  0 | 59 ·7659 |      | 1 ·7764535 |    |
|       |          | 1155 |          | 7932 |       |          | 1194 |          | 8687 |
| 35    | 1726     |      | 05290    |      |  5    | 6465     |      | 55848    |      |
|       |          | 1156 |          | 7956 |       |          | 1196 |          | 8713 |
| 40    | 0570     |      | 1 ·7997334 |    | 10    | 5269     |      | 47135    |      |
|       |          | 1157 |          | 7979 |       |          | 1197 |          | 8741 |
| 45    | 62 ·9413 |      | 89355    |      | 15    | 4072     |      | 38394    |      |
|       |          | 1159 |          | 8004 |       |          | 1198 |          | 8767 |
| 50    | 8254     |      | 81351    |      | 20    | 2874     |      | 29627    |      |
|       |          | 1160 |          | 8027 |       |          | 1199 |          | 8795 |
| 55    | 7094     |      | 73324    |      | 25    | 1675     |      | 20832    |      |
|       |          | 1162 |          | 8053 |       |          | 1201 |          | 8822 |
| 52  0 | 62 ·5932 |      | 1 ·7965271 |    | 30    | 0474     |      | 12010    |      |
|       |          | 1163 |          | 8076 |       |          | 1202 |          | 8849 |
|  5    | 4769     |      | 57195    |      | 35    | 58 ·9272 |      | 03161    |      |
|       |          | 1164 |          | 8100 |       |          | 1203 |          | 8875 |
| 10    | 3605     |      | 49095    |      | 40    | 8069     |      | 1 ·7694286 |    |
|       |          | 1166 |          | 8125 |       |          | 1204 |          | 8904 |
| 15    | 2439     |      | 40970    |      | 45    | 6865     |      | 85382    |      |
|       |          | 1167 |          | 8149 |       |          | 1206 |          | 8931 |
| 20    | 1272     |      | 32821    |      | 50    | 5659     |      | 76451    |      |
|       |          | 1168 |          | 8174 |       |          | 1207 |          | 8959 |
| 25    | 0104     |      | 24647    |      | 55    | 4452     |      | 67492    |      |
|       |          | 1169 |          | 8200 |       |          | 1208 |          | 8987 |

and its Logarithm, measured along Parallels of Latitude.

| LATI-TUDE. | Length in feet. | Diff. | Logarithm. | Diff. | LATI-TUDE. | Length in feet. | Diff. | Logarithm. | Diff. |
|---|---|---|---|---|---|---|---|---|---|
| ° ′ | | | | | ° ′ | | | | |
| 55 0 | 58·3244 | | 1·7658505 | | 57 30 | 54·6431 | | 1·7375354 | |
| | | 1209 | | 9015 | | | 1246 | | 9912 |
| 5 | 2035 | | 49490 | | 35 | 5185 | | 65442 | |
| | | 1211 | | 9042 | | | 1246 | | 9943 |
| 10 | 0824 | | 40448 | | 40 | 3939 | | 55499 | |
| | | 1212 | | 9071 | | | 1248 | | 9976 |
| 15 | 57·9612 | | 31377 | | 45 | 2691 | | 45523 | |
| | | 1213 | | 9100 | | | 1249 | | 10007 |
| 20 | 8399 | | 22277 | | 50 | 1442 | | 35516 | |
| | | 1214 | | 9127 | | | 1251 | | 10041 |
| 25 | 7185 | | 13150 | | 55 | 0191 | | 25475 | |
| | | 1216 | | 9157 | | | 1251 | | 10073 |
| 30 | 5969 | | 03993 | | 58 0 | 53·8940 | | 1·7315402 | |
| | | 1217 | | 9185 | | | 1253 | | 10105 |
| 35 | 4752 | | 1·7594808 | | 5 | 7687 | | 05297 | |
| | | 1218 | | 9213 | | | 1254 | | 10139 |
| 40 | 3534 | | 85595 | | 10 | 6433 | | 1·7295158 | |
| | | 1219 | | 9243 | | | 1255 | | 10172 |
| 45 | 2315 | | 76352 | | 15 | 5178 | | 84986 | |
| | | 1220 | | 9272 | | | 1256 | | 10205 |
| 50 | 1095 | | 67080 | | 20 | 3922 | | 74781 | |
| | | 1222 | | 9300 | | | 1257 | | 10237 |
| 55 | 56·9873 | | 57780 | | 25 | 2665 | | 64544 | |
| | | 1223 | | 9330 | | | 1258 | | 10272 |
| | | | | | 30 | 1407 | | 54272 | |
| 56 0 | 56·8650 | | 1·7548450 | | | | 1259 | | 10305 |
| | | 1224 | | 9360 | 35 | 0148 | | 43967 | |
| 5 | 7426 | | 39090 | | | | 1261 | | 10340 |
| | | 1226 | | 9389 | 40 | 52·8887 | | 33627 | |
| 10 | 6200 | | 29701 | | | | 1262 | | 10373 |
| | | 1226 | | 9418 | 45 | 7625 | | 23254 | |
| 15 | 4974 | | 20283 | | | | 1263 | | 10408 |
| | | 1228 | | 9449 | 50 | 6362 | | 12846 | |
| 20 | 3746 | | 10834 | | | | 1264 | | 10440 |
| | | 1229 | | 9478 | 55 | 5098 | | 02406 | |
| 25 | 2517 | | 01356 | | | | 1265 | | 10477 |
| | | 1230 | | 9508 | 59 0 | 52·3833 | | 1·7191929 | |
| 30 | 1287 | | 1·7491848 | | | | 1266 | | 10511 |
| | | 1231 | | 9538 | 5 | 2567 | | 81418 | |
| 35 | 0056 | | 82310 | | | | 1267 | | 10545 |
| | | 1233 | | 9570 | 10 | 1300 | | 70873 | |
| 40 | 55·8823 | | 72740 | | | | 1269 | | 10581 |
| | | 1234 | | 9599 | 15 | 0031 | | 60292 | |
| 45 | 7589 | | 63141 | | | | 1270 | | 10617 |
| | | 1235 | | 9630 | 20 | 51·8761 | | 49675 | |
| 50 | 6354 | | 53511 | | | | 1271 | | 10651 |
| | | 1236 | | 9661 | 25 | 7490 | | 39024 | |
| 55 | 5118 | | 43850 | | | | 1271 | | 10686 |
| | | 1238 | | 9691 | 30 | 6219 | | 1·7128338 | |
| 57 0 | 55·3880 | | 1·7434159 | | | | 1273 | | 10723 |
| | | 1238 | | 9723 | 35 | 4946 | | 17615 | |
| 5 | 2642 | | 24436 | | | | 1274 | | 10758 |
| | | 1240 | | 9754 | 40 | 3672 | | 06857 | |
| 10 | 1402 | | 14682 | | | | 1275 | | 10794 |
| | | 1241 | | 9784 | 45 | 2397 | | 1·7096063 | |
| 15 | 0161 | | 04898 | | | | 1277 | | 10831 |
| | | 1242 | | 9816 | 50 | 1120 | | 85232 | |
| 20 | 54·8919 | | 1·7395082 | | | | 1277 | | 10867 |
| | | 1243 | | 9848 | 55 | 50·9843 | | 74365 | |
| 25 | 7676 | | 85234 | | | | 1278 | | 10903 |
| | | 1245 | | 9880 | 60 0 | 50·8565 | | 1·7063462 | |

## TABLE XIII.

### Spherical Excess.

| Area of Triangle in Square Miles. | Spherical Excess. | Area of Triangle in Square Miles. | Spherical Excess. | Area of Triangle in Square Miles. | Spherical Excess. |
|---|---|---|---|---|---|
| | ″ | | ″ | | ″ |
| 5 | 0·07 | 60 | 0·79 | 400 | 5·27 |
| 10 | ·13 | 70 | ·92 | 500 | 6·59 |
| 15 | ·20 | 80 | 1·05 | 600 | 7·91 |
| 20 | ·26 | 90 | 1·19 | 700 | 9·23 |
| 30 | ·40 | 100 | 1·32 | 800 | 10·54 |
| 40 | ·53 | 200 | 2·64 | 900 | 11·86 |
| 50 | ·66 | 300 | 3·95 | 1000 | 13·18 |

Computed from formula, spherical excess $= \dfrac{A}{r^2 \sin 1''}$, where A = area of triangle, $r$ = mean radius of curvature.

The table has been made out for latitude 30°, but may be used for all latitudes for topographical, tertiary, and secondary work.

## TABLE XIV.—Astronomical Refraction.

The Mean Refraction is for Barometer 30 inches and Temperature 50° Fahr.

| Apparent Altitude. | Mean Refraction. | Diff. for 1' Alt. | Diff. for +1 in. Baro. | Diff. for −1° Temp. | Apparent Altitude. | Mean Refraction. | Diff. for 1' Alt. | Diff. for +1 in. Baro. | Diff. for −1° Temp. |
|---|---|---|---|---|---|---|---|---|---|
| ° ' | ' " | " | " | " | ° ' | ' " | " | " | " |
| 0  0 | 33 51 | 11·7 | 74 | 8·1 | 4  0 | 11 52 | 2·2 | 24·1 | 1·70 |
|    5 | 32 53 | 11·3 | 71 | 7·6 |   10 | 11 30 | 2·1 | 23·4 | 1·64 |
|   10 | 31 58 | 10·9 | 69 | 7·3 |   20 | 11 10 | 2·0 | 22·7 | 1·58 |
|   15 | 31  5 | 10·5 | 67 | 7·0 |   30 | 10 50 | 1·9 | 22·0 | 1·53 |
|   20 | 30 13 | 10·1 | 65 | 6·7 |   40 | 10 32 | 1·8 | 21·3 | 1·48 |
|   25 | 29 24 |  9·7 | 63 | 6·4 |   50 | 10 15 | 1·7 | 20·7 | 1·43 |
|   30 | 28 37 |  9·4 | 61 | 6·1 | 5  0 |  9 58 | 1·6 | 20·1 | 1·38 |
|   35 | 27 51 |  9·0 | 59 | 5·9 |   10 |  9 42 | 1·5 | 19·6 | 1·34 |
|   40 | 27  6 |  8·7 | 58 | 5·6 |   20 |  9 27 | 1·5 | 19·1 | 1·30 |
|   45 | 26 24 |  8·4 | 56 | 5·4 |   30 |  9 11 | 1·4 | 18·6 | 1·26 |
|   50 | 25 43 |  8·0 | 55 | 5·1 |   40 |  8 58 | 1·3 | 18·1 | 1·22 |
|   55 | 25  3 |  7·7 | 53 | 4·9 |   50 |  8 45 | 1·3 | 17·6 | 1·19 |
| 1  0 | 24 25 |  7·4 | 52 | 4·7 | 6  0 |  8 32 | 1·2 | 17·2 | 1·15 |
|    5 | 23 48 |  7·1 | 50 | 4·6 |   10 |  8 20 | 1·2 | 16·8 | 1·11 |
|   10 | 23 13 |  6·9 | 49 | 4·5 |   20 |  8  9 | 1·1 | 16·4 | 1·09 |
|   15 | 22 40 |  6·6 | 48 | 4·4 |   30 |  7 58 | 1·1 | 16·0 | 1·06 |
|   20 | 22  8 |  6·3 | 46 | 4·2 |   40 |  7 47 | 1·0 | 15·7 | 1·03 |
|   25 | 21 37 |  6·1 | 45 | 4·0 |   50 |  7 37 | 1·0 | 15·3 | 1·00 |
|   30 | 21  7 |  5·9 | 44 | 3·9 | 7  0 |  7 27 | 1·0 | 15·0 | 0·98 |
|   35 | 20 38 |  5·7 | 43 | 3·8 |   10 |  7 17 | 0·9 | 14·6 | 95 |
|   40 | 20 10 |  5·5 | 42 | 3·6 |   20 |  7  8 | 9 | 14·3 | 93 |
|   45 | 19 43 |  5·3 | 40 | 3·5 |   30 |  6 59 | 8 | 14·1 | 91 |
|   50 | 19 17 |  5·1 | 39 | 3·4 |   40 |  6 51 | 8 | 13·8 | 89 |
|   55 | 18 52 |  4·9 | 39 | 3·3 |   50 |  6 43 | 8 | 13·5 | 87 |
| 2  0 | 18 29 |  4·8 | 38 | 3·2 | 8  0 |  6 35 | 0·7 | 13·3 | 0·85 |
|    5 | 18  5 |  4·6 | 37 | 3·1 |   10 |  6 28 | 7 | 13·1 | 83 |
|   10 | 17 43 |  4·4 | 36 | 3·0 |   20 |  6 21 | 7 | 12·8 | 82 |
|   15 | 17 21 |  4·3 | 36 | 2·9 |   30 |  6 14 | 7 | 12·6 | 80 |
|   20 | 17  0 |  4·1 | 35 | 2·8 |   40 |  6  7 | 7 | 12·3 | 79 |
|   25 | 16 40 |  4·0 | 34 | 2·8 |   50 |  6  0 | 6 | 12·1 | 77 |
|   30 | 16 21 |  3·9 | 33 | 2·7 | 9  0 |  5 54 | 0·6 | 11·9 | 0·76 |
|   35 | 16  2 |  3·7 | 33 | 2·7 |   10 |  5 47 | 6 | 11·7 | 74 |
|   40 | 15 43 |  3·6 | 32 | 2·6 |   20 |  5 41 | 6 | 11·5 | 73 |
|   45 | 15 25 |  3·5 | 32 | 2·5 |   30 |  5 36 | 6 | 11·3 | 71 |
|   50 | 15  8 |  3·4 | 31 | 2·4 |   40 |  5 30 | 5 | 11·1 | 71 |
|   55 | 14 51 |  3·3 | 30 | 2·3 |   50 |  5 25 | 5 | 11·0 | 70 |
| 3  0 | 14 35 |  3·2 | 30 | 2·3 | 10  0 |  5 20 | 0·5 | 10·8 | 0·69 |
|    5 | 14 19 |  3·1 | 29 | 2·2 |   10 |  5 15 | 5 | 10·6 | 67 |
|   10 | 14  4 |  3·0 | 29 | 2·2 |   20 |  5 10 | 5 | 10·4 | 65 |
|   15 | 13 50 |  2·9 | 28 | 2·1 |   30 |  5  5 | 5 | 10·2 | 64 |
|   20 | 13 35 |  2·8 | 28 | 2·1 |   40 |  5  0 | 5 | 10·1 | 63 |
|   25 | 13 21 |  2·7 | 27 | 2·0 |   50 |  4 56 | 4 |  9·9 | 62 |
|   30 | 13  7 |  2·7 | 27 | 2·0 | 11  0 |  4 51 | 0·4 |  9·8 | 0·60 |
|   35 | 12 53 |  2·6 | 26 | 2·0 |   10 |  4 47 | 4 |  9·6 | 59 |
|   40 | 12 41 |  2·5 | 26 | 1·9 |   20 |  4 43 | 4 |  9·5 | 58 |
|   45 | 12 28 |  2·4 | 25 | 1·9 |   30 |  4 39 | 4 |  9·4 | 57 |
|   50 | 12 16 |  2·4 | 25 | 1·9 |   40 |  4 35 | 4 |  9·2 | 56 |
|   55 | 12  3 |  2·3 | 25 | 1·8 |   50 |  4 31 | 4 |  9·1 | 55 |

NOTE.—If the Barometer is over 30 inches the Barometer correction is *additive* to the Mean Refraction.
If the Thermometer is less than 50° the Thermometer correction is *additive* to the Mean Refraction.
And *vice versa*.

## TABLE XIV.—Astronomical Refraction.

The Mean Refraction is for Barometer 30 inches and Temperature 50° Fahr.

| Apparent Altitude. | | Mean Refraction. | | Diff. for 1′ Alt. | Diff. for + 1 in. Baro. | Diff. for − 1° Temp. | Apparent Altitude. | | Mean Refraction. | | Diff. for 1′ Alt. | Diff. for + 1 in. Baro. | Diff. for − 1° Temp. |
|---|---|---|---|---|---|---|---|---|---|---|---|---|---|
| ° | ′ | ′ | ″ | ″ | ″ | ″ | ° | ′ | ′ | ″ | ″ | ″ | ″ |
| 12 | 0  | 4 | 28·1 | 0·38 | 9·00 | 0·556 | 42 | 0 | 1 |  4·6 | 0·038 | 2·16 | 0·130 |
|    | 10 | 4 | 24·4 |   37 | 8·86 |   548 | 43 | 0 | 1 |  2·4 |   036 | 2·09 |   125 |
|    | 20 | 4 | 20·8 |   36 | 8·74 |   541 | 44 | 0 | 1 |  0·3 |   034 | 2·02 |   120 |
|    | 30 | 4 | 17·3 |   35 | 8·63 |   533 | 45 | 0 | 0 | 58·1 |   034 | 1·94 |   117 |
|    | 40 | 4 | 13·9 |   33 | 8·51 |   524 | 46 | 0 |   | 56·1 |   033 | 1·88 |   112 |
|    | 50 | 4 | 10·7 |   32 | 8·41 |   517 | 47 | 0 |   | 54·2 |   032 | 1·81 |   108 |
| 13 | 0  | 4 |  7·5 | 0·31 | 8·30 | 0·509 | 48 | 0 | 0 | 52·3 | 0·031 | 1·75 | 0·104 |
|    | 10 | 4 |  4·4 |   31 | 8·20 |   503 | 49 | 0 |   | 50·5 |   030 | 1·69 |   101 |
|    | 20 | 4 |  1·4 |   30 | 8·10 |   496 | 50 | 0 |   | 48·8 |   029 | 1·63 |   097 |
|    | 30 | 3 | 58·4 |   30 | 8·00 |   490 | 51 | 0 |   | 47·1 |   028 | 1·58 |   094 |
|    | 40 | 3 | 55·5 |   29 | 7·89 |   482 | 52 | 0 |   | 45·4 |   027 | 1·52 |   090 |
|    | 50 | 3 | 52·6 |   29 | 7·79 |   476 | 53 | 0 |   | 43·8 |   026 | 1·47 |   088 |
| 14 | 0  | 3 | 49·9 | 0·28 | 7·70 | 0·469 | 54 | 0 | 0 | 42·2 | 0·026 | 1·41 | 0·085 |
|    | 10 | 3 | 47·1 |   28 | 7·61 |   464 | 55 | 0 |   | 40·6 |   025 | 1·36 |   082 |
|    | 20 | 3 | 44·4 |   27 | 7·52 |   458 | 56 | 0 |   | 39·3 |   025 | 1·31 |   079 |
|    | 30 | 3 | 41·8 |   26 | 7·43 |   453 | 57 | 0 |   | 37·8 |   025 | 1·26 |   076 |
|    | 40 | 3 | 39·2 |   26 | 7·34 |   448 | 58 | 0 |   | 36·4 |   024 | 1·22 |   073 |
|    | 50 | 3 | 36·7 |   25 | 7·26 |   444 | 59 | 0 |   | 35·0 |   024 | 1·17 |   070 |
| 15 | 0  | 3 | 34·3 | 0·24 | 7·18 | 0·439 | 60 | 0 |   | 33·6 | 0·023 | 1·12 | 0·067 |
|    | 30 | 3 | 27·3 |   22 | 6·95 |   424 | 61 | 0 |   | 32·3 |   022 | 1·08 |   065 |
| 16 | 0  | 3 | 20·6 |   21 | 6·73 |   411 | 62 | 0 |   | 31·0 |   022 | 1·04 |   062 |
|    | 30 | 3 | 14·4 |   20 | 6·51 |   399 | 63 | 0 |   | 29·7 |   021 | 0·99 |   060 |
| 17 | 0  | 3 |  8·5 |   19 | 6·31 |   386 | 64 | 0 |   | 28·4 |   021 |   95 |   057 |
|    | 30 | 3 |  2·9 |   18 | 6·12 |   374 | 65 | 0 |   | 27·2 |   020 |   91 |   055 |
| 18 | 0  | 2 | 57·6 | 0·17 | 5·98 | 0·362 | 66 | 0 |   | 25·9 | 0·020 | 0·87 | 0·052 |
| 19 | 0  | 2 | 47·7 |   16 | 5·61 |   340 | 67 | 0 |   | 24·7 |   020 |   83 |   050 |
| 20 | 0  | 2 | 38·7 |   15 | 5·31 |   322 | 68 | 0 |   | 23·5 |   020 |   79 |   047 |
| 21 | 0  | 2 | 30·5 |   13 | 5·04 |   305 | 69 | 0 |   | 22·4 |   020 |   75 |   045 |
| 22 | 0  | 2 | 23·2 |   12 | 4·79 |   290 | 70 | 0 |   | 21·2 |   020 |   71 |   043 |
| 23 | 0  | 2 | 16·5 |   11 | 4·57 |   276 | 71 | 0 |   | 19·9 |   020 |   67 |   040 |
| 24 | 0  | 2 | 10·1 | 0·10 | 4·35 | 0·264 | 72 | 0 |   | 18·8 | 0·019 | 0·63 | 0·038 |
| 25 | 0  | 2 |  4·2 |   09 | 4·16 |   252 | 73 | 0 |   | 17·7 |   018 |   59 |   036 |
| 26 | 0  | 1 | 58·8 |   09 | 3·97 |   241 | 74 | 0 |   | 16·6 |   018 |   56 |   033 |
| 27 | 0  | 1 | 53·8 |   08 | 3·81 |   230 | 75 | 0 |   | 15·5 |   018 |   52 |   031 |
| 28 | 0  | 1 | 49·1 |   08 | 3·65 |   219 | 76 | 0 |   | 14·4 |   018 |   48 |   029 |
| 29 | 0  | 1 | 44·7 |   07 | 3·50 |   209 | 77 | 0 |   | 13·4 |   017 |   45 |   027 |
| 30 | 0  | 1 | 40·5 | 0·07 | 3·36 | 0·201 | 78 | 0 |   | 12·3 | 0·017 | 0·41 | 0·025 |
| 31 | 0  | 1 | 36·6 |   06 | 3·23 |   193 | 79 | 0 |   | 11·2 |   017 |   38 |   023 |
| 32 | 0  | 1 | 33·0 |   06 | 3·11 |   186 | 80 | 0 |   | 10·2 |   017 |   34 |   021 |
| 33 | 0  | 1 | 29·5 |   06 | 2·99 |   179 | 81 | 0 |   |  9·2 |   017 |   31 |   018 |
| 34 | 0  | 1 | 26·1 |   05 | 2·88 |   173 | 82 | 0 |   |  8·2 |   017 |   27 |   016 |
| 35 | 0  | 1 | 23·0 |   05 | 2·78 |   167 | 83 | 0 |   |  7·1 |   017 |   24 |   014 |
| 36 | 0  | 1 | 20·0 | 0·05 | 2·68 | 0·161 | 84 | 0 |   |  6·1 | 0·017 | 0·20 | 0·012 |
| 37 | 0  | 1 | 17·1 |   05 | 2·58 |   155 | 85 | 0 |   |  5·1 |   017 |   17 |   010 |
| 38 | 0  | 1 | 14·4 |   05 | 2·49 |   149 | 86 | 0 |   |  4·1 |   017 |   14 |   008 |
| 39 | 0  | 1 | 11·8 |   04 | 2·40 |   144 | 87 | 0 |   |  3·1 |   017 |   10 |   006 |
| 40 | 0  | 1 |  9·3 |   04 | 2·32 |   139 | 88 | 0 |   |  2·0 |   017 |   07 |   004 |
| 41 | 0  | 1 |  6·9 |   04 | 2·24 |   134 | 89 | 0 |   |  1·0 |   017 |   03 |   002 |

NOTE.—If the Barometer is over 30 inches the Barometer correction is *additive* to the Mean Refraction.
If the Thermometer is less than 50° the Thermometer correction is *additive* to the Mean Refraction.
And *vice versa*.

## TABLE XV.

### Parallax of the Sun.

| Altitude. | Parallax. | Altitude. | Parallax. |
|---|---|---|---|
| ° | ″ | ° | ″ |
| 0 | 8·8 | 50 | 5·7 |
| 5 | 8·8 | 55 | 5·1 |
| 10 | 8·7 | 60 | 4·4 |
| 15 | 8·5 | 65 | 3·7 |
| 20 | 8·3 | 70 | 3·0 |
| 25 | 8·0 | 75 | 2·3 |
| 30 | 7·7 | 80 | 1·5 |
| 35 | 7·2 | 85 | 0·8 |
| 40 | 6·8 | 90 | 0·0 |
| 45 | 6·2 | | |

Computed from the formula

Parallax in altitude = horizontal parallax × cos altitude.

The sun's horizontal parallax varies during the year from 8″·95 to 8″·65.
The above is a mean value, sufficiently exact for all topographical purposes.

## TABLE XVI.
### Latitude by Circum-Meridian Observations.
### Reduction to Meridian.

$$m = \frac{2\sin^2\frac{t}{2}}{\sin 1''}$$

| t | m. | 1m. | 2m. | 3m | 4m. | 5m. | 6m. | 7m. | 8m. | 9m. | 10m. | 11m. | 12m. | 13m. | 14m. | 15m. | 16m. | 17m. | 18m. | 19m. |
|---|---|---|---|---|---|---|---|---|---|---|---|---|---|---|---|---|---|---|---|---|
| s. | " | " | " | " | " | " | " | " | " | " | " | " | " | " | " | " | " | " | " | " |
| 0 | 0·0 | 2·0 | 7·8 | 17·7 | 31·4 | 49·1 | 70·7 | 96·2 | 125·7 | 159·0 | 196·3 | 237·5 | 282·7 | 331·7 | 384·7 | 441·5 | 502·5 | 567·2 | 635·9 | 708·4 |
| 1 | 0·0 | 2·0 | 8·0 | 17·9 | 31·7 | 49·4 | 71·1 | 96·7 | 126·2 | 159·6 | 197·0 | 238·3 | 284·5 | 332·6 | 385·6 | 442·5 | 503·5 | 568·3 | 637·0 | 709·7 |
| 2 | 0·0 | 2·1 | 8·1 | 18·1 | 31·9 | 49·7 | 71·5 | 97·1 | 126·7 | 160·2 | 197·6 | 239·0 | 283·4 | 333·6 | 386·6 | 443·5 | 504·6 | 569·4 | 638·2 | 710·9 |
| 3 | 0·0 | 2·1 | 8·2 | 18·3 | 32·2 | 50·1 | 71·9 | 97·6 | 127·2 | 160·8 | 198·2 | 239·6 | 284·3 | 334·3 | 387·5 | 444·5 | 505·6 | 570·5 | 639·4 | 712·1 |
| 4 | 0·0 | 2·2 | 8·4 | 18·5 | 32·5 | 50·4 | 72·3 | 98·0 | 127·8 | 161·4 | 198·9 | 240·4 | 285·8 | 335·2 | 388·8 | 445·6 | 506·6 | 571·5 | 640·6 | 713·4 |
| 5 | 0·0 | 2·3 | 8·5 | 18·7 | 32·7 | 50·7 | 72·7 | 98·5 | 128·3 | 162·0 | 199·6 | 241·1 | 286·6 | 336·0 | 389·6 | 446·6 | 507·7 | 572·8 | 641·7 | 714·8 |
| 6 | 0·0 | 2·4 | 8·7 | 19·1 | 33·0 | 51·1 | 73·1 | 99·0 | 128·8 | 162·5 | 200·3 | 241·9 | 287·4 | 336·8 | 390·2 | 447·5 | 508·8 | 573·9 | 642·9 | 715·9 |
| 7 | 0·0 | 2·4 | 8·8 | 19·3 | 33·3 | 51·4 | 73·5 | 99·4 | 129·3 | 163·2 | 200·9 | 242·5 | 288·2 | 337·6 | 391·2 | 448·5 | 509·8 | 575·0 | 644·1 | 717·1 |
| 8 | 0·0 | 2·5 | 8·9 | 19·3 | 33·5 | 51·7 | 73·9 | 99·9 | 129·9 | 163·8 | 201·6 | 243·3 | 289·0 | 338·6 | 392·1 | 449·5 | 510·9 | 576·1 | 645·3 | 718·4 |
| 9 | 0·0 | 2·6 | 9·1 | 19·5 | 33·8 | 52·1 | 74·3 | 100·4 | 130·4 | 164·4 | 202·2 | 244·1 | 289·8 | 339·4 | 393·0 | 450·5 | 511·9 | 577·2 | 646·5 | 719·6 |
| 10 | 0·1 | 2·7 | 9·2 | 19·7 | 34·1 | 52·4 | 74·7 | 100·8 | 130·9 | 165·0 | 202·9 | 244·8 | 290·6 | 340·3 | 393·9 | 451·5 | 513·0 | 578·4 | 647·7 | 720·9 |
| 11 | 0·1 | 2·7 | 9·4 | 19·9 | 34·3 | 52·7 | 75·1 | 101·3 | 131·5 | 165·5 | 203·5 | 245·5 | 291·4 | 341·2 | 394·8 | 452·5 | 514·0 | 579·5 | 648·9 | 722·1 |
| 12 | 0·1 | 2·8 | 9·5 | 20·1 | 34·6 | 53·1 | 75·5 | 101·8 | 132·0 | 166·1 | 204·2 | 246·3 | 292·2 | 342·0 | 395·8 | 453·5 | 515·1 | 580·7 | 650·0 | 723·4 |
| 13 | 0·1 | 2·9 | 9·6 | 20·3 | 34·9 | 53·4 | 75·9 | 102·3 | 132·5 | 166·8 | 204·9 | 247·0 | 293·0 | 343·0 | 396·7 | 454·5 | 516·1 | 581·7 | 651·2 | 724·6 |
| 14 | 0·1 | 3·0 | 9·8 | 20·5 | 35·2 | 53·8 | 76·3 | 102·7 | 133·1 | 167·4 | 205·6 | 247·7 | 293·8 | 343·7 | 397·6 | 455·5 | 517·2 | 582·9 | 652·4 | 725·9 |
| 15 | 0·1 | 3·1 | 9·9 | 20·7 | 35·5 | 54·1 | 76·7 | 103·2 | 133·6 | 168·0 | 206·3 | 248·5 | 294·6 | 344·6 | 398·6 | 456·5 | 518·3 | 584·0 | 653·3 | 727·2 |
| 16 | 0·1 | 3·1 | 10·1 | 20·9 | 35·7 | 54·5 | 77·1 | 103·7 | 134·2 | 168·6 | 206·9 | 249·2 | 295·5 | 345·6 | 399·6 | 457·5 | 519·3 | 585·1 | 654·5 | 728·5 |
| 17 | 0·2 | 3·2 | 10·2 | 21·1 | 36·0 | 54·8 | 77·5 | 104·2 | 134·7 | 169·2 | 207·6 | 249·9 | 296·3 | 346·4 | 400·6 | 458·5 | 520·4 | 586·3 | 655·7 | 729·7 |
| 18 | 0·2 | 3·3 | 10·4 | 21·4 | 36·3 | 55·1 | 77·9 | 104·6 | 135·3 | 169·8 | 208·3 | 250·7 | 297·1 | 347·2 | 401·6 | 459·5 | 521·5 | 587·4 | 656·9 | 730·9 |
| 19 | 0·2 | 3·4 | 10·5 | 21·6 | 36·6 | 55·5 | 78·3 | 105·1 | 135·8 | 170·4 | 208·9 | 251·4 | 297·8 | 348·1 | 402·3 | 460·5 | 522·5 | 588·5 | 658·4 | 732·2 |
| 20 | 0·2 | 3·5 | 10·7 | 21·8 | 36·9 | 55·8 | 78·8 | 105·6 | 136·3 | 171·0 | 209·3 | 253·2 | 298·6 | 349·0 | 403·6 | 461·5 | 523·6 | 589·6 | 659·6 | 733·5 |
| 21 | 0·2 | 3·6 | 10·8 | 22·0 | 37·2 | 56·2 | 79·2 | 106·1 | 136·9 | 171·6 | 210·3 | 252·9 | 299·4 | 349·8 | 404·2 | 462·5 | 524·7 | 590·8 | 660·8 | 734·7 |
| 22 | 0·3 | 3·7 | 11·0 | 22·3 | 37·4 | 56·5 | 79·6 | 106·6 | 137·4 | 172·2 | 211·0 | 253·6 | 300·2 | 350·7 | 405·1 | 463·5 | 525·7 | 591·9 | 662·0 | 736·0 |
| 23 | 0·3 | 3·8 | 11·2 | 22·5 | 37·7 | 56·9 | 80·0 | 107·0 | 138·0 | 172·9 | 211·7 | 254·4 | 301·0 | 351·6 | 406·1 | 464·5 | 526·8 | 593·0 | 663·2 | 737·3 |
| 24 | 0·3 | 3·8 | 11·3 | 22·7 | 38·0 | 57·3 | 80·4 | 107·5 | 138·5 | 173·5 | 212·3 | 255·1 | 301·8 | 352·5 | 407·0 | 465·5 | 527·9 | 594·2 | 664·4 | 738·5 |

| | | | | | | | | | | | | | | | | |
|---|---|---|---|---|---|---|---|---|---|---|---|---|---|---|---|---|
| 25 | 0·3 | 3·9 | 11·5 | 22·9 | 38·3 | 57·6 | 80·8 | 108·0 | 139·1 | 174·1 | 213·0 | 255·9 | 302·6 | 353·3 | 408·0 | 466·5 | 529·9 | 595·3 | 665·6 | 739·8 |
| 26 | 0·3 | 4·0 | 11·6 | 23·1 | 38·6 | 58·0 | 81·3 | 108·5 | 139·6 | 174·7 | 213·7 | 256·6 | 303·5 | 354·2 | 408·8 | 467·5 | 530·0 | 595·5 | 666·8 | 741·1 |
| 27 | 0·4 | 4·1 | 11·8 | 23·4 | 38·9 | 58·3 | 81·7 | 109·0 | 140·2 | 175·3 | 214·4 | 257·4 | 304·3 | 355·0 | 409·8 | 468·5 | 531·1 | 596·5 | 668·0 | 742·5 |
| 28 | 0·4 | 4·2 | 11·8 | 23·6 | 39·2 | 58·7 | 82·1 | 109·5 | 140·7 | 175·9 | 215·1 | 258·1 | 305·0 | 355·8 | 410·8 | 469·5 | 532·2 | 597·8 | 669·2 | 743·8 |
| 29 | 0·5 | 4·3 | 12·1 | 23·8 | 39·5 | 59·0 | 82·5 | 110·0 | 141·3 | 176·6 | 215·8 | 258·8 | 305·8 | 356·6 | 411·7 | 470·5 | 533·3 | 599·9 | 670·4 | 744·9 |
| 30 | 0·5 | 4·4 | 12·3 | 24·0 | 39·8 | 59·4 | 83·0 | 110·4 | 141·8 | 177·2 | 216·4 | 259·6 | 306·7 | 357·7 | 412·7 | 471·5 | 534·3 | 601·0 | 671·6 | 746·2 |
| 31 | 0·5 | 4·5 | 12·4 | 24·3 | 40·1 | 59·8 | 83·4 | 110·9 | 142·2 | 177·7 | 217·1 | 260·4 | 307·5 | 358·6 | 413·6 | 472·6 | 535·4 | 602·2 | 672·8 | 747·4 |
| 32 | 0·5 | 4·6 | 12·6 | 24·5 | 40·6 | 60·1 | 83·8 | 111·4 | 142·7 | 178·4 | 217·8 | 261·1 | 308·4 | 359·5 | 414·6 | 473·6 | 536·5 | 603·3 | 674·1 | 748·7 |
| 33 | 0·6 | 4·7 | 12·8 | 24·7 | 40·8 | 60·5 | 84·2 | 111·9 | 143·3 | 178·8 | 218·5 | 261·9 | 309·2 | 360·3 | 415·5 | 474·6 | 537·7 | 604·3 | 675·3 | 750·0 |
| 34 | 0·6 | 4·8 | 12·8 | 25·0 | 40·9 | 60·8 | 84·7 | 112·4 | 144·1 | 179·5 | 219·2 | 262·6 | 310·0 | 361·1 | 416·5 | 475·6 | 538·7 | 605·6 | 676·5 | 751·3 |
| 35 | 0·5 | 4·9 | 13·2 | 25·2 | 41·2 | 61·2 | 85·1 | 112·9 | 144·3 | 180·3 | 219·9 | 263·4 | 310·8 | 362·2 | 417·4 | 476·6 | 539·7 | 606·8 | 677·7 | 752·6 |
| 36 | 0·5 | 5·0 | 13·4 | 25·4 | 41·5 | 61·6 | 85·5 | 113·4 | 145·1 | 180·7 | 220·6 | 264·1 | 311·7 | 363·1 | 418·4 | 477·7 | 540·8 | 607·9 | 678·8 | 753·8 |
| 37 | 0·6 | 5·1 | 13·6 | 25·7 | 41·8 | 61·9 | 86·0 | 113·8 | 145·6 | 181·2 | 221·3 | 264·9 | 312·5 | 364·0 | 419·4 | 478·7 | 541·9 | 609·1 | 680·1 | 755·0 |
| 38 | 0·6 | 5·2 | 13·8 | 25·9 | 42·1 | 62·3 | 86·4 | 114·3 | 146·3 | 182·2 | 221·8 | 265·7 | 313·3 | 364·8 | 420·3 | 479·7 | 543·0 | 610·2 | 681·3 | 756·4 |
| 39 | 0·6 | 5·3 | 13·8 | 26·2 | 42·5 | 62·7 | 86·8 | 114·9 | 146·9 | 182·8 | 222·7 | 266·4 | 314·1 | 365·7 | 421·3 | 480·7 | 544·1 | 611·4 | 682·5 | 757·7 |
| 40 | 0·9 | 5·4 | 14·0 | 26·4 | 42·8 | 63·0 | 87·2 | 115·4 | 147·5 | 183·5 | 223·4 | 267·2 | 315·0 | 366·6 | 422·2 | 481·7 | 545·2 | 612·5 | 683·6 | 758·9 |
| 41 | 0·9 | 5·5 | 14·1 | 26·6 | 43·1 | 63·4 | 87·7 | 115·9 | 148·0 | 184·1 | 224·1 | 267·9 | 315·8 | 367·4 | 423·2 | 482·7 | 546·3 | 613·7 | 684·9 | 760·2 |
| 42 | 0·9 | 5·6 | 14·3 | 26·9 | 43·4 | 63·8 | 88·1 | 116·3 | 148·6 | 184·7 | 224·9 | 268·7 | 316·6 | 368·4 | 424·2 | 483·8 | 547·4 | 614·8 | 686·1 | 761·5 |
| 43 | 1·0 | 5·7 | 14·5 | 27·1 | 43·7 | 64·2 | 88·6 | 116·8 | 149·2 | 185·4 | 225·5 | 269·4 | 317·4 | 369·3 | 425·1 | 484·8 | 548·4 | 616·0 | 687·3 | 762·8 |
| 44 | 1·0 | 5·8 | 14·5 | 27·4 | 44·0 | 64·5 | 88·9 | 117·3 | 149·7 | 185·9 | 226·2 | 270·1 | 318·2 | 370·2 | 426·1 | 485·8 | 549·5 | 617·4 | 688·7 | 764·0 |
| 45 | 1·1 | 5·4 | 14·8 | 27·6 | 44·3 | 64·9 | 89·5 | 117·9 | 150·3 | 186·5 | 226·8 | 270·9 | 319·1 | 371·1 | 427·0 | 486·8 | 550·6 | 618·5 | 689·9 | 765·4 |
| 46 | 1·2 | 5·5 | 14·9 | 28·0 | 44·6 | 65·3 | 89·9 | 118·4 | 150·8 | 187·1 | 227·5 | 271·6 | 319·9 | 372·0 | 428·0 | 487·9 | 551·7 | 619·6 | 691·2 | 766·5 |
| 47 | 1·2 | 5·7 | 15·1 | 28·2 | 45·0 | 65·7 | 90·4 | 118·9 | 151·5 | 187·7 | 228·3 | 272·3 | 320·7 | 372·9 | 429·0 | 488·9 | 552·8 | 620·7 | 692·4 | 767·8 |
| 48 | 1·3 | 5·8 | 15·3 | 28·5 | 45·3 | 66·1 | 90·8 | 119·4 | 152·0 | 188·4 | 229·0 | 273·0 | 321·5 | 373·8 | 429·9 | 489·9 | 553·9 | 621·8 | 693·6 | 769·0 |
| 49 | 1·3 | 5·9 | 15·5 | 28·7 | 45·6 | 66·4 | 91·2 | 120·0 | 152·6 | 189·1 | 229·7 | 273·7 | 322·4 | 374·7 | 430·9 | 491·0 | 555·0 | 623·0 | 694·8 | 770·3 |
| 50 | 1·4 | 6·0 | 14·8 | 29·1 | 45·9 | 66·8 | 91·7 | 120·5 | 153·1 | 189·9 | 230·5 | 274·4 | 323·3 | 375·6 | 431·9 | 492·0 | 556·0 | 624·1 | 696·0 | 771·5 |
| 51 | 1·4 | 6·1 | 15·9 | 29·4 | 46·2 | 67·2 | 92·1 | 120·6 | 153·8 | 190·5 | 231·2 | 275·2 | 324·1 | 376·5 | 432·8 | 493·1 | 557·2 | 625·2 | 697·3 | 772·7 |
| 52 | 1·5 | 6·2 | 16·1 | 29·6 | 46·5 | 67·6 | 92·5 | 121·5 | 154·3 | 191·1 | 231·8 | 275·8 | 325·0 | 377·4 | 433·8 | 494·1 | 558·3 | 627·6 | 698·5 | 773·9 |
| 53 | 1·5 | 6·3 | 16·2 | 29·9 | 46·8 | 68·0 | 93·0 | 122·0 | 154·8 | 191·8 | 232·5 | 276·6 | 325·8 | 378·3 | 434·8 | 495·1 | 559·4 | 627·6 | 699·7 | 775·1 |
| 54 | 1·6 | 6·4 | 16·5 | 30·2 | 47·1 | 68·3 | 93·5 | 122·5 | 155·5 | 192·4 | 233·2 | 277·3 | 326·7 | 379·2 | 435·8 | 496·2 | 560·5 | 628·8 | 701·0 | 776·3 |
| 55 | 1·6 | 7·2 | 16·7 | 30·4 | 47·5 | 68·8 | 93·9 | 123·1 | 156·1 | 193·1 | 234·0 | 278·2 | 327·5 | 380·2 | 436·7 | 497·2 | 561·6 | 629·8 | 702·2 | 778·4 |
| 56 | 1·7 | 7·3 | 16·9 | 30·6 | 47·8 | 69·1 | 94·4 | 123·6 | 156·7 | 193·7 | 234·7 | 278·9 | 328·4 | 381·1 | 437·8 | 498·3 | 562·8 | 631·2 | 703·5 | 779·7 |
| 57 | 1·7 | 7·5 | 17·1 | 30·9 | 48·1 | 69·5 | 94·8 | 124·1 | 157·2 | 194·5 | 235·4 | 279·6 | 329·2 | 382·2 | 438·7 | 499·3 | 563·9 | 632·3 | 704·7 | 781·0 |
| 58 | 1·8 | 7·6 | 17·3 | 31·1 | 48·4 | 69·9 | 95·3 | 124·6 | 157·8 | 195·0 | 236·1 | 280·3 | 330·0 | 383·2 | 439·9 | 500·4 | 565·0 | 633·5 | 705·9 | 782·3 |
| 59 | 1·9 | 7·7 | 17·5 | 31·2 | 48·8 | 70·3 | 95·7 | 125·1 | 158·4 | 195·7 | 236·8 | 281·3 | 330·9 | 383·8 | 440·6 | 501·4 | 566·2 | 634·7 | 707·1 | 783·6 |

## TABLE XVII.

### Longitude by Occultations.

| Lat. | Correction of the Moon's Equatorial Parallax for the figure of the Earth. (Compression $\frac{1}{294\cdot 26}$.) ||||| 
| --- | --- | --- | --- | --- | --- |
| | Horizontal Parallax. ||||| 
| | 54′ | 56′ | 58′ | 60′ | 62′ |
| ° | ″ | ″ | ″ | ″ | ″ |
| 0 | 0 | 0 | 0 | 0 | 0 |
| 8 | ·2 | ·2 | ·2 | ·2 | ·2 |
| 16 | ·8 | ·9 | ·9 | ·9 | ·9 |
| 20 | 1·3 | 1·3 | 1·4 | 1·4 | 1·5 |
| 24 | 1·8 | 1·9 | 2·0 | 2·0 | 2·0 |
| 28 | 2·5 | 2·5 | 2·6 | 2·7 | 2·8 |
| 32 | 3·1 | 3·2 | 3·3 | 3·4 | 3·5 |
| 36 | 3·8 | 4·0 | 4·1 | 4·3 | 4·4 |
| 40 | 4·6 | 4·7 | 4·9 | 5·1 | 5·2 |
| 44 | 5·3 | 5·5 | 5·7 | 5·9 | 6·1 |
| 48 | 6·4 | 6·7 | 7·0 | 7·2 | 7·4 |
| 52 | 6·9 | 7·1 | 7·4 | 7·6 | 7·9 |
| 56 | 7·6 | 7·9 | 8·2 | 8·5 | 8·7 |
| 60 | 8·3 | 8·6 | 8·9 | 9·2 | 9·6 |
| 64 | 9·0 | 9·3 | 9·6 | 10·0 | 10·3 |
| 68 | 9·5 | 9·9 | 10·3 | 10·6 | 11·0 |
| 72 | 10·0 | 10·4 | 10·8 | 11·2 | 11·6 |
| 76 | 10·5 | 10·9 | 11·2 | 11·6 | 12·0 |
| 80 | 10·8 | 11·2 | 11·6 | 12·0 | 12·4 |

## TABLE XVIII.

### Longitude by Occultations.

REDUCTION OF LATITUDE.
Compression $\frac{1}{294\cdot26}$

| Lat. | Red. | Lat. | Red. | Lat. | Red. |
|---|---|---|---|---|---|
| ° | ′ ″ | ° | ′ ″ | ° | ′ ″ |
| 0 | 0 0 | 30 | 10 7 | 60 | 10 9 |
| 1 | 0 24 | 31 | 10 19 | 61 | 9 56 |
| 2 | 0 49 | 32 | 10 30 | 62 | 9 43 |
| 3 | 1 13 | 33 | 10 40 | 63 | 9 29 |
| 4 | 1 37 | 34 | 10 50 | 64 | 9 14 |
| 5 | 2 1 | 35 | 10 59 | 65 | 8 59 |
| 6 | 2 25 | 36 | 11 7 | 66 | 8 43 |
| 7 | 2 49 | 37 | 11 14 | 67 | 8 26 |
| 8 | 3 13 | 38 | 11 21 | 68 | 8 9 |
| 9 | 3 36 | 39 | 11 26 | 69 | 7 51 |
| 10 | 3 59 | 40 | 11 31 | 70 | 7 32 |
| 11 | 4 22 | 41 | 11 35 | 71 | 7 13 |
| 12 | 4 45 | 42 | 11 38 | 72 | 6 54 |
| 13 | 5 7 | 43 | 11 40 | 73 | 6 34 |
| 14 | 5 28 | 44 | 11 41 | 74 | 6 13 |
| 15 | 5 50 | 45 | 11 42 | 75 | 5 52 |
| 16 | 6 11 | 46 | 11 42 | 76 | 5 31 |
| 17 | 6 31 | 47 | 11 41 | 77 | 5 9 |
| 18 | 6 51 | 48 | 11 38 | 78 | 4 46 |
| 19 | 7 11 | 49 | 11 36 | 79 | 4 24 |
| 20 | 7 30 | 50 | 11 32 | 80 | 4 1 |
| 21 | 7 49 | 51 | 11 27 | 81 | 3 38 |
| 22 | 8 8 | 52 | 11 22 | 82 | 3 14 |
| 23 | 8 24 | 53 | 11 15 | 83 | 2 50 |
| 24 | 8 41 | 54 | 11 8 | 84 | 2 26 |
| 25 | 8 57 | 55 | 11 0 | 85 | 2 2 |
| 26 | 9 12 | 56 | 10 52 | 86 | 1 38 |
| 27 | 9 27 | 57 | 10 42 | 87 | 1 14 |
| 28 | 9 41 | 58 | 10 32 | 88 | 0 49 |
| 29 | 9 54 | 59 | 10 21 | 89 | 0 24 |

Formula :

$$\phi - \phi' = \frac{e^2}{(2 - e^2) \sin 1''} \sin 2\phi - \left\{ \frac{e^2}{2 - e^2} \right\}^2 \frac{\sin 4\phi}{2 \sin 1''},$$
$$e = \sqrt{2c - c^2},$$
$$c = \frac{1}{294\cdot26}.$$

## APPENDIX I.

### TABLE I.—LATITUDES, LONGITUDES AND REVERSE AZIMUTHS.

This Table gives the logarithms of the quantities, P, Q, R, S, T used in the forms employed for the computation of Terrestrial Latitudes, Longitudes, and Reverse Azimuths (*vide* Form 2).

Suppose that **A** and **B** are two stations and that we have given—

$\lambda$ = Latitude of **A**,
$L$ = Longitude of **A**,
$A$ = Azimuth of **B** from **A** (measured from S. to W.)
$c$ = Distance in feet at mean sea-level between **A** and **B** ;

and we require

$\lambda_b$ = Latitude of **B**,
$L_b$ = Longitude of **B**,
$B$ = Azimuth of **A** from **B**.

If we put

$$\lambda_b = \lambda + \Delta\lambda; \quad L_b = L + \Delta L; \quad \text{and} \quad B = \pi + A + \Delta A;$$

it will be seen that we require the differences

$$\Delta\lambda, \quad \Delta L \quad \text{and} \quad \Delta A.$$

The values of these quantities are found as follows :—

Putting

$$\Delta\lambda = \delta_1\lambda + \delta_2\lambda,$$
$$\Delta L = \delta_1 L + \delta_2 L,$$
$$\Delta A = \delta_1 A + \delta_2 A,$$

then

$\delta_1\lambda = P \cos A . c,$       $\delta_2\lambda = \delta_1 AR \sin A . c,$
$\delta_1 L = \delta_1\lambda Q \sec \lambda \tan A,$       $\delta_2 L = \delta_2\lambda S \cot A,$
$\delta_1 A = \delta_1 L \sin \lambda,$       $\delta_2 A = \delta_2 LT,$

which are the expressions used in the form for computation.

It will be noticed that each term is found consecutively from the preceding one

The proper signs to be given to the quantities will be found in Tables II (A) and II (B)

For latitudes other than those given in the Table the quantities must be found by interpolation.

## APPENDIX II.

### COMPUTATION OF DISTANCES AND AZIMUTHS.

Given the latitudes and longitudes of two points A and B.

| REQUIRED. | Azimuths **B** from **A** (A) and **A** from **B** ($\pi + A + \Delta A$), and length in feet between Station **A** and Station **B** (c). | |
|---|---|---|
| Data. | Latitude. | Longitude. |
| Station **A** ... ... ... ... | $\lambda_a =$ ° ′ ″ | $L_a =$ ° ′ ″ |
| Station **B** ... ... ... ... | $\lambda_b =$ | $L_b =$ |
| $\therefore \Delta\lambda =$ ± | $\therefore \Delta L =$ ± | |
| $\dfrac{\Delta\lambda}{2} =$ ± | $\dfrac{\Delta L}{2} =$ ± | |
| $\lambda_m =$ Middle Latitude $\left(\lambda_a + \dfrac{\Delta\lambda}{2}\right) =$ | | |

$\Delta\lambda$  When **A** and **B** are both N., or S. of the Equator :

$\Delta\lambda$ is $\dfrac{+}{-}$ when $\lambda_b$ is $\dfrac{\text{greater}}{\text{less}}$ than $\lambda_a$, and $\lambda_m = \frac{1}{2}(\lambda_a + \lambda_b)$.

If the ray crosses the Equator :

$\Delta\lambda$ is always negative and $= \lambda_a + \lambda_b$, and $\lambda_m = \frac{1}{2}$ diff. between $\lambda_a$ and $\lambda_b$.

$\Delta L$  When **A** and **B** are both W. of Greenwich :

$\Delta L$ is $\dfrac{+}{-}$ when $L_b$ is $\dfrac{\text{less}}{\text{greater}}$ than $L_a$.

When **A** and **B** are both E. of Greenwich :

$\Delta L$ is $\dfrac{+}{-}$ when $L_b$ is $\dfrac{\text{greater}}{\text{less}}$ than $L_a$.

If the ray crosses the meridian of Greenwich :—

$\Delta L = L_a + L_b$ and is $\dfrac{+}{-}$ if $L_b$ is $\dfrac{\text{E.}}{\text{W.}}$ of Greenwich.

| Formulæ. | Logarithms. | Results. |
|---|---|---|
| $\dfrac{\Delta A}{2} = \dfrac{\Delta L}{2}\sin\lambda_m$ | To 5 places of Decimals.<br>$\log\dfrac{\Delta L}{2}$ (in secs.) =<br>$\log\sin\lambda_m$ =<br>$\log\dfrac{\Delta A}{2}$ = | $\dfrac{\Delta A}{2} =$ ″<br>[When $\lambda_m$ is $\dfrac{\text{N.}}{\text{S.}}$ of equator $\Delta A$ has the $\dfrac{\text{same}}{\text{opposite}}$ sign as $\Delta L$.] |
| $\log\tan\left(A + \dfrac{\Delta A}{2}\right) =$<br>$\log$ Table XII.† − log Table XI.† + log $\Delta L$ − log $\Delta\lambda$ | log Table XII. =<br>Co-log Table XI. =<br>log $\Delta L$ (in secs.) =<br>Co-log $\Delta\lambda$ =<br>$\log\tan\left(A + \dfrac{\Delta A}{2}\right) =$ | $A + \dfrac{\Delta A}{2} =$ ° ′ ″<br>[Quadrant obtained from Table II. (A) or II. (B) with given signs of $\Delta\lambda$ and $\Delta L$ as above.] |
| $\log c = \log\sec\left(A + \dfrac{\Delta A}{2}\right) + \log$ Table XI.† + log $\Delta\lambda$ | $\log\sec\left(A + \dfrac{\Delta A}{2}\right) =$<br>log Table XI. =<br>log $\Delta\lambda$ (in secs.) =<br>$\log c =$ | $c =$ |
| Required Azimuths.<br>—<br>A<br>$\pi + A + \Delta A$ | $A + \dfrac{\Delta A}{2}$ =<br>$\dfrac{\Delta A}{2}$ − + or −<br>Diff. =<br>$\pi +$ Sum = | ° ′ ″<br><br>= azimuth of B from A.<br>= azimuth of A from B. |

† Values for middle latitude, i.e., for $\lambda + \dfrac{\Delta\lambda}{2}$.

## APPENDIX III.

### ALTERNATIVE FORMULÆ FOR DETERMINING LATITUDE, LONGITUDE, AND REVERSE AZIMUTH IN TRIANGULATION; ALSO THOSE FOR THE REVERSE PROCESS.

*Direct Process :—*

$$\frac{\Delta A}{2} = \frac{\Delta L}{2}\left(\sin \lambda + \frac{\Delta \lambda}{2}\right) \quad \ldots \ldots \ldots \quad (1),$$

$$\Delta \lambda = -\frac{c}{\rho} \cos\left(A + \frac{\Delta A}{2}\right) \operatorname{cosec} 1'' \quad \ldots \ldots \ldots \quad (2),$$

$$\Delta L = -\frac{c}{\nu} \cdot \frac{\sin\left(A + \frac{\Delta A}{2}\right)}{\cos\left(\lambda + \frac{\Delta \lambda}{2}\right)} \operatorname{cosec} 1'' \quad \ldots \ldots \ldots \quad (3),$$

where A, λ, L are the initial azimuth, latitude, and longitude, and ΔA, Δλ, ΔL, are the increments to be added to the initial values to obtain those of the distant station at the end of the line, of which $c$ is the length in feet.

Now
$$\log \frac{\operatorname{cosec} 1''}{\rho} = P \text{ (Table I.)}$$

and
$$\log \frac{\operatorname{cosec} 1''}{\nu} = P + Q \text{ (Table I.)}.$$

To obtain a value for $\frac{\Delta A}{2}$, the approximate position of the end of the line must be known by measurement from a plane-table or approximate computation, and the rough values for ΔL and Δλ thus obtained are substituted in (1). The rest is obvious.

*The Reverse Process,* i.e., the latitudes and longitudes of the two stations being known, to compute the azimuths and distance :—

Here we know ΔL, λ, and Δλ, and can therefore compute the value of ΔA from (1).

Dividing (3) by (2),
$$\frac{\Delta L}{\Delta \lambda} = \frac{\rho}{\nu} \tan\left(A + \frac{\Delta A}{2}\right) \sec\left(\lambda + \frac{\Delta \lambda}{2}\right) \quad \ldots \ldots \ldots \quad (4),$$

where
$$\log \frac{\rho}{\nu} = Q \text{ (Table I.)}.$$

From this, $\left(A + \frac{\Delta A}{2}\right)$ can be computed, and then from (2) $c$ can be found.

NOTE.—Both in the Direct Process and in the Reverse Process the values of P and Q to be used are those for the *mean latitude* $\left(\lambda + \frac{\Delta \lambda}{2}\right)$

## APPENDIX IV.

### TABLE IV.—COMPUTATION OF HEIGHTS.

This table is to facilitate finding the subtended angle when only one angle has been observed—

$\left.\begin{array}{l}D_a \\ E_a\end{array}\right\} = $ angle of $\left\{\begin{array}{l}\text{Depression} \\ \text{Elevation}\end{array}\right\}$ at **A**.

$\left.\begin{array}{l}D_b \\ E_b\end{array}\right\} = $ do. do. at **B**.

$c$ = distance in feet between **A** and **B**.

*1st case.*—When reciprocal angles have been observed—

(a) If both angles are depressions, take half the difference.
(b) If one angle only is an elevation, take half the sum.

The result is an auxiliary angle called the subtended angle (S)—

$$i.e., S = \frac{D_b - D_a}{2} \text{ or } \frac{E_a + D_b}{2}.$$

*2nd case.*—When one angle only ($D_a$ or $E_a$) has been observed, we must first find the subtended angle—

*Rule.*—Subtract the quantity in Table IV. from log $c$. The natural number corresponding to the resulting logarithm is an angle K in seconds of arc, then—

$$S = K - D, \text{ or } K + E.$$

Having now found S in either case, the difference in height in feet ($h$) between **A** and **B** is obtained from the formula—

$$h = c \tan S.$$

EXAMPLE OF 2ND CASE—

Given log $c = 4 \cdot 8907$
and $D = 0°\ 14'\ 25''$,

to find S in lat. 20° with coefficient of refraction $\cdot 07$—

log $c = 4 \cdot 8907$

$\left.\begin{array}{l}\text{Number in Table IV.} \\ \text{under } \cdot 07 \text{ for lat. } 20°\end{array}\right\} = 2 \cdot 3716$

log K = $2 \cdot 5191$

$\therefore$ K = $330'' = 0°\ 5'\ 30''$
D $\phantom{= 330''} = 0\ 14\ 35$

Hence S = K - D = 0 9 5

*See* Form II., p. 31.

## APPENDIX V.

### THE SAG OF STEEL TAPES.

Let $l$ be the length of the tape between the supports $= 2s$.
    $z$  ,,  ,,  half the chord.
    $c$  ,,  ,,  tape of which the weight = horizontal tension.
    $w$  ,,  weight of the tape between the supports.
    $t$  ,,  tension applied.
The supports are supposed to be on the same level.

Then, $s = \dfrac{c}{2}(e^{\frac{z}{c}} - e^{-\frac{z}{c}})$. (See any elementary text-book.)

Expanding $e^{\frac{z}{c}}$ and $e^{-\frac{z}{c}}$,

(i.) $$s = z + \frac{z^3}{c^2\lfloor 3} + \frac{z^5}{c^4\lfloor 5} + \ldots,$$

a rapidly converging series if $z$ is small compared with $c$.

The term $\dfrac{z^5}{c^4\lfloor 5}$ is very small; for instance, let the tape be 100 feet long, and let it weigh 1 lb.; let the tension applied be 20 lbs. Then this term is less than $\dfrac{1}{2,000,000}$.

So that for light tapes and the tensions ordinarily used:—

(ii.) $$\text{Sag} = 2(s - z) = \frac{2z^3}{c^2\lfloor 3},$$

*i.e.*, **the sag varies approximately as the cube of the span.**

Since $z$ approximately equals $s$, and horizontal tension nearly equals $t$, we have from (ii.),

(iii.) $$\text{sag} = \frac{2\left(\dfrac{l}{2}\right)^3}{\dfrac{t^2 l^2 \lfloor 3}{w^2}} = \frac{lw^2}{24t^2}, \text{ approximately.}$$

This last expression will usually be sufficiently accurate for topographical bases.

If greater accuracy is required, find the true value of $c$, thus

$$c = \left(t^2 - \frac{w^2}{4}\right)^{\frac{1}{2}} \times \frac{2s}{w},$$

and having obtained an approximate value of sag from (iii.), put $z = s - \tfrac{1}{2}$ sag, then

$$\text{sag} = \frac{2z^3}{c^2\lfloor 3} + \frac{2z^5}{c^4\lfloor 5}.$$

## APPENDIX VI.

### LIST OF INSTRUMENTS, TOOLS, AND CAMP EQUIPMENT.

The variety of conditions under which topographical and geographical surveys may be carried out is so great that no practical benefit would result from the compilation of an ideal outfit of instruments and stores for a survey party or expedition. But the following list may be useful for selecting from. It will rarely happen that all the instruments therein mentioned are required.

In fitting out an expedition it is also necessary to consider what *stationery* and *technical books* will be wanted.

### INSTRUMENTS.

Abney levels.
Arrows.
Barometers, aneroid (with Kew certificate).
Boards, drawing.
Clinometers, Indian pattern.
Chronometers, box.
Compasses, beam.
 „ prismatic.
 „ trough (for plane-tables).
 „ proportional.
Field glasses.
Heliographs.
Heliotropes or Heliostats.
Hypsometer.
Instruments, drawing, in leather rolls.
Lamps, bicycle.
 „ bull's-eye.
 „ reflecting.
Leather cases, carrying.
Level's, Cooke's.
 „ "Y."
Log line.
Pantagraph.
Parallel rulers.
Perambulators.
Protractors, brass.
 „ ivory, wood or paper.
Plane-table, 18" × 24", with tripod and cover.
 „ portable, 18" × 18", and cover.
Scales, plotting.

Scales, Marquois.
 „ 36-inch, brass or steel, diagonal.
Sight rule, 16-inch, for plane-tables.
Straight-edge, steel, 36-inch.
Subtense bars.
Tacheometers.
Tapes, steel or invar, 100 feet.
 „ steel or invar, 300 feet or 100 meters.
Telescopes.
Thermometers, in lined brass tubes.
Theodolites, 5-inch transit, micrometer microscope.
Theodolites, 6-inch transit, micrometer microscope.
Theodolites, Vernier (3-inch, 5-inch or 6-inch).
 „ subtense.
Tin tubes for maps.
Tripods for prismatic compasses.
T-squares.
Watches, half chronometer, mean time.
 „ half chronometer, siderial time.

### TOOLS, &c.

Axes, felling, with spare helves.
Carpenter's tools, rolls of.
Crowbars.
Matchets.
Nails.
Pliers.

Rope, 3-inch.
Saws.
Shovels.

These should be supplemented by the tools in use in the country.

## CAMP EQUIPMENT.

Baths, canvas.
Beds, camp.
Bedding and valises.
Boxes for small stores.
Buckets.
Canteens and messing kit.
Candle lamps.
Chairs, camp.
Curtains, mosquito.
Hurricane lamps.
Indiarubber sheets.
Lamps, Lord's.
Mess boxes and cooking utensils.
Stoves, camp and cooking (for cold climates)
Tables, camp.
Tents, Willesden canvas, double roof, ridge.
Washing kit.

## APPENDIX VII.

This account of sketching in dense forest has been extracted from the "Manual of Field Sketching." The method described is often useful in exploratory expeditions, and may sometimes be used in deliberate work for filling in minor tracks between points more rigorously fixed.

### SKETCHING IN DENSE FOREST.*

In many countries, parts of Africa especially, it is impossible, on account of dense bush, jungle, or grass, to survey any features other than the line of the road or track. As the paths wind about very much, the bearings have to be taken at short intervals, and such surveys being generally made whilst the sketcher is marching with a column or caravan, it is inconvenient and would occasion too great a loss of time to plot the bearings and distances on the march. It is preferable, therefore, to enter them in a note-book and to plot them on arrival in camp. The bearings are better taken with a prismatic compass, if available, but if not, with an ordinary pocket compass. Distances are best arrived at by noting the times at which each bearing is taken or object passed, and estimating the rate of marching during the interval. A convenient form of note-book is formed by ruling four lines about half an inch apart down the centre of each page (see fig. 70). Supposing the day's traverse is being commenced from Yandahu village, start by obtaining a back bearing to the last village. This is entered as B. B. (back bearing) to village, 208°, and a forward bearing to the next village in front entered as F. B. village (forward bearing) Yolu 71°. In the thickest bush country a native has generally a very good idea of the relative positions of the villages round his own, and these bearings from village to village are very useful as checks on the direction of the traverse.

If a distant view is obtainable it is advisable to take back readings to any marked physical feature as far away as possible and near which the track has passed, also to any points in advance near which it appears probable it may pass. These long bearings of points, say 30 or 40 miles apart, are more useful than those from village to village in checking and correcting the direction of the traverse.

In the example (fig. 70) bearings were obtained to two hills about 30 miles ahead. The bearing of the track immediately ahead is then taken and entered in the centre column, and the time in the first or left-hand column.

It is often impossible to see the general direction of the track, which probably disappears in the bush a few yards ahead; in this case it is necessary to have a native to show the direction, but constant watching is necessary, as many natives are very slow in understanding what is required, and seem unable to strike a mean between the direction of the path just a few yards in front and its position two or three miles ahead. To get a general direction of about half a mile of road, so as to take a bearing every ten minutes, is a very convenient mean, and will give fair results. On arrival at the point where the direction changes and a new bearing has to be taken the estimated rate of marching is entered in the third column.

If it be impossible, from a want of guides or other cause, to get satisfactorily the forward direction of the track, the best way of proceeding is to take the bearing upon which one happens to be marching at fixed intervals, say every five minutes of time, which may be

---

* The method here described is also best for long continuous sketches from horse or camel in hot climates. If the bearings are long it is better to dismount from the horse when taking them (not necessarily from the camel). The plotting is done on arrival in camp.

shortened or lengthened according as the result is desired to be more or less accurate. Sometimes it is convenient to watch the compass swinging when held in the hand and take the approximate mean bearing as judged every two or three minutes.

It is necessary to watch carefully to see when the bearing of the path changes. If the

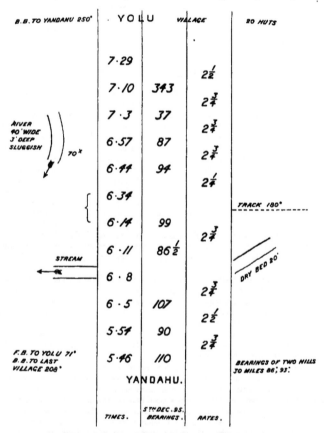

Fig. 70.—SPECIMEN PAGE OF COMPASS TRAVERSE.

sun be shining, one's shadow is a good guide; in the early morning, before the sun rises, it is very difficult to judge. In this case take a fresh bearing every five minutes.

The time occupied in taking the bearing and booking it is so short as not to require any special consideration; these intervals can be allowed for in the estimation at the rate of marching. When, however, an appreciable delay is occasioned it is desirable to note it and make allowance for it in the plotting. In the example, 6.14 is the time of arrival at the spot and 6.34 that of starting off again. It is sufficient to bracket the two times together to show what is intended.

The distances are plotted as straight lines; it is, therefore, necessary to allow in the "rate of marching" for the windings of the path. The maximum rate entered is $2\frac{3}{4}$ miles an

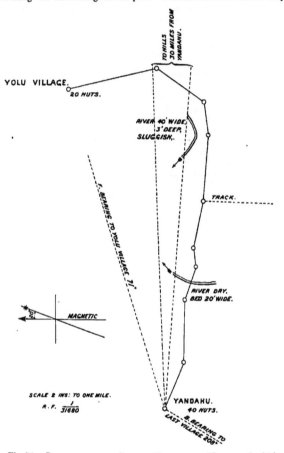

Fig. 71.—PLOTTING OF THE COMPASS TRAVERSE. (Shown on fig. 70.)

hour, and this is as much as most caravans of porters will accomplish on an African bush track. If the track goes up and down hill, or is very stony or otherwise bad, it is necessary to reduce the rate entered accordingly.

After a little practice the error in length of traverses of this nature done with ordinary care should not exceed about eight miles in a hundred, and this is caused generally by a persistent over- or under-estimation of the rate of marching. In direction the result is, as a rule, very good ; it is wonderful how the number of small errors in bearings neutralise one another.

The track included in fig. 70 is plotted on fig. 71.

For plotting it is convenient to make one or two scales showing the distances to be plotted in minutes of marching at the different rates.

Try to start and end up traverses of this nature on points of which the positions are known.

When marching with a column or caravan it is a good thing to work, if possible, for a mile or so beyond the camping ground at the end of the march, so as to be able in the morning to start off ahead and clear of the caravan.

It is advisable to carry the watch by some means which enables the time to be readily noted. A leather case, of the pattern used in South Africa, on the belt, or a band on the wrist, are both convenient methods.

It is a good thing to practise at first in a country of which there exists an accurate survey, as this gives a means of checking the work, and shows the topographer how much reliance he can place upon it.

Excellent traverses have been executed in a similar way during canoe voyages down rivers in Africa and elsewhere.

## APPENDIX VIII.

### PHOTOGRAPHIC SURVEYING.

A map of a tract of country may be constructed from photographs of it, and, under exceptional conditions, such a process may be a useful and economical one. But suitable conditions are rare, and the resulting map is necessarily less accurate than one made by the methods which have been described in this book.

Photographic surveying is practically equivalent to constructing a map by intersection only, avoiding the use of interpolation, traversing, and local measurement. As a survey system it has two fundamental defects: the ground is not walked over, and the map is not made in the field but plotted in an office.

The conditions favourable to the use of photo-topographic methods are:—

(i.) Inaccessibility of the tract of country to be surveyed, which may be, for instance, the upper portion of a high snow range.

(ii.) Well marked features. The method would, it is obvious, be useless on the rolling veldt, on plains, or in forests.

(iii.) A very short field season, such as may occur in sub-arctic regions.

(iv.) The want of skilled topographers.

The most successful instance of the use of photo-topography is to be found in the Canadian surveys of the Rocky Mountains.

Many treatises have been published on the subject; of those in English, the reader is referred to:—

'Photographic Surveying,' by E. Deville, Surveyor General of Dominion Lands, Ottawa. Government Printing Bureau, 1895.

'Photographic Methods and Instruments,' by J. A. Flemer, in the United States Coast and Geodetic Survey Report for 1897.

"Photographic Surveying," by J. Bridges Lee, in the Royal Geographical Society's 'Hints to Travellers.' Eighth Edition, 1901.

The use of photo-topography is so rarely necessary that it is not considered desirable to increase the bulk of this book by the inclusion of a detailed description.

## APPENDIX IX.

### USEFUL TRIGONOMETRICAL FORMULÆ.

**(i.) Plane Triangles—**

Given three sides, $a, b, c$,

$$\sin \frac{A}{2} = \sqrt{\frac{(s-b)(s-c)}{bc}},$$

$$\cos \frac{A}{2} = \sqrt{\frac{s(s-a)}{bc}},$$

where

$$s = \tfrac{1}{2}(a+b+c).$$

Given two sides and included angle $a, b, C$,

$$\tan \frac{A-B}{2} = \frac{a-b}{a+b} \cot \frac{C}{2}$$

Given angles and a side, A, B, C, $a$,

$$\frac{a}{\sin A} = \frac{b}{\sin B} = \frac{c}{\sin C}.$$

Given two sides and angle opposite one of them, $a, b$, A,

$$\sin B = \frac{b}{a} \sin A.$$

**(ii.) Spherical Triangles—**

Given three sides, $a, b, c$,

$$\sin \frac{A}{2} = \sqrt{\frac{\sin(s-b)\sin(s-c)}{\sin b \sin c}},$$

$$\tan \frac{A}{2} = \sqrt{\frac{\sin(s-b)\sin(s-c)}{\sin s \sin(s-a)}},$$

where

$$s = \tfrac{1}{2}(a+b+c).$$

Given two sides and included angle, $a, b$, C,

$$\tan \frac{A+B}{2} = \frac{\cos \tfrac{1}{2}(a-b)}{\cos \tfrac{1}{2}(a+b)} \cot \frac{C}{2},$$

$$\tan \frac{A-B}{2} = \frac{\sin \tfrac{1}{2}(a-b)}{\sin \tfrac{1}{2}(a+b)} \cot \frac{C}{2}.$$

Given two sides and angle opposite one of them, $a, b$, A,

let

$$\tan \theta = \tan b \cos A,$$

$$\cos \theta' = \frac{\cos \theta \cos a}{\cos b},$$

then

$$c = \theta \pm \theta' \text{ (two solutions)}.$$

Given three angles, A, B, C,

$$\cos a = \frac{\sin (A + \theta)}{\sin B \sin C \sin \theta},$$

where

$$\cot \theta = \frac{\cos B \cos C}{\sin A}.$$

Given two angles and interjacent side, A, B, c,

$$\cos c = \frac{\cos A \sin (B - \theta)}{\sin \theta},$$

where

$$\cot \theta = \cos c \tan A;$$

or,

$$\tan \frac{a + b}{2} = \frac{\cos \frac{A - B}{2}}{\cos \frac{A + B}{2}} \tan \frac{c}{2},$$

$$\tan \frac{a - b}{2} = \frac{\sin \frac{A - B}{2}}{\sin \frac{A + B}{2}} \tan \frac{c}{2}.$$

Given two angles and side opposite one of them, A, B, a,

let

$$\cot \theta = \tan B \cos a,$$

$$\sin \theta' = \frac{\sin \theta \cos A}{\cos B} \text{ (two solutions)},$$

then

$$c = \theta \pm \theta'.$$

## APPENDIX X.

### LENGTH OF THE METRE.

In a book published by the Ordnance Survey, entitled, "Comparison of Standards of Length," by Captain A. R. Clarke, R.E., F.R.S., 1866, will be found an elaborate account of the methods employed to determine the values of the principal national units of length in terms of each other. The most important relation is the following:—

THE METRE = 1·093623 yard.
= 39·37043 inches.

THE YARD = 0·914392 metre.

## REEVES'S TANGENT-MICROMETER.

A micrometer which can be readily fitted to any theodolite by means of which angles can be read with much greater accuracy than by a Vernier, has been recently designed by Mr. E. A. Reeves of the Royal Geographical Society.

The Tangent-micrometer will be valuable when circumstances prevent the use of micrometer-microscopes (and sometimes the extra bulk and large size of the boxes is a difficulty). A theodolite fitted with tangent-micrometers is cheaper than one fitted with micrometer-microscopes.

By means of the addition of a micrometer drum combined with an indicator, a carefully constructed tangent screw serves also as a micrometer, and enables readings of the arc to be made with considerable accuracy, whilst the arrangement adds practically nothing to the size or weight of the instrument.

Casella, 147, Holborn Bars, London, E.C., is the sole maker of this micrometer.

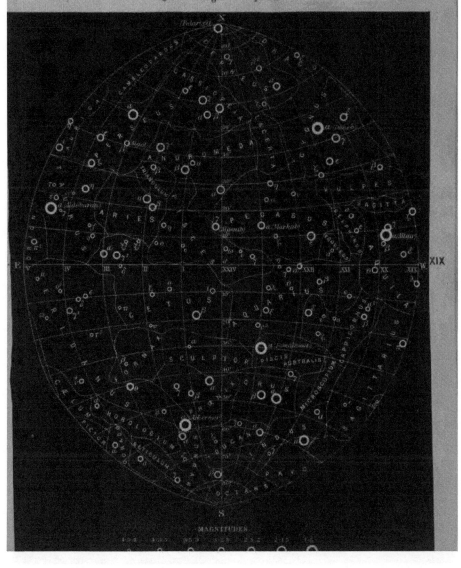

Plate XXX.

## STAR CHART
### SHOWING ALL N.A. STARS (1904)
*up to 4.5 mag. in four plates*

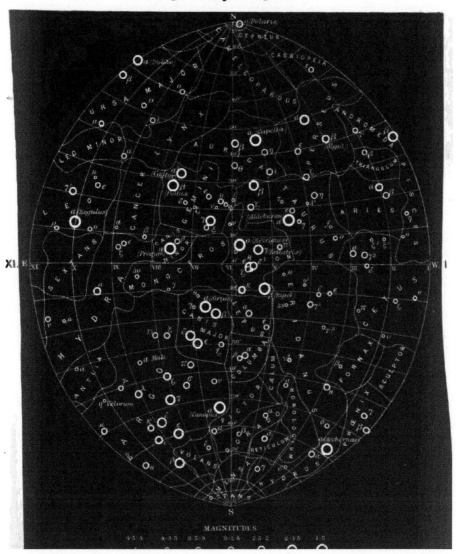

# STAR CHART
### SHOWING ALL N.A. STARS (1904)
*up to 4·5 mag. in four plates*

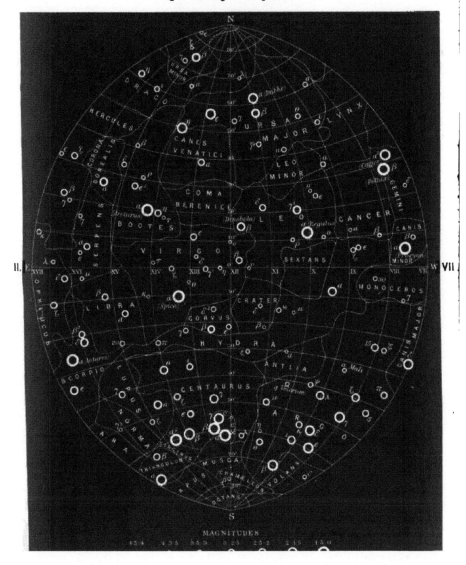

## STAR CHART
### SHOWING ALL N.A. STARS (1904)
#### up to 4·5 mag. in four plates

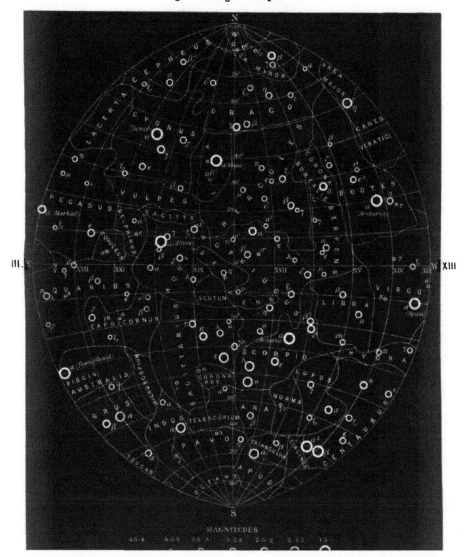

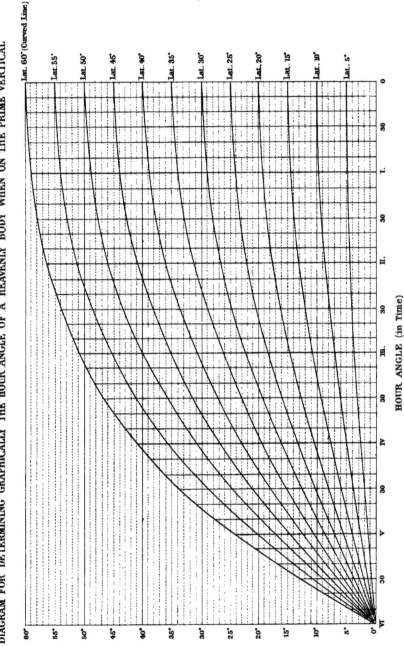

Plate XXXIII.

DIAGRAM FOR DETERMINING GRAPHICALLY THE HOUR ANGLE OF A HEAVENLY BODY WHEN ON THE PRIME VERTICAL

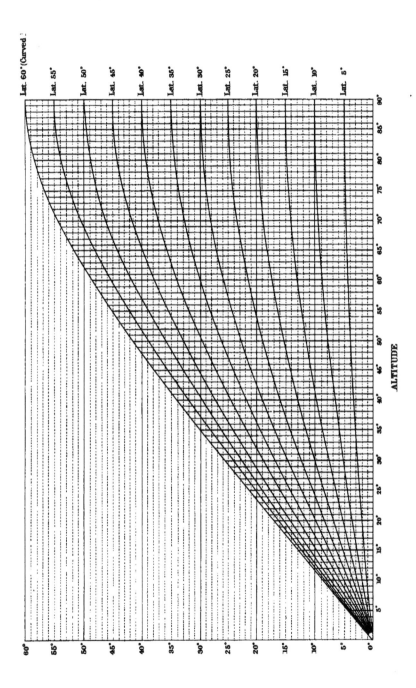

# INDEX.

### A.

Accuracy of various methods of determining heights, 86.
Accuracy of various methods of determining longitude, 203.
Active Service, surveys on, 63.
Adams, Mr. C. W., measurements with steel tapes, 8.
Adjustment of errors in long traverses, 207.
Adjustments of levels, 75, 78.
Adjustment of long traverse, approximate, 50.
,, for run, 14.
,, of theodolite, 15.
Afghan Boundary Commission, 71.
Airy's projection, 97, 106.
Alidade or sight vane, 57.
Alternative formulæ for latitude, longitude and reverse azimuth, 270.
Altitudes observed, corrections to be applied to, 140.
Altitude of star, 128.
Albers's projection, 97, 103.
Aneroid barometer, determination of heights by, 86.
Apparent day, 131.
,, places of stars, 139.
Appendix I.—Explanation of Table I. and Topographical Form 2, 267.
,, II.—Computation of distances and azimuths, 268.
,, III.—Alternative formulæ for determining latitudes, longitudes and reverse azimuths, 270.
,, IV.—Computation of heights, explanation of Table IV., 271.
,, V.—The sag of steel tapes, 272.
,, VI.—List of instruments, tools and camp equipment, 273.
,, VII.—Sketching in dense forest, 275.
,, VIII.—Photographic surveying, 279.
,, IX.—Useful trigonometrical formulæ, 280.
,, X.—Length of the metre, 282.
Apse Line, 124.
Arc or zero, 21.
Astronomical computations, 169.
,, co-ordinates, 128.
,, stations in long traverses, 43.
,, triangle, 130, 143.

Axis of earth, definition, 127.
Azimuth, definition of, 129.
,, determination of, 147.
,, stations in long traverses, 44.
,, for trigonometrical computation, 26.

### B.

Barometer heights, 86.
Bartholomew's half-inch maps, 122.
Bases, 5.
Base computations, 6.
,, length of, 5.
,, line, figures, 8.
,, measurements, rapid, 64.
,, of precision, 8.
,, probable errors of, 5.
,, site for, 5.
,, standard for, 5.
,, terminal points, 5.
,, of verification, 8.
Baskets, signal, 10.
Bemrose paper, 117.
Bench marks, 74, 84.
Boiling-point thermometer, determination of height by, 87.
,, ,, tables, 226.
Bonne's projection, 97, 102.
Boundary Commissions, 42, 43, 71, 73, 184, 190.
Box chronometer, 141.

### C.

Cadastral plan, 1.
Camp equipment, 274.
Canadian surveys, photo-surveying in, 279.
Cassini's projection, 98, 111.
Catenary, 8, 272.
Celestial sphere, 123.
,, ,, definition of, 127.
Chain of triangles, 11.
Chainage errors, 41.
Check levelling, 84.
Chronometers, comparison of, 146.
Chronometer, error, 144.
,, watches, 141.

Clarke, Colonel, 43, 105, 282.
Clarke's figure of the earth, 209.
" projection, 97, 105.
Clearings for long traverses, 43.
Clinometer, Indian pattern, 61.
Coefficient of refraction, 29.
" expansion of steel, 6.
Collignon's projection, 97, 104.
Collimation of levels, adjustments for, 76, 78.
" theodolite, adjustments for, 16.
Colour lithography, 115.
Comparison of chronometers, 146.
" watches, 146.
Compass traverses, 34, 276.
" trough, 57.
Compression of earth, 209.
Computations, astronomical, 169.
Computation of co-ordinates of traverse, 35.
Computations. Lengths of sides, 23, 24.
Computation, trigonometrical heights, 28, 31, 223, 271.
" for rapid triangulation, 68.
" of rectangular co-ordinates, 25.
" trigonometrical latitudes and longitudes, 26, 267, 270.
Conical projections, 96, 97, 107, 138.
Constellations, 125.
Contour, definition of a, 89.
" interval, 91.
" intervals in British and Foreign topographical maps, 121.
Contouring, 61, 89.
" degrees of accuracy in, 89.
" methods of, 89.
" with aneroid barometer, 91.
" " Indian clinometer and plane-table, 90.
" " level or theodolite, 89.
" " water level, 90.
Conventional projections, 97.
Conversion of mean and apparent time, 133.
" " sidereal time, 134, 135.
Cooke's Level, 75.
Cotton, Mr. E. P., steel tape measurements by, 8.
Co-ordinates, rectangular, 25.
Culmination (moon), accuracy of, 203.
" of stars, definition of, 128.
Curvature of the earth, 80.
" and refraction amount and formula for, 81.
Cylindrical projections, 96, 98, 110.

D.

Darwin, Major, pamphlet, 110.
Datum level, 30, 74.
Declination, definition of, 129.
Determination of time, 143.
Diaphragm of theodolite, 16, 17.
Distances and azimuths, 268.
Drawing on stone, 115.

E.

Ecliptic, definition of, 127.
Elastic extension, 6.
Elgin, local attraction at, 43.
Elongation of star, 137.
Error, angular of traverse, 37.
" definition of, 204.
" law of propagation, 206.
Errors, adjustment of in long traverses, 207.
" of compass traverses, 278.
" frequency of, 204.
" in a network of long traverses, adjustment of, 207.
" in long traverses, 46.
" probable, 205.
" residual, 205.
Equal altitudes for time, 145.
Equal-area projection, 96, 101.
Equation of time, 139.
" " definition of, 131.
" " variation in, 132.
Equator, definition of, 127.
Equinoxes, 127.

F.

Face, 21.
Farquharson, Col. Sir J., 119.
Field book form for traverses, 36.
" projections, 95.
Focus, theodolite. adjustment, 15.
" level adjustment, 75.
Forest, compass sketching in, 275.
Form-lines, 91.
Framework of a map, 2, 4.

G.

Gauss's method of collimating, 17, 79.
Geodetic survey, 1.
" triangulation definition of, 3.
Germain, on projections, 98.
Geographical surveys, 3.
Gill, Sir David, 208.
Glareanus, treatise of, 106.
Globular projections, 98, 110.
Gnomonic projection, 97, 104.
Gold Coast Boundary, 43-45.
Grant, Colonel, diagram for occultations, 189.
Graticule, definition of, 92.
Graticules, tables for constructing, 227, 232.
Guillaume, Mr., Nickel-Steel alloys, 8.

H.

Height above sea, correction for, to base measurement, 7.

Heights, by barometer, Tables V., VI., 224.
,, boiling-point thermometer, Tables VII., VIII., 226.
,, trigonometrical, 28, 31, 223, 271.
,, degrees of accuracy in methods of determining, 86.
Hill sketching, 89, 91.
Hills, Major, 201.
Hobday, Colonel, 203.
Horizon, definition of, 127.
Horizontal angles, method of observing, 20.
,, equivalent, 62.
,, parallax of sun, 140, 261.
Hour angle, definition of, 129.
Hypsometer, determination of heights by, 86, 87, 226.

### I.

Identification of trig. stations, 66, 68.
Indian clinometer, 61.
Ink, lithographic, 113.
Instruments, tools and equipment, list of, 273.
Interpolation, trigonometrical, 30, 32, 33, 69, 70, 72.
Intersected points, 11, 68.
Invar, 8.
,, tapes, 44.
Impressions, number of, obtainable by zincography and lithography, 116.

### J.

James's projection, 97, 105.

### K.

Kepler's laws, 124.

### L.

Layer system, 122.
Lays, lithographic, 115.
Layvange's projection, 97, 101.
La Hire's ,, 97, 106.
Lambert's ,, 97, 100, 101.
Latitude by Pole Star, 139, 156.
,, ,, meridian altitude, 152.
,, ,, circum-meridian altitudes, 152.
,, determination of, 151.
,, definition of, 127.
,, linear value of quarter degrees of, 232.
,, ,, ,, ,, second of arc along meridian, 234.
,, longitude and reverse azimuth, alternative formulæ, 270.
Latitudes, longitudes and reverse azimuths, 267.
,, ,, ,, ,, ,, Table I., 309.

Latitudes, reduction of Table XVIII. for occultations, 265.
Length of one second along meridian, Table XI., 234.
,, of one second along parallel, Table XII., 246.
,, of sights when levelling, 82.
,, of quarter degrees, Table X., 232.
Level, various patterns of, 75.
,, Cooke's reversible, description and adjustment of, 75.
,, Dumpy, adjustments of, 78.
,, "Y" pattern, adjustments of, 78.
Levelling, 74.
,, check, 84.
,, definition and principle of, 74.
,, ordinary or engineering, 74.
,, instruments, 74.
,, field-book form for, 84.
,, geodetic or precise, 74.
,, staves, description of, 80.
,, procedure, 81.
,, with the theodolite, 85.
,, a theodolite, 18.
Level readings, "Forward," "Back," and "Intermediate," 82.
,, scale of theodolite, value of, 19.
Liberian Boundary Commission, 42.
Lithography, 113.
Lithographic equipment in the field, 112.
,, drawing and printing, rate of, 113.
,, printing, 114.
,, transfers, 113.
,, work in the field, personnel required for, 113.
Local attraction, 43.
Logarithm tables, Shortrede's, 142.
Longitude, definition of, 127.
,, example of rapidly carrying forward, 71.
,, by altitudes of moon, 203.
,, ,, chronometer, 186.
,, ,, eclipses of Jupiter's satellites, 203.
,, linear value of one second of arc along parallel, 246.
,, linear value of quarter degree of, 232.
,, by lunar distances, 203.
,, ,, moon-culminating stars, 201.
,, ,, ,, occultations, 187.
,, ,, ,, photographs, 201.
,, ,, telegraph, 185.
,, ,, triangulation, 184.
,, relative accuracy of different methods of determining, 203.
"Long" traverses, 42.
,, ,, astronomical stations for, 43.
,, ,, azimuth stations in, 44.
,, ,, computations of, 45.
,, ,, errors in, 44.
,, ,, reduction to sea level, 46.
,, traverse, computation form, 48.

## M.

Magnitudes of stars, 126.
Map projections, not necessarily geometrical, 92.
" " general discussion, 95.
" " list of, 96.
Maps, British and Foreign Government, topographical (see also topographical maps), 118.
Mapping *versus* sketching, 118.
Marching, rate of, in forest, 277.
Marks and mark stones, 10.
Mean sea level, 30.
" time, definition of, 131.
Mercator's projection, 97, 100.
Meridian, definition of, 127.
Metre, in terms of yard and inch, 282.
Micrometer microscope theodolites, 13, 141.
" taking runs of, 14.
Mollweide's projection, 97, 104.
Moon, 125.
" altitudes, longitude by, 203.
" culminations, longitude by, 201.
" distance from earth, 123.
" occultations, longitude by, 187.
" photographs for longitude, 201.
Moon's altitude in occultation, 200.
" parallax, correction for, Table XVII., 264.
" vertex, angles from in occultation, 298.
Mounting a plane-table, 57.

## N.

Nautical almanac, 138.
Nigeria, Boundary Commission, 190.
Note-book for compass sketching, 275.
Nyasa, Boundary Commission, 184.

## O.

Observations for time, 143.
Occultation of star by the moon, 187.
" example of prediction of, 196.
" illumination of moon's limb, 198.
Offset, lithographic, 116.
Ordnance survey, 5, 8, 118.
" " signal, 10.
Orthographic projection, 97, 104.
Orthomorphic " 96, 97, 100.

## P.

Pamirs, Boundary Commission, 73.
Paper, Bemrose, 117.
" lithographic, 115.
" transfer, 113.
Parallax, adjustment for, 15, 75.
" effects of in occultation, 187.
" of sun, Table XV., 261.
" moon's correction for Table XVII., 264.
" of stars, 126.
" sun's diurnal, 140.

Perspective projections, 96.
Photographs of moon for longitude, 201.
Photographic surveying, 279.
Pillars, isolated, 15.
Planets, 123, 124.
Plane-table, description of, 57.
" " identification of points by, 66.
" " interpolation with, 58.
" " intersection with, 58.
" " mounting a, 57.
" " resection from two points with, 60.
" " rules for interpolation with, 59.
" " surveying with the, 57.
" " traversing with, 60.
Plane-tabling, 57, 66.
Plotting a graticule, 93.
Polar distance, 129.
Pole star, 126.
" and brush, 10.
Polyconical projections, 97, 109, 110.
Prediction of occultation, 196.
Prime vertical, definition of, 127.
" sun or star on, 138, 144.
Printing, lithographic, 114.
" sun, 116, 117.
" zincographic, 116.
Prints from Bemrose and other papers, 117.
Probable error, 205.
" " of arithmetic mean, formula for, 205.
Projection, conical, for South Africa, 109.
Projections, list of, 96, 97, 98.
" of some important maps, 99.
" the choice of, 98.
Proper motion of stars, 126.

## R.

Rapid triangulation, 63, 65, 71.
" " computations for, 68.
Rate of marching in forest, 277.
Reduction to meridian for latitude observations, Table XVI., 262.
Referring object, 20.
Refraction, astronomical, 140, 259.
" terrestrial, amount and effects of, 81.
" " co-efficient of, 29, 223.
Register, lithographic, 116.
Reproduction of maps in the field, 112.
Residuals, 205.
Reverse azimuths, 26, 267, 270.
Right ascension, definition of, 130.
" angled spherical triangles, 137.
Roller, lithographic, 116.
Roy, General, 118.
Runs (micrometer), 14.
Ruysch, projection by, 107.

Sag of steel tapes, 6, 8, 272.
Satellite stations, 22.
Scales of foreign topographical maps, list of, 120.
„ „ topographical maps, 1.
„ for „ „ 122.
Scheme for traverse survey, 46.
Sea level, reduction to, 46.
Sections of ground, drawing, 82, 85.
Selection of stars for time, 144.
Semi-diameter, sun's, 140.
Sidereal day, 131.
Sight vane or allidade, 57.
„ „ for traverses, 35.
Signs for latitudes, longitudes and reverse azimuths, Table II., 222.
Signals, trigonometrical, 10.
„ for traverses, 35.
Sinusoidal projection, 97, 102.
Sketch contouring, 91.
Sketches, futility of compiling to form maps, 118.
Sketching in dense forest, 275.
Solar system, 123.
Solstices, 127.
South Africa, maps of, 119.
Spherical excess, 23, 258.
„ trigonometry, application to Field Astronomy, 136.
Stations, selection of, 10.
Stars, apparent places of, 139.
„ distance of, 123.
„ elongation of, 137.
„ magnitude of, 126.
„ names of, 126.
„ observation to, 142.
„ on prime vertical, 138.
„ selection of for azimuth, 148.
„ „ „ latitude, 153.
„ „ „ time, 144.
Steel, expansion of, 6.
„ tapes, sag of, 272.
Stereographic projection, 97, 105.
Stone, lithographic printing in colours, 115.
„ „ preparation of, 113, 115.
„ „ transfer to, 114.
Subtense method, 51, 52.
„ station, 42.
„ traverses, 41, 53.
Surveying, definition of, 1.
Survey Departments of British and Foreign Governments, 118.
„ of India, The, 119.
Sun, distance from earth, 123.
„ observations to, 142.
„ printing and apparatus for same, 116, 117.
Sun's diurnal parallax, 140, 261.
Swing, 20.

T.

Tables, 209.
Table I.—For computing latitude, longitude and reverse azimuths, 209, 270.
„ II.—For applying signs to computations obtained by Table I., 222.
„ III.—Computation of heights from vertical angles, 223, 271.
„ IV.—Computation of heights, to find subtended angle, 223, 271.
„ V. and VI.—For determining relative heights by barometer, 224.
„ VII. and VIII.—For determining heights with boiling-point thermometer, 226.
„ IX.—For facilitating drawing graticules of maps, 227.
„ X.—Length of quarter degree of latitude and longitude, 232.
„ XI.—Linear value of one second of arc along meridians, 234.
„ XII.—Linear value of one second of arc along parallels, 246.
„ XIII.—Spherical excess, 258.
„ XIV.—Astronomical refraction, 259.
„ XV.—Parallax of the sun, 261.
„ XVI.—Reduction to meridian (for latitude), 262.
„ XVII.—Correction of moon's equatorial parallax, 264.
„ XVIII.—Reduction of latitude (for occultations), 265.
Tacheometry, 51, 53.
Tapes, steel, 5, 272.
Theodolite, 11.
„ adjustments of, 15.
„ „ „ collimation, 16.
„ „ „ horizontality of wire, 17.
„ „ „ levelling, 18.
„ „ „ focus and parallax, 15.
„ „ „ transit axis, 18.
„ „ „ verticality of wire, 17.
„ examination of, 14.
„ micrometer, 12.
„ vernier, 12.
„ observations, 68.
„ setting up a, 15.
„ traverses, 34.
„ use of in field astronomy, 141.
Theodolites, portable, 63.
„ prices of, 141.
„ weights of, 141.
Theory of errors, brief notes on, 204.
„ „ „ examples of use, 206.
Time, determination of, 143.
„ mean and sidereal conversion of, 134, 135.
Togo-Gold Coast Boundary Commission, 44-47.
Topography, definition of, 1.
Topographical surveying, definition of, 1.
„ map, 1.

Topographical maps of South Africa, 119.
" " " Austria, 119.
" " " Belgium, 119.
" " " France, 119, 120.
" " " Germany, 120.
" " " Italy, 120.
" " " Russia, 120.
" " " Spain, 120.
" " " Switzerland, 120.
" " " United States, 120.
" " " the Survey of India, 119.
" " " " Ordnance Survey, 118.
Transit axis level, 18.
Traverses between trigonometrical points, 34.
" errors of, 34.
" linear measurement of, 36.
" observations for, 35.
" procedure for carrying out, 35.
Traverse, definition of, 34.
" angular error of, 37.
" book, 36.
" compass, in forest, 275.
" computation, 35, 42, 45.
" " of co-ordinates of, 35.
" long, computation form, 48.
" field book, form for, 36.
" different classes of, 34.
" stations, 34.
" subtense, 41, 53.
Transfer, lithographic, 113.
Triangulation, 10.
" topographical, definition, 3.
" classes of, 3.
" probable errors of, 207.
" rapid, 63, 65, 71.
" principal, definition, 3.
" secondary, definition, 3.
" tertiary, definition, 3.
Triangles, size of, 10.
" length of sides, 3.

Trigonometrical formulæ, 280.
" heights, tables, 223.
" interpolation, 30, 69, 70, 72.
" signals, 10.
Troughton and Simms, Messrs., 14.

U.

Unmapped area, 2.
Uses of maps in war, 121.

V.

Verniers, 12.
Vertical angles, best time for, 28.
" " methods of observing, 21.
" interval of contours, 62.

W.

Wahab, Colonel, 61.
Water level, 90.
Watches, comparison of, 146.
Watson, General, 118.
Werner's projection, 97, 102.

Y.

Yard in terms of metre, 282.
Year, length of, 134.

Z.

Zero or arc, 21.
Zincography, 113, 116.
Zenith, definition of, 127.
" distance, 128.
Zenithal projections, 96, 106.

**DATE DUE**

| | | |
|---|---|---|
| FEB ?  | | |
| | | |
| FEB 27 1989 | | |
| | | |
| MAR 28 1989 | | |
| | | |
| MAY 22 | | |
| | | |
| DEC 15 2001 | | |
| JAN 17 2002 | | |

DEMCO 38-297

ACME
BOOK? ? ? INC.

NOV 6 1984

100 CAMBRIDGE STREET
CHARLESTOWN, MASS.

Ingram Content Group UK Ltd.
Milton Keynes UK
UKHW022321050623
422929UK00005B/456